Lecture Notes in Computer Science 7300

Commenced Publication in 1973
Founding and Former Series Editors:
Gerhard Goos, Juris Hartmanis, and Jal

Henning Bordihn   Martin Kutrib
Bianca Truthe (Eds.)

Languages Alive

Essays Dedicated to Jürgen Dassow
on the Occasion of His 65th Birthday

 Springer

Volume Editors

Henning Bordihn
Universität Potsdam, Institut für Informatik
August-Bebel-Straße 89
14482 Potsdam, Germany
E-mail: henning@cs.uni-potsdam.de

Martin Kutrib
Universität Giessen, Institut für Informatik
Arndtstraße 2
35392 Giessen, Germany
E-mail: kutrib@informatik.uni-giessen.de

Bianca Truthe
Otto-von-Guericke-Universität Magdeburg, Fakultät für Informatik
Universitätsplatz 2
39106 Magdeburg, Germany
E-mail: truthe@iws.cs.uni-magdeburg.de

Cover illustration by Bianca Truthe
© Bianca Truthe

ISSN 0302-9743   e-ISSN 1611-3349
ISBN 978-3-642-31643-2   e-ISBN 978-3-642-31644-9
DOI 10.1007/978-3-642-31644-9
Springer Heidelberg Dordrecht London New York

Library of Congress Control Number: 2012940987

CR Subject Classification (1998): F.1, F.2, F.4, G.2.3, I.1.3

LNCS Sublibrary: SL 1 – Theoretical Computer Science and General Issues

Typesetting: Camera-ready by author, data conversion by Scientific Publishing Services, Chennai, India

Printed on acid-free paper

Springer is part of Springer Science+Business Media (www.springer.com)

Jürgen Dassow
(painted by his wife Christine)

Preface

This Festschrift is dedicated to Professor Jürgen Dassow on the occasion of his 65th birthday on July 11, 2012. The volume contains articles on recent research in the theory of automata, formal languages, biologically inspired computations, and related topics. The contributing authors – leading researchers, colleagues, and friends – honor Jürgen Dassow with papers covering a wide range of topics in the field in which he has been very active for many years.

Jürgen Dassow studied mathematics at the University of Rostock. He obtained his doctoral degree (Dr. rer. nat.) with the dissertation *Über das Verhalten von lokalen Ringen bei monoidalen Transformationen* in 1972 and his habilitation (Dr. rer. nat. habil.) with the habilitationsschrift *Ein modifizierter Vollständigkeitsbegriff in einer Algebra von Automatenabbildungen* in 1978, both from the University of Rostock. He started his academic career as a research associate in Rostock. In 1980, Jürgen Dassow took a position as a lecturer in Magdeburg, where he later became a professor of mathematics in 1987 and a professor of computer science in 1992.

As a scientist, he is particularly known for his work on *regulated rewriting* and on *cooperating distributed grammar systems*. Further main areas of his research are *grammatical picture generation*, *Lindenmayer systems*, and *biologically inspired formalisms*. These five topics have also been chosen for the illustration appearing on the cover of this book. The green tree in the center of the picture was obtained after graphically interpreting a string generated by a Lindenmayer system. The speech balloons emerging from birds residing in the tree symbolize the connection between natural and formal languages. The picture in the upper right corner shows symbolically a part of a DNA strand that stands for the work inspired by biology, while the one in the upper left corner represents a chain code picture that emblematizes the work on grammatical picture generation. The speech balloon in the lower left corner contains a derivation tree from which an unwanted part is to be cut – regulating a rewriting process. The picture in the lower right corner symbolizes a cooperating distributed grammar system which in turn models some kind of 'blackboard architecture' where different agents perform actions toward a common solution. Further interpretations are welcome and left to the reader . . .

To date, Jürgen Dassow has been the author/coauthor of more than 200 scientific papers, four monographs, and 18 proceedings. He has given at least 100 lectures at conferences and universities in 18 countries. Moreover, he has been a very active organizer of scientific exchange. He is the Editor-in-Chief of the *Journal of Automata, Languages and Combinatorics* (JALC), the successor of the renowned East German *Journal of Information Processing and Cybernetics* (EIK), and has organized several international conferences and workshops, for instance the First International Workshop on Descriptional Complexity of

Automata, Grammars and Related Structures (DCAGRS), one of the two predecessors of the workshop Descriptional Complexity of Formal Systems (DCFS), and the Second International Conference on Developments in Language Theory (DLT). In addition, Jürgen Dassow renders outstanding organizing services within the German community. In 1992, he was one of the founders of the special interest group on Automata and Formal Languages of the German Association of Informatics (Gesellschaft für Informatik). For 14 years, he served as a member of its Steering Committee, from 1993 to 2003 as Chairman. The *Theorietag Automaten und Formale Sprachen*, the annual meeting of the special interest group, was organized by Jürgen Dassow in 1991, 2001, and 2011. Moreover, he was the scientific host to five researchers who spent one or two years at the University of Magdeburg with a Fellowship from the Alexander von Humboldt Foundation of Germany. Many other guests were supervised during short-term visits.

Beside his research activities, Jürgen Dassow took responsibility in the administration of the University of Magdeburg. He served as the rector and prorector of the university, and as dean of the Departments of Mathematics and Computer Science. Last but not least, he has supervised several diploma and PhD theses.

The various authors and their contributions to this Festschrift cover many areas of automata and language theory and reflect the international reputation of Jürgen Dassow. This volume contains papers on picture languages, cooperating distributed systems of automata, quantum automata, grammar systems, online computation, word equations, biologically motivated formal systems, controlled derivations, descriptional complexity, as well as on 'classic' topics of automata and language theory. The contributions have been refereed by at least two reviewers according to usual standards.

We would like to thank all those who helped to realize this Festschrift, the authors and the referees for their contributions and for timely cooperation, as well as Alfred Hofmann and Anna Kramer from Springer for their friendly collaboration and help in the preparation of this volume.

Dear Professor Dassow, dear Jürgen, we are grateful for all the ideas and inspiring work you shared with us. We wish you many more years full of pleasure. Happy Birthday!

April 2012

Henning Bordihn
Martin Kutrib
Bianca Truthe

Table of Contents

Peptide Computers

M. Sakthi Balan[1] and Helmut Jürgensen[2]

[1] ECOM Research Lab, Education and Research, Infosys Limited,
Bangalore - 560100, India
`sakthi_muthiah@infosys.com`
[2] Department of Computer Science, The University of Western Ontario,
London, Ontario, Canada, N6A 5B7
`hjj@csd.uwo.ca`

Abstract. A peptide computer is a formal model for computations based on peptide-antibody interactions. We provide a rigorous detailed formal model and prove that this model leads to a well-defined computational behaviour. We review existing results concerning the power and limitations of peptide computers and the types of non-determinism arising in such computers on the basis of this formal model.

1 Introduction

To consider the interactions between peptides and antibodies as a model of computation was proposed in 2001 by Hug and Schuler [30]. Like a DNA computer, a peptide computer utilizes the massive parallelism of chemical reactions. Some of the potential advantages of a peptide computer compared to other biologically inspired models of computing could be the larger number of basic building blocks, the different binding affinities, and a potentially greater tolerance to experimental conditions [30].

The classical model of computation, often referred to as the Turing model, relies on the intuition of writing symbols on a sheet of paper. Several models relying on other perceived modes of computation have been formulated, for example, models relying on biological or physical or chemical phenomena. The fundamental discussion regarding the essence of computing is far from settled. Even shortly after the formulation of Church's or Turing's theses, there were well-founded formal mathematical and philosophical arguments stating that the concept of Turing computability was too wide (Péter [40]) or too narrow (Kalmár [36]). With our increased, but still and likely ever incomplete, understanding of natural processes, many aspects of computability keep undergoing scrutiny (e.g. [29, 41]).

Natural processes can be modelled as computations at different levels. One can work at a behavioural level, a phenomenological level or at the actual level of the respective science involved. For any computing model, some abstraction is needed.

A formal mathematical model of a peptide computer was introduced by Balan in 2004 [3]. This model involved an 'agent' – like a human in a laboratory – to supervise the process and to evaluate the respective situation. This model

H. Bordihn, M. Kutrib, and B. Truthe (Eds.): Dassow Festschrift 2012, LNCS 7300, pp. 1–29, 2012.

is referred to as *look-and-do* by us in [8]. A more rigorous, but less intuitive alternative definition, which does not need such an agent, was indicated by us briefly in [8] and also in [4]. Certain essential details were omitted there because of space limitations. Below we expand this definition by providing all details and we augment it by explanations.

Several special results by Balan, Krithivasan and Sivasubramanyam [2, 10–12] published between 2001 and 2004 concern the simulation of Boolean functions by and the computational power of, peptide computers[1]. Much of this is summarized in the survey [9]. Depending on the details of the definition, a peptide computer is at least as powerful as a Turing machine [7, 8, 10]. In view of the ongoing discussions regarding the foundations of computability, one could envisage a model of peptide computing which is even more powerful than Turing machines.

When studying computing models inspired by natural processes one may have several complementary goals, for instance:

- Understand the natural processes.
- Model the processes for simulation.
- Use the processes to perform computations.
- Use the model to understand aspects of computation.
- Use the model to build computers.
- Simulate the model to build computers.

In this respect, peptide computers are similar to other models of computation inspired by natural phenomena. The mathematical model of peptide computing is an abstraction of the bio-chemical processes, which attempts to capture essential properties and ignores issues which appear to be less important. As is common in the literature on computing based on biological, chemical or other processes occurring in nature, we focus on the mathematical aspects. It is obvious that, at some point, in-vitro experiments would be needed to validate the practicability of the proposed computations.

Our focus is on the mathematical model and its potential. For the bio-chemical background we refer to [17]. In the book *Immunological Computation* [21] one finds a much simplified introduction and a huge and useful, but not quite representative, list of references. Also several interesting applications including fault-tolerant computing and intrusion detection are presented there.

---

[1] There seems to be a recent book *Peptide Computing,* which we have not seen yet [42], ISBN 9786131428623. We have not found this book in any library search across the world. According to the publishers the book "exclusively contains articles from the free Wikipedia Encyclopedia, which means that we are not able to send a review copy of the book because the content is available freely on the internet ..." (email of 9 December, 2011 from l.sabjan at vdm-vsg.de). The book is announced to have 112 pages. On the internet we found one Wikipedia article on *Peptide Computing* as a stub (end of November 2011) and about five additional pages on the topic which are not copyrighted. According to the information available, this book is not a relevant source and will be ignored in the sequel.

Our conclusions in the present paper refer to the mathematical model. One hopes that they translate into the reality of the bio-chemical world. If the abstraction leads to unrealistic conclusions, the model will have to be adjusted. The intention is, however, that the conclusions are not only realistic, but lead to insights into the natural processes.

As is the case with other models of computing based on natural phenomena – like reaction systems [23], DNA computing [38], membrane computing [16, 39], physarum machines [1] or soliton automata [14, 22] – the following fundamental problems – among many other ones – need to be solved for peptide computing:

- How should one encode the input in a natural and efficient fashion? How can this be programmed?
- How can one automate the computing procedure? How does one obtain the required substances and the right amounts?
- How can one decode the system state to obtain the result of a computation?
- How does one deal with inherent non-determinism?

Several aspects of the first three issues are addressed in [5] and [13]. For computations requiring huge bounded or even unbounded resources, an incremental strategy is proposed in [13] to provide additional molecules when needed[2]. Automation or programming of the computation control is studied in [5]. For example, an instruction set for peptide computing based on Head's framework for aqueous computing architecture [27] is proposed.

Computational processes are often non-deterministic, depending on the level of detail at which they are observed. They may be non-deterministic at a micro-level so that state changes are achieved non-deterministically, while at a macroscopic level the state changes are deterministic; they may also be truly non-deterministic at the macroscopic level, meaning that a state may have multiple successor states. The issue of non-determinism in peptide computing is briefly addressed in [4]. Non-determinism is inherent to processes when the state-transition is *absolutely* unclear. One knows the potential next states, but does not know anything about which one it will be. Using probabilities in such a situation does not capture the problem. In natural processes, non-determinism can express itself in various ways: (1) as a transient uncertainty about the next state[3]; (2) as an uncertainty about when the next state is reached; (3) as an uncertainty as to how the next state is reached; (4) as an uncertainty about the next state. One attempts to exploit some of these uncertainties by parallelism: the process is run in parallel; the states reached are evaluated. If parallel computing does not cost anything, exponential time reduces to polynomial-time.

A peptide, a sequence of amino acids attached by covalent bonds called peptide bonds, consists of recognition sites, called *epitopes,* for the antibodies. A peptide can contain more than one epitope for the same or different antibodies.

---

[2] This is similar to a Turing machine being given additional symbols and additional tape space when needed; for peptide computing (or DNA computing), the details are significantly more complicated, however.

[3] A similar phenomenon is encountered in asynchronous circuits.

With each antibody, which attaches to a specific epitope, a binding power is associated, called its *affinity*. When antibodies compete for recognition sites – which may overlap in the given peptide – then the antibodies with greater affinity have higher priority. For further information regarding the bio-chemical processes themselves we refer to, for example, [17]. Dynamic global computing models for the immune system are presented in [31, 43].

In [30] it was shown how to solve the satisfiability problem using peptide computing and in a subsequent paper [12] it was shown how to solve two further NP-complete problems – *Hamiltonian circuit* and *exact cover by 3-set*. Moreover in [12], a simulation of a Turing machine by peptide computing is presented to show that peptide computing is computationally complete in the sense of Turing computability. Towards formalizing peptide computing, a model of a peptide computer was proposed in [7, 8]. A formal peptide computer defines the notion of a step and it was also shown in [8] that such a computer can be simulated by a Turing machine under some conditions. A survey on peptide computing as investigated to this point is presented in [9]. The model used in these articles is formal enough to convey the ideas of the processes involved, but not rigorous enough to lend itself to formal proofs. In this paper, we remedy this situation by providing a mathematically accurate definition of a peptide computer, a peptide program and related notions.

We model molecules as words over an appropriate alphabet; epitopes are infixes[4] of such words. Epitopes can overlap. One could ignore the internal structure of epitopes and their mutual arrangements and thus arrive at a computing model solely based on finding optimal matchings between sets of complementary entities with pairwise matching preferences. This model is introduced informally in [26, 32, 33] as *matching automaton*. The idea is illustrated using the *Stable Marriage Problem* (see, for example, [24, 25, 37]). The computational power of this more abstract automaton model, also based on antibody reactions, has not yet been explored.

Also other models of computation, inspired by natural processes, rely on matching strings or patterns. For DNA computing string matching is one of the key features. Using quite different basic concepts, both peptide computing and DNA computing can simulate universal Turing machines. As an automaton theoretic abstraction of DNA computing the model of Watson-Crick automata has been studied (see the survey [20]), including parallel communicating Watson-Crick automaton systems [18, 19]. In terms of language acceptance such systems are at most as powerful as linearly bounded Turing machines.

This paper is organized as follows: After this introductory section, we summarize the background information in Section 2. Section 3 is the centre part of this article. We offer a rigorous definition of the concepts needed for the microstructure of computing based on peptide-antibody interactions, and we prove that these definitions lead to a solid basis for an automaton-theoretic interpretation of processes using such interactions. We discuss the timing of such processes.

---

[4] A word u is an *infix* of a word w if there are words x and y such that $xuy = w$. Infixes or often also referred to as *factors* or *subwords*.

In Section 4, we indicate how, using our rigorous model, the solution of problems by peptide computing could be formulated. We also address complexity issues there. We review ideas about non-determinism briefly in Section 5. Originally, we planned the present article to be devoted to the sources of non-determinism exclusively. This issue required the rigorous formulation of peptide computing as it is now presented in Section 3. To keep this paper within limits, we decided to defer the detailed treatment of non-determinism to another article. The main point of this paper is to achieve a definition of peptide computation which is both meaningful and manageable.

2   Preliminaries

For a set S, $|S|$ denotes the cardinality of S. When S is a singleton set, $S = \{x\}$ say, we often omit the set brackets, that is, we write x instead of $\{x\}$. For sets S and T, consider a relation $\varrho \subseteq S \times T$. Then ϱ^{-1} is the relation $\varrho^{-1} = \{(t, s) \mid (s, t) \in \varrho\}$ and, for $s \in S$, $\varrho(s) = \{t \mid (s, t) \in \varrho\}$ and $\operatorname{dom} \varrho = \{s \mid s \in S, \varrho(s) \neq \emptyset\}$. We use the notation $\varrho : S \xrightarrow{\circ} T$ to denote a partial mapping of S into T. In that case $\operatorname{dom} \varrho$ is the subset of S on which ϱ is defined. The notation $\varrho : S \to T$ means that ϱ is a total mapping of S into T, hence $\operatorname{dom} \varrho = S$ in this case.

Let S be a non-empty set. A *multiset* on S is a pair $M = (I, \iota)$ where I is a set, the *index set*, and ι is a mapping of I into S, the *index mapping*. A multiset M is non-empty, if I is non-empty; it is finite if I is finite. For $s \in S$, the number $\operatorname{mult}_M(s) = |\{i \mid i \in I, \iota(i) = s\}|$ is the *multiplicity* of s. When I is countable, we write $M = \{m_i \mid i \in I\}$ where $m_i = \iota(i)$ is implied. With this notation, it is possible that $m_i = m_j$ while $i \neq j$ for $i, j \in I$. We use modified standard symbols for set theoretic operations also for multisets: However, on multisets, union ($\cup_{\mathrm{m}}$) is disjoint union and both intersection ($\cap_{\mathrm{m}}$) and difference ($\setminus_{\mathrm{m}}$) take multiplicities into account. We also need the operation $\times_{\mathrm{m}}$ of direct (or Cartesian) product of multisets. We write $s \in_{\mathrm{m}} M$ if $\operatorname{mult}_M(s) > 0$. Inclusion of multisets is denoted by $\subseteq_{\mathrm{m}}$. Formally this can be handled by appropriate operations on the index sets[5]. Multisets as defined above are also called *families* in the literature. The usual definition of a multiset as a set $\{(s, \operatorname{mult}(s)) \mid s \in S\}$ of pairs is adequate only, if all multiplicities are finite and if different instances of the same element need not be distinguished. In several parts of this paper, the latter point is extremely important. In those cases, which we have indicated, it is crucial that the reader consider distinguishable instances of the same object rather than its multiplicity.

By $\mathbb{N}$ and $\mathbb{N}_0$ we denote the sets of positive integers and of non-negative integers, respectively. The set $\mathbb{B} = \{0, 1\}$ represents the set of Boolean values. By $\mathbb{R}$ we denote the set of real numbers, and $\mathbb{R}_+ = \{r \mid r \in \mathbb{R}, r \geq 0\}$. For $i, j \in \mathbb{N}_0$ with $i \leq j$, the *interval* $[i, j]$ is the set $\{i, i+1, i+2, \ldots, j\}$.

An alphabet is a non-empty set. Let X be an alphabet. Then X^* is the set of all words over X including the empty word λ, and $X^+ = X^* \setminus \{\lambda\}$.

---

[5] For a rigorous treatment of multisets see [15, 34].

For a word $w \in X^*$, $|w|$ is its length. Let $w = x_0 x_1 \cdots x_{n-1}$ with $n = |w|$ and $x_0, x_1, \ldots, x_{n-1} \in X$. For i and j with $0 \le i \le j < n$, define $w[i, j] = x_i x_{i+1} \cdots x_j$. Any word $u \in X^*$ with $w \in uX^*$ is a prefix of w; let $\mathsf{Pref}(w)$ be the set of prefixes of w; the words in $\mathsf{Pref}_+(w) = \{u \mid u \in X^+, w \in uX^+\}$ are the proper prefixes of w. One defines suffixes by left-right duality. A word $u \in X^*$ with $w \in X^*u$ is a suffix of w; let $\mathsf{Suff}(w)$ be the set of suffixes of w; the words in $\mathsf{Suff}_+(w) = \{u \mid u \in X^+, w \in X^+u\}$ are the proper suffixes of w. Similarly, a word $u \in X^*$ with $w \in X^*uX^*$ is an infix of w, $\mathsf{Inf}(w)$ is the set of infixes of w and $\mathsf{Inf}_+(w) = \{u \mid u \in X^+, u \in \mathsf{Inf}(w), u \ne w\}$ is the set of proper infixes of w. A language over X is a subset of X^*. For a language L over X and $Y \in \{\mathsf{Pref}, \mathsf{Pref}_+, \mathsf{Inf}, \mathsf{Inf}_+\}$, $Y(L) = \bigcup_{w \in L} Y(w)$. Finally two words u and v overlap if u has a proper prefix which is a proper suffix of v or vice versa.

In this paper we use the following simplifying terminology: Two words u and v *are in conflict,* if any one of the following conditions holds true:

1. $u \in \mathsf{Pref}_+(v)$ or vice versa;
2. $u \in \mathsf{Suff}_+(v)$ or vice versa;
3. $u \in \mathsf{Inf}_+(v)$ or vice versa;
4. u and v overlap.

Let L be a language over X and $w \in X^*$. An L-*decomposition* of w is a pair of sequences $(u_0, u_1, \ldots, u_k)$, $(v_0, v_1, \ldots, v_{k-1})$ of words in X^* such that $u_0 v_0 u_1 v_1 \cdots v_{k-1} u_k = w$, $v_0, v_1, \ldots, v_{k-1} \in L$ and $u_0, u_1, \ldots, u_k \notin X^*LX^*$. A language in X^+ such that every word has a unique L-decomposition is called a *solid code* [35]. Consider $w \in X^+$ of length n, say $w = x_0 x_1 \cdots x_{n-1}$ with $x_i \in X$ for $i = 0, 1, \ldots, n-1$. An L-decomposition of w as above can be specified by a set of pairs $\{(i_l, j_l) \mid l = 0, 1, \ldots, k-1\}$ such that, for $l = 0, 1, \ldots, k-1$, $v_l = w[i_l, j_l]$. Let $\partial_L(w)$ be the set of L-decompositions when represented in this way. Let $\mathcal{D}(L) = \{(w, d) \mid w \in X^*, d \in \partial_L(w)\}$ be the set of words together with all their L-decompositions.

To illustrate the concept of L-decomposition, consider the language $L = \{ab, bb, aba\}$ over the alphabet $X = \{a, b\}$ and the word $w = ababbaaba$. The word w has six L-decompositions as follows:

$$
\begin{array}{ll}
(\lambda, \lambda, ba, \lambda) & (ab, ab, aba) \\
(\lambda, \lambda, ba, a) & (ab, ab, ab) \\
(\lambda, a, a, \lambda) & (ab, bb, aba) \\
(\lambda, a, a, a) & (ab, bb, ab) \\
(\lambda, \lambda, a, \lambda) & (aba, bb, aba) \\
(\lambda, \lambda, a, a) & (aba, bb, ab)
\end{array}
$$

Hence $\partial_L(w)$ consists of the following six sets:

$$
\begin{array}{ll}
\{[0, 1], [2, 3], [6, 8]\}, & \{[0, 1], [2, 3], [6, 7]\}, \\
\{[0, 1], [3, 4], [6, 8]\}, & \{[0, 1], [3, 4], [6, 7]\}, \\
\{[0, 2], [3, 4], [6, 8]\}, & \{[0, 2], [3, 4], [6, 7]\}.
\end{array}
$$

The words ab and bb overlap; so do ab and aba. Moreover, ab is also a proper prefix of aba. The words bb and aba each overlap themselves. Obviously, L is not a solid code.

3 Peptide Computing – The Model

We provide the formal definition of peptide computing as proposed in Section 5 of [8], but modified and expanded to greater detail. We also include intuitive explanations.

Definition 1. *A peptide computer is a quintuple $\mathcal{P} = (X, E, A, \alpha, \beta)$ with the following properties:*

1. *X is a finite alphabet;*
2. *$E \subseteq X^+$ is a language;*
3. *A is a countable alphabet with $A \cap X^* = \emptyset$;*
4. *$\alpha \subseteq E \times A$ is a relation;*
5. *$\beta : E \times A \to \mathbb{R}_+$ is a mapping such that $\beta(e, a) > 0$ if and only if $(e, a) \in \alpha$.*

The components of the quintuple $\mathcal{P} = (X, E, A, \alpha, \beta)$, which define a peptide computer, have the following intuitive meanings:

1. The symbols in X are the basic building units, like molecules, of which larger units are made.
2. The words in E represent molecules which can serve as epitopes.
3. The symbols in A represent antibodies. Their internal structure is not relevant for the model.
4. The relation α states which antibodies can be attached to which epitopes. Thus $(e, a) \in \alpha$ means that antibody a can be attached to the epitope e.
5. The value of $\beta(e, a)$ denotes the affinity between epitope e and antibody a.

In the rest of this paper we frequently use the terms *antibodies* or *epitopes* instead of *symbols* or *words* or *sequences* etc. This is intended to help the intuition, but should always be understood in the rigorous sense of Definition 1.

We now define how antibodies are attached to epitopes.

Definition 2. *Let $w \in X^+$, where $w = x_0 x_1 \cdots x_n$ with $x_0, x_1, \ldots, x_n \in X$.*

1. *Let $d \in \partial_E(w)$ and $\tau : d \overset{\circ}{\to} A$. The partial mapping τ is legal if, for all $(i, j) \in \mathsf{dom}\,\tau$, one has*
$$\bigl(w[i, j], \tau(i, j)\bigr) \in \alpha.$$

2. *An A-attachment to w is a pair $\vartheta = (d, \tau)$ with $d \in \partial_E(w)$ and $\tau : d \overset{\circ}{\to} A$ such that τ is legal.*

Consider a word $w \in X^+$ as above, $d \in \partial_E(w)$ and a partial mapping $\tau : d \overset{\circ}{\to} A$. Suppose $d = \{(i_l, j_l) \mid l = 0, 1, \ldots, k - 1\}$. Then $\vartheta = (d, \tau)$ defines a word $w_\vartheta \in \bigl(X \cup (E \times A)\bigr)^*$ as follows:

> For all $l = 0, 1, \ldots, k-1$, if $(i_l, j_l) \in \mathsf{dom}\,\tau$, then replace the infix $w[i_l, j_l]$
> by the pair $\big(w[i_l, j_l], \tau(i_l, j_l)\big)$ in w.

If τ is legal, then $w_\vartheta \in (X \cup \alpha)^*$ and ϑ is an *A-attachment to w*. The situation is illustrated in Fig. 1. Here the set $X \cup \alpha$ is treated as the alphabet consisting of the symbols of the alphabet X and the pairs in the relation α. Words over the alphabet $X \cup \alpha$ are thus strings of symbols in X and pairs in α. The former represent symbols to which nothing has been attached. The latter represent epitopes to which antibodies have been attached. This formalism retains the information about the string forming the epitope even when an antibody is attached to it. If at some stage such an antibody becomes unattached, the epitope is again represented by the corresponding string over the alphabet X.

Remark 1. For each $z \in (X \cup \alpha)^+$ there is a word $w \in X^+$ and an *A*-attachment ϑ to w such that $z = w_\vartheta$. Moreover, w is unique with this property.

Proof. The word w is obtained by applying, to z, the homomorphism, which maps every $x \in X$ to itself and every pair $(e, a) \in \alpha$ to e. The *A*-attachment ϑ is the inverse of this application of the homomorphism. □

In Remark 1, the word w is unique. However, ϑ need not be unique as the *A*-attachment may result in z for several different *E*-decompositions of w. The word w is obtained from z by *stripping* all attachments away from z, $\mathsf{strip}\,(z) = w$. Moreover, with $\vartheta = (d, \tau)$ as above, we define the multisets $\mathsf{att}\,(z)$ and $\mathsf{epi}\,(z)$ on A and E, respectively, by

$$\mathsf{mult}_{\mathsf{att}(z)}(a) = \big|\{(i,j) \mid (i,j) \in \mathsf{dom}\,\tau \wedge \tau(i,j) = a\}\big|$$

and

$$\mathsf{mult}_{\mathsf{epi}(z)}(e) = \big|\{(i,j) \mid (i,j) \in d \wedge (\mathsf{strip}\,(z))[i,j] = e\}\big|$$

for $a \in A$ and $e \in E$. The multiset $\mathsf{att}\,(z)$ consists of all symbols in A which are attached in z. The multiset $\mathsf{epi}\,(z)$ consists of all words in E which occur in the current *E*-decomposition d of $w = \mathsf{strip}\,(z)$.

When several epitopes and antibodies are present, conflicts can arise between the choice of both, the epitopes and the antibodies. Epitopes in different *E*-decompositions may overlap. Different antibodies may attach to the same epitopes. Some of these conflicts are partially resolved using affinity values. However, some non-determinism remains. The case of conflicting epitopes is illustrated in Fig. 2.

Our next goal is to define *basic reactions*. We achieve this in several steps. First we define the situation when an antibody dominates in a certain "area" of epitopes. Then we explain the resulting reaction in such a case. Finally, we define the basic reactions between words in $(X \cup \alpha)^+$.

Definition 3. Let $z \in (X \cup \alpha)^+$ and $a \in A$. Let $w \in X^+$, and let $\vartheta = (d, \tau)$ be such that $z = w_\vartheta$. Let $d' \in \partial_E(w)$ and $(i,j) \in d'$.

w with $d \in \partial_E(w)$; $v_i \in E$ for $i = 0, 1, \ldots, 5$

w with dom τ indicated

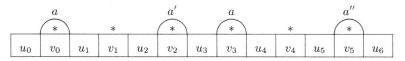

$z = w_\vartheta$ with $\vartheta = (d, \tau)$

Fig. 1. An A-attachment to w resulting in z. Potential binding sites are indicated by the symbol $*$; here $\mathsf{att}(z) = \{a, a', a, a''\}$.

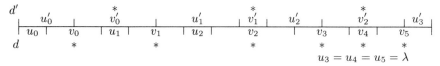

Fig. 2. Conflicting epitopes in E-decompositions d and d' of a word w. The epitopes are indicated by the symbol $*$.

1. The symbol a is said to dominate (i, j) in z if
$$\beta\big(w[i, j], a\big) > \beta\big(w[i', j'], \tau(i', j')\big)$$
   for all $(i', j') \in \mathsf{dom}\, \tau$ with $[i', j'] \cap [i, j] \neq \emptyset$.
2. If a dominates (i, j) in z, then all pairs (i', j') with $(i', j') \in \mathsf{dom}\, \tau$ and $[i', j'] \cap [i, j] \neq \emptyset$ are said to be affected (by a at (i, j)).

If an antibody a dominates an epitope e at a specific position, then this implies that all antibodies bound to epitopes which overlap the specific instance of e have lower affinities for their current bindings. As a result, the bindings at the affected sites are broken, and the antibody a attaches to that occurrence of the epitope e. This intuition guides the definition of a basic reaction.

Definition 4. Let $z \in (X \cup \alpha)^+$ and $a \in A$. Let $w \in X^+$, and let $\vartheta = (d, \tau)$ be such that $z = w_\vartheta$.

1. The pair (z, a) is said to be a critical pair, if there are $d' \in \partial_E(w)$ and $(i, j) \in d'$ such that a dominates (i, j) in z. In that case, the quadruple (z, a, i, j) is called a critical constellation.

2. *Let (z, a, i, j) be a critical constellation. A basic reaction $\mathsf{react}_{(i,j)}(z, a)$ between z and a at (i, j) consists of the following steps, and results in the multiset $\mathsf{Res}_{(i,j)}(z, a)$ and the set $\mathsf{Out}_{(i,j)}(z, a)$.*

   (a) *Initially, $\mathsf{Res}_{(i,j)}(z, a)$ and $\mathsf{Out}_{(i,j)}(z, a)$ are empty.*

   (b) *For each affected pair (i', j'), a copy of $\tau(i', j')$ is put into $\mathsf{Out}_{(i,j)}(z, a)$, and the multiplicity $\mathsf{mult}_{\mathsf{Res}_{(i,j)}(z,a)}(\tau(i', j'))$ of $\tau(i', j')$ in $\mathsf{Res}_{(i,j)}(z, a)$ is increased by one.*

   (c) *Let $Y \subseteq \mathsf{dom}\,\tau$ be the set of pairs which are not affected and let $d'' \in \partial_E(w)$ be such that $Y \cup (i, j) \subseteq d''$. Let $\bar{\vartheta} = (d'', \tau'')$ be the A-attachment with $\mathsf{dom}\,\tau'' = Y \cup (i, j)$ and*

$$\tau''(p) = \begin{cases} \tau(p), & \text{if } p \in Y, \\ a, & \text{if } p = (i, j). \end{cases}$$

   Insert a copy of $w_{\bar{\vartheta}}$ into $\mathsf{Res}_{(i,j)}(z, a)$, that is, let $\mathsf{mult}_{\mathsf{Res}_{(i,j)}(z,a)}(w_{\bar{\vartheta}}) = 1$.

In the reaction $\mathsf{react}_{(i,j)}(z, a)$, the antibody a, which dominates (i, j) in z, attaches to the epitope at (i, j). Simultaneously, the antibodies previously attached to the epitopes affected by a at (i, j), if any, are released. Together with the new word $w_{\bar{\vartheta}}$ obtained from z in this fashion, they form the multiset $\mathsf{Res}_{(i,j)}(z, a)$. The set (not multiset) of antibodies released in this reaction is $\mathsf{Out}_{(i,j)}(z, a)$.

Example 1. Consider the situation in Fig. 2. Let a_3 and a_5 be attached to v_3 and v_5, respectively. Nothing is attached to v_4, nor to v_2'. Let z be the result of these attachments to w. The positions of the various parts of w are given by

$$w[i_3, j_3] = v_3, \ w[i_4, j_4] = v_4, \ w[i_5, j_5] = v_5, \ w[i_2', j_2'] = v_2'$$

where, by assumption,

$$i_3 < i_2' < j_3 < i_4 = j_3 + 1 \le j_4 < i_5 = j_4 + 1 < j_2' < j_5.$$

Thus, $\{(i_3, j_3), (i_5, j_5)\} \subseteq \mathsf{dom}\,\tau$. The details of this situation are shown in Fig. 3(a).

Let $a \in A$, and suppose that $\beta(v_3, a_3) = 1$, $\beta(v_5, a_5) = 4$ and $\beta(v_2', a) = 5$. Then a dominates (i_2', j_2') in z. The three pairs (i_3, j_3), (i_4, j_4) and (i_5, j_5) are affected.

Thus (z, a) is a critical pair and (z, a, i_2', j_2') is a critical constellation. The basic reaction $\mathsf{react}_{(i_2', j_2')}(z, a)$ achieves the following: After Step (b) $\mathsf{Out}_{(i_2', j_2')}(z, a)$ is the *set* $\{a_3\} \cup \{a_5\}$ and $\mathsf{Res}_{(i_2', j_2')}(z, a)$ is the *multiset* $\{a_3\} \cup_m \{a_5\}$. In Step (c) the word $w_{\bar{\vartheta}}$ is added to $\mathsf{Res}_{(i_2', j_2')}(z, a)$. Here $\bar{\vartheta}$ is obtained from ϑ by removing a_3 from v_3 and a_5 from v_5, and by attaching a to v_2'. The relevant part of the resulting word $w_{\bar{\vartheta}}$ is shown in Fig. 3(b).

A reaction $\mathsf{react}_{(i,j)}(z, a)$ is modelled as a single step. In reality it consists of a non-deterministic sequence of substeps in which antibodies are released and attached. For the result, the order in which such substeps occur is not important.

We also need to consider basic reactions between (instances of) words $z_1, z_2 \in (X \cup \alpha)^+$. The words z_1 and z_2 need not be distinct.

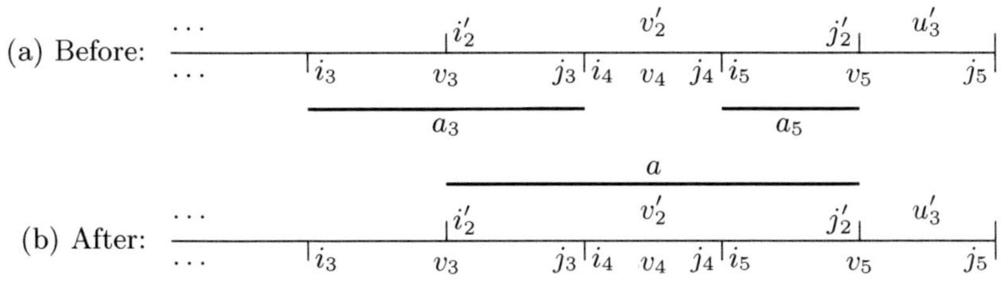

Fig. 3. The situation of Example 1 before and after the basic reaction

Definition 5. *For $h = 1, 2$, let $z_h \in (X \cup a)^+$. Let $w_h \in X^+$, and let $\vartheta_h = (d_h, \tau_h)$ be such that $z_h = w_{h, \vartheta_h}$.*

1. *The pair (z_1, z_2) is said to be a* critical pair, *if there exist $(i_2, j_2) \in \operatorname{dom} \tau_2$, $d' \in \partial_E(w_1)$ and $(i_1, j_1) \in d'$ such that $\tau_2(i_2, j_2)$ dominates (i_1, j_1) in z_1 and*

$$\beta\big(w_1[i_1, j_1], \tau_2(i_2, j_2)\big) > \beta\big(w_2[i_2, j_2], \tau_2(i_2, j_2)\big).$$

In that case, the sixtuple $(z_1, z_2, i_2, j_2, i_1, j_1)$ is said to be a critical constellation.

2. *Let $(z_1, z_2, i_2, j_2, i_1, j_1)$ be a critical constellation and let $a = \tau_2(i_2, j_2)$. A basic reaction* $\operatorname{react}_{(i_2, j_2), (i_1, j_1)}(z_1, z_2)$ *between z_1 and z_2 from (i_2, j_2) to (i_1, j_1) consists of the following steps and results in the multiset* $\operatorname{Res}_{(i_2, j_2), (i_1, j_1)}(z_1, z_2)$ *and the set* $\operatorname{Out}_{(i_2, j_2), (i_1, j_1)}(z_1, z_2)$.

   (a) *In the separation step* $\operatorname{sep}_{(i_2, j_2)}(z_1, z_2)$ *the multiset* $\operatorname{Sep}_{(i_2, j_2)}(z_1, z_2)$ *is formed as follows:*

      i. *Initially,* $\operatorname{Sep}_{(i_2, j_2)}(z_1, z_2)$ *contains only z_1 and z_2. If $z_1 \neq z_2$ then each has multiplicity one. If z_1 and z_2 are different instances of the same word z, then this word has multiplicity two. If z_1 and z_2 are the same instance of the word z, then this word has multiplicity one.*

      ii. *Decrease the multiplicity of z_2 in* $\operatorname{Sep}_{(i_2, j_2)}(z_1, z_2)$ *by one.*

      iii. *Let τ_2' be the restriction of τ_2 to $\operatorname{dom} \tau_2 \setminus (i_2, j_2)$. Let $\vartheta_2' = (d_2, \tau_2')$. Let $z_2' = w_{2, \vartheta_2'}$.*

      iv. *Add a and z_2' to* $\operatorname{Sep}_{(i_2, j_2)}(z_1, z_2)$, *that is, increase their multiplicities by one.*

   (b) *If $z_1 \in \operatorname{Sep}_{(i_2, j_2)}(z_1, z_2)$, continue as follows:*

      i. *Perform* $\operatorname{react}_{(i_1, j_1)}(z_1, a)$ *to form* $\operatorname{Res}_{(i_1, j_1)}(z_1, a)$ *and* $\operatorname{Out}_{(i_1, j_1)}(z_1, a)$.

      ii. *Remove one occurrence of z_1 and one occurrence of a from* $\operatorname{Sep}_{(i_2, j_2)}(z_1, z_2)$.

      iii. *Let* $\operatorname{Res}_{(i_2, j_2), (i_1, j_1)}(z_1, z_2)$ *be the multiset union of* $\operatorname{Sep}_{(i_2, j_2)}(z_1, z_2)$ *and* $\operatorname{Res}_{(i_1, j_1)}(z_1, a)$.

      iv. *Let* $\operatorname{Out}_{(i_2, j_2), (i_1, j_1)}(z_1, z_2) = \operatorname{Out}_{(i_1, j_1)}(z_1, a)$.

(c) *Otherwise: Let*

$$\mathsf{Res}_{(i_2,j_2),(i_1,j_1)}(z_1,z_2) = \mathsf{Res}_{(i_1,j_1)}(z_2',a)$$

and let

$$\mathsf{Out}_{(i_2,j_2),(i_1,j_1)}(z_1,z_2) = \mathsf{Out}_{(i_1,j_1)}(z_2',a).$$

In the reaction $\mathsf{react}_{(i_2,j_2),(i_1,j_1)}(z_1,z_2)$ one starts with the following situation: The antibody a is bound to an epitope $w_2[i_2,j_2]$ in z_2. In z_1 there is an epitope $w_1[i_1,j_1]$ to which the symbol a has a greater affinity. Then the bond in z_2 is broken, that is, the antibody a is released from z_2. The resulting word is z_2'. After that, a attaches to $w_1[i_1,j_1]$ releasing antibodies from affected epitopes. In the end, one has the word z_2', the word z_1' formed in the reaction $\mathsf{react}_{(i_1,j_1)}(z_1,a)$ and the antibodies released in that reaction.

A special case arises when z_1 and z_2 refer to the same occurrence of the same word z. Let $w = w_1 = w_2$. In this case, a is initially attached to $w[i_2,j_2]$ and dominates (i_1,j_1) in z. It follows that $(i_1,j_1) \neq (i_2,j_2)$. After the separation step, $\mathsf{Sep}_{(i_2,j_2)}(z_1,z_2)$ consists of $z' = z_2'$ and a, each with multiplicity one. Then a attaches to z_2' at (i_1,j_1) releasing the antibodies from the affected epitopes.

Again, a reaction $\mathsf{react}_{(i_2,j_2),(i_1,j_1)}(z_1,z_2)$ is modelled as a single step. For both types of basic reactions, the result is a multiset consisting of the words after the changes and the antibodies which have been released. The output is the set of those antibodies. Reactions occur when words in $(X \cup \alpha)^+$ and elements of A are present and when, in addition, dominance conditions as specified in Definitions 4 and 5 prevail. This intuition is expressed formally in the next definition.

Definition 6. *Let $\mathcal{P}$ be a peptide computer.*

1. *A peptide configuration is a finite multiset of words in $(X \cup \alpha)^+ \cup A$.*
2. *A peptide configuration Q contains a critical pair, if one of the following conditions holds:*
   (a) *there are $z, a \in Q$ and i, j such that (z, a, i, j) is a critical constellation or*
   (b) *there are $z_1, z_2 \in Q$ and i_1, j_1, i_2, j_2 such that $(z_1, z_2, i_2, j_2, i_1, j_1)$ is a critical constellation.*
3. *A peptide configuration is said to be stable if it does not contain any critical pair.*
4. *The base of a peptide configuration Q is the multiset $\mathsf{base}(Q)$ of symbols in A and words in X^+ with multiplicities as follows:*

$$\mathsf{mult}_{\mathsf{base}(Q)}(a) = \mathsf{mult}_Q(a) + \sum_{z \in (X \cup \alpha)^+} \mathsf{mult}_Q(z) \cdot \mathsf{mult}_{\mathsf{att}(z)}(a)$$

and

$$\mathsf{mult}_{\mathsf{base}(Q)}(w) = \sum_{\substack{z \in (X \cup \alpha)^+ \\ w = \mathsf{strip}(z)}} \mathsf{mult}_Q(z)$$

for $a \in A$ and $w \in X^+$.

Peptide configurations denote the states of the computer $\mathcal{P}$. If Q is a peptide configuration, then also base (Q) is a peptide configuration. If a Q is stable, then there are no critical pairs which could be involved in basic reactions. On the other hand, if there is a critical pair, a reaction will occur leading to a new configuration. This process continues non-deterministically without any input from the environment; it stops when a stable configuration is reached.

We say that a peptide configuration contains a critical constellation (z, a, i, j) or $(z_1, z_2, i_2, j_2, i_1, j_1)$ if it contains (z, a) or (z_1, z_2) as a critical pair, respectively, and the constellation is critical.

Definition 7. *Let $\mathcal{P}$ be a peptide computer and let Q be a peptide configuration.*

1. *The* immediate successors *of Q are defined as follows:*

    (a) *If Q contains a critical constellation (z, a, i, j) then the immediate successor* $\mathsf{succ}_{z,a,(i,j)}(Q)$ *of Q is the multiset*

    $$\left(Q \setminus_{\mathrm{m}} \{z, a\}\right) \cup_{\mathrm{m}} \mathsf{Res}_{(i,j)}(z, a).$$

    (b) *If Q contains a critical constellation $(z_1, z_2, i_2, j_2, i_1, j_1)$ then the immediate successor* $\mathsf{succ}_{z_1,z_2,(i_2,j_2),(i_1,j_1)}(Q)$ *of Q is the multiset[6]*

    $$\left(Q \setminus_{\mathrm{m}} \{z_1, z_2\}\right) \cup_{\mathrm{m}} \mathsf{Res}_{(i_2,j_2),(i_1,j_1)}(z_1, z_2).$$

    (c) *If Q is stable its only immediate successor is Q.*

2. *Let* $\mathsf{succ}(Q)$ *be the set (not multiset) of all immediate successors of Q. For a set $\mathcal{Q}$ of configurations, let*

$$\mathsf{succ}(\mathcal{Q}) = \bigcup_{Q \in \mathcal{Q}} \mathsf{succ}(Q).$$

Let $\mathcal{Q}$ be a set of configurations of $\mathcal{P}$. We consider the transformation of the configurations in $\mathcal{Q}$ under successive basic reactions. Those configurations in $\mathcal{Q}$, which are stable, will not change. The unstable ones may change and give rise to sets of configurations. Define

$$\mathsf{succ}^n(\mathcal{Q}) = \begin{cases} \mathcal{Q}, & \text{if } n = 0, \\ \mathsf{succ}(\mathcal{Q}), & \text{if } n = 1, \\ \mathsf{succ}\left(\mathsf{succ}^{n-1}(\mathcal{Q})\right), & \text{if } n > 1, \end{cases}$$

for $n \in \mathbb{N}_0$. A configuration Q is said to be *reachable from* $\mathcal{Q}$ if $Q \in \mathsf{succ}^n(\mathcal{Q})$ for some $n \in \mathbb{N}_0$.

Remark 2. Let Q be a peptide configuration. Then $Q \in \mathsf{succ}^n(\mathsf{base}(Q))$ for some $n \in \mathbb{N}_0$, that is, Q is reachable from $\mathsf{base}(Q)$.

---

[6] Here $\{z_1, z_2\}$ is considered as a multiset as follows: If $z_1 \neq z_2$, each word occurs with multiplicity one. If $z_1 = z_2 = z$ then z occurs with multiplicity one or two, depending on whether the same or different occurrences of z are considered in the critical pair.

Proof. To obtain Q from $\mathsf{base}(Q)$, a sequence of basic reactions of the form $\mathsf{react}_{(i,j)}(z,a)$ is sufficient. If Q contains n instances of epitopes to which antibodies are attached, then such a sequence of basic reactions has length n. By the definition of $\mathsf{base}(Q)$, there are enough words and antibodies of the required kinds available. As the basic reactions occur non-deterministically, such a sequence is possible. Hence $Q \in \mathsf{succ}^n(\mathsf{base}(Q))$. □

Lemma 1. *Let Q be a peptide configuration. Then $\mathsf{base}(Q') = \mathsf{base}(Q)$ for every $Q' \in \mathsf{succ}(Q)$.*

Proof. Each basic reaction changes only the places where symbols from A are located. It does not change the number of their occurrences nor the number of occurrences of words $w \in X^+$ which are stripped versions of words $z \in (X \cup \alpha)^+$. □

As a consequence of Lemma 1, there is only a bounded number of different peptide configurations which can result from a sequence of basic reactions applied to a given peptide configuration Q. Hence, the sequence $\{\mathsf{succ}^n(Q)\}_{n \in \mathbb{N}_0}$ is ultimately periodic. Below, we show that the sequence actually converges.

Theorem 1. *Let $\mathcal{P}$ be a peptide computer, and let Q be a configuration of $\mathcal{P}$. The configuration Q is stable if and only if $\mathsf{succ}(Q) = \{Q\}$.*

Proof. If Q is stable, $\mathsf{succ}(Q) = \{Q\}$ by definition.

Assume that $\mathsf{succ}(Q) = \{Q\}$ and that Q is not stable. Then Q contains a critical constellation. We distinguish two cases depending on the type of the critical constellation.

1. *The critical constellation has the form (z, a, i, j). As a dominates (i, j) in z, $a \notin \mathsf{Res}_{(i,j)}(z,a)$. Hence*

$$\mathsf{mult}_{\mathsf{succ}_{z,a,(i,j)}(Q)}\, a < \mathsf{mult}_Q\, a$$

   and, thus, $\mathsf{succ}_{z,a,(i,j)}(Q) \neq Q$.

2. *The critical constellation has the form $(z_1, z_2, i_2, j_2, i_1, j_1)$. We distinguish two cases:*

   (a) First, assume that z_1 and z_2 do not refer to the same occurrence of the same word. The basic reaction results[7] in the multiset EC which contains one occurrence of z_2', and the multiset $\mathsf{Res}_{(i_1,j_1)}(z_1,a)$. Let z_1' be the word obtained in $\mathsf{react}_{(i_1,j_1)}(z_1,a)$. As $Q = \mathsf{succ}_{z_1,z_2,(i_2,j_2),(i_1,j_1)}(Q)$ and $z_2' \neq z_2$, one has $z_1 = z_2'$, $z_2 = z_1'$ and $\mathsf{Out}_{(i_2,j_2),(i_1,j_1)}(z_1,z_2) = \emptyset$. It follows that $w_1 = w_2 = xe_1ye_2z$ for some words $x, y, z \in (X \cup \alpha)^*$ and $e_1, e_2 \in E$ such that $e_1 = w[i_1, j_1]$ and $e_2 = w[i_2, j_2]$ or vice versa. The epitopes cannot overlap as the output set is empty. As these cases are analogous, we consider only the former. Thus a is attached to e_2 in z_2

---

[7] Throughout, we use the symbols with the same meanings as in the respective definitions.

and, later, to e_1 in z'_1. Therefore, a is attached to both e_1 and e_2 in z_2. This implies that a is attached to e_1 in $z'_2 = z_1$ leading to $\mathsf{Res}_{(i_1,j_1)}(z_1, a)$ containing an occurrence of a, a contradiction!

(b) Next assume that z_1 and z_2 refer to the same occurrence of the same word z. As $Q = \mathsf{succ}_{z_1,z_2,(i_2,j_2),(i_1,j_1)}(Q)$, the output set is empty. The resulting word in $\mathsf{Res}_{(i_1,i_2)}(z'_2, a)$ is different from z. Thus

$$\mathsf{Res}_{(i_2,j_2),(i_1,j_1)}(z_1, z_2) \neq Q,$$

again a contradiction!

This completes the proof.   □

Theorem 2. *Let $\mathcal{P}$ be a peptide computer, and let Q be a configuration of $\mathcal{P}$. If $Q \in \mathsf{succ}(Q)$, then Q is stable and, hence, $\mathsf{succ}(Q) = \{Q\}$.*

Proof. In every basic reaction, an antibody a attaches to an epitope where either nothing was attached or antibodies with lower affinity where attached. Thus, if Q is a successor of itself, Q does not contain any critical constellations. Hence, Q is stable and, by Theorem 1, Q is its only successor.   □

On the basis of Theorem 2 one is inclined to conjecture that every configuration in a set $\mathcal{Q}$ of configurations is stable if and only if $\mathsf{succ}(\mathcal{Q}) = \mathcal{Q}$. Assuming this, one would have a simple test for when all configurations in such a set $\mathcal{Q}$ are stable. Proving this claim turned out to be unexpectedly difficult. We state and prove this claim further below as Theorem 5.

Let $\mathcal{P}$ be a peptide computer, and let Q be a configuration of $\mathcal{P}$. Let $\mathsf{stable}(Q)$ be the set of all stable configurations which are reachable from Q.

Lemma 2. *Let $\mathcal{P}$ be a peptide computer and let Q be a configuration of $\mathcal{P}$. The following statements hold true:*

1. *If $Q' \in \mathsf{succ}(Q)$ then $\mathsf{stable}(Q') \subseteq \mathsf{stable}(Q)$.*
2. *$\mathsf{stable}(Q) \subseteq \mathsf{stable}(\mathsf{base}(Q))$.*

Proof. The first statement is true because reachability is a transitive relation. The second statement follows from Lemma 1 and the fact that Q is reachable from $\mathsf{base}(Q)$.   □

Our next goal is to derive a constructive description of the set $\mathsf{stable}(B)$ of all stable peptide configurations having the same base B. If Q is any configuration with base B, then, by Lemma 2, all stable configurations, which are reachable from Q, are contained in that set. The construction reveals a layered structure of the stable configurations. The existence of these layers is then used to prove that the sequence of reactions originating at any configuration Q will ultimately stabilize. This is stated formally in Theorem 4 below. Therefore, we describe a construction of the set $\mathsf{stable}(\mathsf{base}(Q))$ of all stable configurations Q' satisfying $\mathsf{base}(Q') = \mathsf{base}(Q)$. To do so we need some auxiliary notions and notation as follows:

Let epitopes (Q) and antibodies (Q) be the multisets of occurrences of epitopes and of antibodies in Q, respectively. Here it is important that one distinguishes different instances of the same element and does not just count its multiplicity. Thus we assume the existence of appropriate index sets M and N for epitopes (Q) and antibodies (Q), respectively. Then e_μ with $\mu \in M$ is a specific instance of the epitope $e \in E$ occurring in epitopes (Q), and a_ν with $\nu \in N$ is a specific instance of the antibody $a \in A$ occurring in antibodies (Q). In the sequel we do not mention the index sets M and N explicitly.

Consider the multiset

$$\mathfrak{B} = \text{epitopes} \, (Q) \times_m \text{antibodies} \, (Q)$$

of pairs (e_μ, a_ν). Intuitively, such a pair denotes a potential attachment (binding) of a_ν to e_μ. The multiset $\mathfrak{B}$ represents all potential attachments between epitopes and antibodies using the elements of base (Q). Attachments may exclude each other. This is explained below.

Each pair in $\mathfrak{B}$ has an affinity value $\beta(e_\mu, a_\nu) = \beta(e, a)$. As Q is finite, β has only finitely different values on $\mathfrak{B}$. Let these be

$$0 \le \beta_0 < \beta_1 < \cdots < \beta_k.$$

For $h = 0, 1, \ldots, k$, let B_h be the submultiset of those pairs $(e_\mu, a_\nu) \in_m \mathfrak{B}$ which have the affinity value β_h. The family $\{B_h \mid h = 0, 1, \ldots, k\}$ forms a partition of $\mathfrak{B}$.

Consider two pairs $p_1 = (e_{\mu_1}, a_{\nu_1})$ and $p_2 = (e_{\mu_2}, a_{\nu_2})$ in $\mathfrak{B}$. These *pairs are in conflict* in the following two situations:

1. e_{μ_1} and e_{μ_2} are distinct (as instances, $\mu_1 \ne \mu_2$) and in conflict;
2. e_{μ_1} and e_{μ_2} are the same (as instances, $\mu_1 = \mu_2$), but a_{ν_1} and a_{ν_2} are different (as instances, $\nu_1 \ne \nu_2$).

A submultiset of $\mathfrak{B}$ is conflict-free, if it contains no pair in conflict. Consider a conflict-free multiset $B \subseteq_m \mathfrak{B}$ and an arbitrary multiset $\mathfrak{B}' \subseteq_m \mathfrak{B}$. We define ConflictFree $(B, \mathfrak{B}')$ to be the set of maximal submultisets B' of $\mathfrak{B}'$ such that, for each of these multisets B', the multiset $B \cup_m B'$ is conflict-free.

A conflict-free submultiset B of $\mathfrak{B}$ defines a unique configuration config (B) as follows: Starting from base (Q), for each pair $(e_\mu, a_\nu) \in_m B$, the antibody a_ν is attached to the epitope e_μ.

We now start constructing stable configurations with base (Q) as their basis.

Construction 1. *For $h = k, k-1, \ldots, 1$, we build sets $\mathfrak{B}_{h-1}(C_h)$ of multisets C_{h-1}. From these we define conflict-free submultisets $\mathfrak{C}_{h-1}(C_{h-1})$ of $\mathfrak{B}$ to be used in the next step.*

1. *One starts with $\mathcal{B}_k$ defined as the set of maximal conflict-free multisets contained in B_k. For $C_k \in \mathcal{B}_k$, let $\mathfrak{C}_k(C_k) = \text{ConflictFree} \, (C_k, \mathfrak{B} \setminus_m B_k)$.*
2. *Now let $\mathcal{B}_{k-1}(C_k)$ be the set of multisets of the form $C_k \cup_m B'$ where B' is a maximal submultiset of $\mathfrak{C}_k(C_k) \cap_m B_{k-1}$. For $C_{k-1} \in \mathcal{B}_{k-1}(C_k)$, let*

$$\mathfrak{C}_{k-1}(C_{k-1}) = \text{ConflictFree} \, (C_{k-1}, \mathfrak{B} \setminus_m (B_k \cup_m B_{k-1})) \, .$$

3. *Continuing this, for $h > 0$, we obtain $\mathcal{B}_{h-1}(C_h)$ as the set of multisets of the form $C_h \cup_m B'$ where B' is a maximal submultiset of $\mathfrak{C}_h(C_h) \cap_m B_{h-1}$. For $C_{h-1} \in \mathcal{B}_{h-1}(C_h)$, let*

$$\mathfrak{C}_{h-1}(C_{h-1}) = \mathsf{ConflictFree}\left(C_{h-1}, \mathfrak{B} \setminus_m \bigcup_m^{h-1 \leq i \leq k} B_i \right).$$

4. *When $h = 1$, stop the construction. The result of the construction is the set union $\mathfrak{B}_0$ of all the sets $\mathcal{B}_0(C_1)$ of multisets obtained in this way.*

Theorem 3. *Let $\mathcal{P}$ be a peptide computer, and let Q be a configuration of $\mathcal{P}$. Let $\mathfrak{B}_0(Q)$ be the result of Construction 1. Then*

$$\{\mathsf{config}\,(B) \mid B \in \mathfrak{B}_0(Q)\} = \mathsf{stable}\,(\mathsf{base}\,(Q))\,.$$

Proof. By construction, each configuration $\mathsf{config}\,(B)$ with $B \in \mathfrak{B}_0(Q)$ is stable and has the same base as Q. Moreover, each configuration without critical constellation is found by the construction. □

Theorem 4. *Let $\mathcal{P}$ be a peptide computer, and let $\mathcal{Q}$ be a finite set of configurations of $\mathcal{P}$. There is an $n \in \mathbb{N}_0$ such that every configuration in $\mathsf{succ}^n\,(\mathcal{Q})$ is stable.*

Proof. Suppose, the statement is true for every $Q \in \mathcal{Q}$, that is, there is an $n_Q \in \mathbb{N}_0$ such that every configuration in $\mathsf{succ}^{n_Q}\,(Q)$ is stable. Then it is true for $\mathcal{Q}$ with $n = \max\{n_Q \mid Q \in \mathcal{Q}\}$. Therefore, it is sufficient to prove the statement for the case of $|\mathcal{Q}| = 1$. Let $\mathcal{Q} = \{Q\}$.

In each basic reaction a specific instance of an antibody improves its situation: It was either not attached and now attaches to an epitope; or it was attached, but changes to an epitope with greater affinity. It can only be set free again if another antibody attaches to a conflicting site with an even greater affinity. Eventually, each antibody reaches a situation which cannot be improved and, if is attached, from which it cannot be removed again. The former happens, because there will be no epitopes to which it can attach with a greater affinity; the latter happens, because there will be no antibodies which could attach to a conflicting site with a greater affinity. Globally, each basic reaction improves the situation for at least one instance of an antibody. That antibody may, of course, become unattached again later, but only because an instance of a different antibody replaced it through greater affinity. As Q is finite, after finitely many steps, all epitopes and antibodies will have reached a stable situation as in Construction 1. □

The basic reactions occurring in a configuration can be considered as steps of a hill-climbing algorithm. The reachable stable configurations correspond to the local maxima in the hill-climbing setting. Because of the non-deterministic choice of the basic reactions, all local maxima are potentially reached. It is possible that our proof could be replaced by one using results on matching automata and the Stable Marriage Problem [24–26, 32, 33, 37]. Our present direct approach reveals an essential part of the rôle of non-determinism.

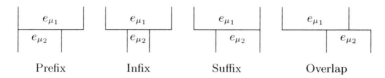

Fig. 4. Four cases of conflicts for pairs $p_1 = (e_{\mu_1}, a_{\nu_1})$ and $p_2 = (e_{\mu_2}, a_{\nu_2})$ in $\mathfrak{B}$. Further cases arise from exchanging the rôles of e_{μ_1} and e_{μ_2}.

Corollary 1. *Let Q be a finite set of configurations, and let $n \in \mathbb{N}_0$ be such that all configurations in $\mathsf{succ}^n(Q)$ are stable. Then $\mathsf{succ}(\mathsf{succ}^n(Q)) = \mathsf{succ}^n(Q)$.*

In view of Corollary 1, we define $\mathsf{succ}^*(Q) = \lim_{n \to \infty} \mathsf{succ}^n(Q)$. By Theorem 4, the limit is reached at some finite n. The value of n can be very large. However, in many cases, it may be sufficient to stop at an earlier situation $\mathsf{succ}^m(Q)$ with $m < n$, but $\mathsf{stable}(Q) \subseteq \mathsf{succ}^m(Q)$. The operation $\mathsf{succ}(Q)$ describes non-deterministic *micro-steps* of the peptide computer $\mathcal{P}$. The operation $\mathsf{succ}^*(Q)$ describes its *steps*. The duration of such steps defines the *internal time* of the peptide computer when considered as a system: it functions without an external clock, and without synchronization. Time or the "clock" advances when the reactions have stopped.

As mentioned above, one can also prove the converse of Corollary 1. This leads to a simple criterion to determine whether a set Q of configurations consists only of stable configurations.

Theorem 5. *Let $\mathcal{P}$ be a peptide computer, and let Q be a finite set of configurations of $\mathcal{P}$. Every configuration in Q is stable if and only if $\mathsf{succ}(Q) = Q$.*

Proof. If every configuration in Q is stable then $\mathsf{succ}(Q) = Q$ by Corollary 1. For the converse statement, observe that $\mathsf{succ}(Q) = Q$ implies $\mathsf{succ}^k(Q) = Q$ for all $k \in \mathbb{N}$. By Theorem 4, $\mathsf{succ}^*(Q) = Q$ and all configurations in Q are stable. □

The following (rather artificial) small example illustrates how non-determinism arises in unessential and essential ways:

Example 2. The components of the peptide computer $\mathcal{P}$ are as follows:

- $X = \{x, y, z\}$.
- $A = \{a, b\}$.
- $E = \{x, y, z, xy, yz\}$.
- β is given by the following table:

$e \in E$	$\beta(e, \cdot)$	
	a	b
x	0	1
y	0	1
z	0	1
xy	2	0
yz	2	0

The relation α is inferred from the definition of β.

Consider the configuration Q with $\mathsf{mult}_Q\, xyz = 1$, $\mathsf{mult}_Q\, a = 1$ and $\mathsf{mult}_Q\, b = 2$. The successor of Q, $\mathsf{succ}\,(Q)$, consists of the following multisets:

$$\mathsf{succ}\,(Q) = \Big\{ \{(xy, a)z, b, b\},\ \{x(yz, a), b, b\},$$
$$\{(x, b)yz, a, b\},\ \{x(y, b)z, a, b\},\ \{xy(z, b), a, b\} \Big\}.$$

Continuing the computation of successors, one obtains:

$$\mathsf{succ}^2\,(Q) = \Big\{ \{(xy, a)(z, b), b\},\ \{(x, b)(yz, a), b\},\ \{(xy, a)z, b, b\},$$
$$\{(x, b)(y, b)z, a\},\ \{(x, b)y(z, b), a\},$$
$$\{x(y, b)(z, b), a\},\ \{x(yz, a), b, b\} \Big\}$$

$$\mathsf{succ}^3\,(Q) = \Big\{ \{(xy, a)(z, b), b\},\ \{(x, b)(yz, a), b\},$$
$$\{(xy, a)z, b, b\},\ \{(x(yz, a), b, b\} \Big\}$$

and

$$\mathsf{succ}^4\,(Q) = \Big\{ \{(xy, a)(z, b), b\},\ \{(x, b)(yz, a), b\} \Big\} = \mathsf{succ}^*\,(Q).$$

Non-determinism in the succesors of Q results from the fact that the order of basic reactions is non-deterministic. The non-determinism in $\mathsf{succ}^*\,(Q)$ is inherent to the process. It is a consequence of the fact that certain reactions may be irreversible.

To define peptide computations, we need the notions of *state*, *instruction*, and *program* for peptide computers.

Definition 8. *Let $\mathcal{P}$ be a peptide computer.*

1. *Let $\Gamma(\mathcal{P})$ be the set of stable configurations of $\mathcal{P}$. The set $\Gamma(\mathcal{P})$ is called the set of states of $\mathcal{P}$.*
2. *A (peptide) instruction has the form $+Q$ or $-Q$, where Q is a peptide configuration of $\mathcal{P}$. If Q' is a peptide configuration and I is a peptide instruction, then*
$$I(Q') = \begin{cases} Q' \cup_m Q, & \text{if } I = +Q, \\ Q' \setminus_m Q, & \text{if } I = -Q, \end{cases}$$
is the result of I applied to Q'.
3. *Let Q' be peptide configuration. An instruction $-Q$ is called a flushing instruction for Q', if Q contains only antibodies and, for all $a \in A$, $\mathsf{mult}_Q\, a \geq \mathsf{mult}_{Q'}\, a$.*
4. *A peptide program is a pair $(\mathfrak{P}, \chi)$ where $\mathfrak{P}$ is a mapping from Γ^* into the set of peptide instructions and χ is a function $\chi : \Gamma \to \mathbb{B}$.*

In the sequel we write Γ instead of $\Gamma(\mathcal{P})$. The set Γ is infinite. Only stable configurations are considered as states. Instructions add or remove elements from configurations. If $-Q$ is a flushing instruction for Q', then $-Q(Q')$ results in the removal of all the symbols in A, which are not bound to any sequence in X^+, from the configuration Q'. In a peptide program $(\mathfrak{P}, \chi)$, the mapping $\mathfrak{P}$ defines the set of potential instructions to be used for the next step, taking into account the past sequence of states; the mapping χ determines a halting condition. Details are provided in the next definition.

Definition 9. *Let $\mathcal{P}$ be a peptide computer and let $(\mathfrak{P}, \chi)$ be a peptide program for $\mathcal{P}$. A peptide computation according to $(\mathfrak{P}, \chi)$ is a word $c = Q_0 Q_1 \cdots Q_t \in \Gamma^*$ with $Q_0, Q_1, \ldots, Q_t \in \Gamma$ and $Q_0 = \emptyset$ such that*

$$Q_i \in \mathsf{succ}^* \left(\left(\mathfrak{P}(Q_0 Q_1 \cdots Q_{i-1})(Q_{i-1}) \right) \right)$$

for $i = 1, 2, \ldots, t$.

A computation as above starts with the empty multiset Q_0 as the initial state and ends when $\chi(Q_i) = 1$ for the first time. Then Q_i is a final state.

Our definition of a peptide program incorporates non-determinism and takes into account the full computational history (similar to recursion over all previous values). The computation starts with the configuration $Q_0 = \emptyset$ using the instruction $I_0 = \mathfrak{P}(\lambda)$. Basic reactions lead to the next state $Q_1 \in \mathsf{succ}^* (I_0(\emptyset))$. Note that nothing happens, if I_0 has the form $-Q$ for some configuration Q. On the other hand, if $I_0 = +Q$, then $Q_1 \in \mathsf{succ}^* (Q)$. Now suppose, that the state sequence so far is given by the word $Q_0 Q_1 \cdots Q_{i-1} \in \Gamma^*$. The next instruction to be used is $I_i = \mathfrak{P}(Q_0 Q_1 \cdots Q_{i-1})$, and it is applied to Q_{i-1}. By applying this instruction, a configuration $Q'_i = I_i(Q_{i-1})$ is obtained, which need not be stable. Reactions will result in a stable configuration $Q_i \in \mathsf{succ}^* (Q'_i)$. If $\chi(Q_i) = 1$, the computation stops. Otherwise the computation continues with the next instruction.

Our model is non-deterministic. Natural processes seem to be inherently non-deterministic unless one makes a special effort to force determinism. On the other hand, when observed, natural processes often *seem* to be deterministic. This discrepancy could result from our inability to observe or model all the relevant parameters. For the mathematical model we accept the non-determinism. We identify and discuss the sources of non-determinism further below.

No model of computation is usable unless the interface with the environment is specified. This issue is often ignored. In the classical models, like the Turing machine, it is somehow, but not explicitly, and certainly not adequately built into the model. In nature-based models it is usually hidden in a pre-processing and a post-processing phase, which often absorb the actual computational complexity of the problem at hand. We pinpoint this issue without providing a solution by defining input and output decodings, that is, by clarifying how a peptide computer could be considered as a Moore (or Mealy) automaton.

To encode inputs we need a mapping γ from inputs to peptide programs, an input encoding; we also need an output decoding, that is, a mapping δ from Γ to outputs. It is an open question how such encodings and decodings would be implemented. This is an unsolved problem – not just for peptide computing, but for many of the nature-inspired computing models.

Definition 10. *A binary relation f from inputs to outputs is* peptide computable *if there is a peptide computer $\mathcal{P}$, a computable input encoding γ of inputs
into the set of peptide programs for $\mathcal{P}$ and a computable decoding δ of Γ into
outputs such that the following condition is satisfied for every $x \in \text{dom } f$:*
    *Let $(\mathfrak{P}_x, \chi_x) = \gamma(x)$. One has $y \in f(x)$ if and only if there is a peptide compu-
tation $Q_0 Q_1 \cdots Q_t$ according to $(\mathfrak{P}_x, \chi_x)$ such that $\chi(Q_t) = 1$ and $\delta(Q_t) = y$.*

We have not imposed any restrictions on how the functions γ and δ are defined.
As mentioned before, the issue of how define the interfaces γ and δ is usually not
treated rigorously in computability and complexity theory. While this may not
always be obvious (obscured by intuition), this is an inherent problem. By Definition 10, we relativize the issue to the choice of the input encoding and the output
decoding.

Using the input encoding γ, the input is considered as a program. This idea
permits the computer $\mathcal{P}$ to access the input in many different ways:

1. as a whole with a single "input" instruction $+Q$;
2. as a sequence of such "input" instructions;
3. as a sequence of inputs in which certain inputs are re-visited;
4. as a sequence of inputs with intermediate calculations.

In the extreme, $\gamma(x)$ could be the same program for all $x \in \text{dom } f$. Then the
same program would be used for all x. This would imply that $f(x) = f(x')$ for
all $x, x' \in \text{dom } f$.

By definition, a partial function f is peptide computable if it is peptide computable as a relation: For every $x \in \text{dom } f$, every computation according to $\gamma(x)$
results in the same output, that is, $f(x)$.

4 Solving Problems by Peptide Computers

In this section we briefly review and discuss known techniques for solving problems with peptide computers.

We start with an example adapted from a proposal due to Hug and Schuler [30]:
Given two sets S_1 and S_2, determine whether $|S_1| < |S_2|$.

Example 3. First we define the components of the peptide computer $\mathcal{P}$. Let
$X = \{x, y, z\}$, $E = \{x, y, z, xy, yz\}$, $A = \{a_1, a_2, b, c_1, c_2, d\}$ and β be defined
according to following table:

$e \in E$	$\beta(e, a)$ for $a = \ldots$					
	a_1	a_2	b	c_1	c_2	d
x	0	0	0	3	0	0
y	0	0	0	0	2	1
z	0	0	4	0	0	0
xy	1	0	0	0	0	0
yz	0	3	0	0	0	0

The relation α can be derived from this table. Consider the multisets C_1 and C_2 of the symbols c_1 and c_2, respectively, satisfying

$$\mathsf{mult}_{C_1}\, c_1 = |S_1| \text{ and } \mathsf{mult}_{C_2}\, c_2 = |S_2|.$$

Let $(\mathfrak{P}, \chi)$ be the encoding $\gamma(S_1, S_2)$. Let R be a multiset consisting of a very large number of occurrences of the word xyz. Let A_1, A_2, B and D be at least equally large multisets consisting of occurrences of only the symbols a_1, a_2, b and d, respectively.

The instructions in $\mathfrak{P}$ are as follows, writing P_i instead of $\mathfrak{P}(Q_0 Q_1 \cdots Q_i)$:

1. $\mathfrak{P}(\lambda)(\emptyset) = +R$. This results in $Q_0 = \mathsf{succ}^*(R) = R$.
2. $P_0(Q_0) = +A_2$. The symbols a_2 attach to the epitopes yz, resulting in Q_1.
3. $P_1(Q_1) = +C_1$. The symbols c_1 attach to the epitopes x, resulting in Q_2. Depending on the relative sizes of R and C_1, there may be unattached occurrences of c_1 or free epitopes x in Q_2.
4. $P_2(Q_2) = +B$. As $\beta(yz, a_2) < \beta(z, b)$, the symbols b attach to the epitopes z. The symbols a_2 are unattached. The resulting state is Q_3.
5. $P_3(Q_3) = +A_1$. The resulting state is Q_4.
6. $P_4(Q_4) = +C_2$. The resulting state Q_5 consists of the following objects:
   (a) Words xyz with antibodies attached:
      i. an antibody of type b is attached every epitope of type z;
      ii. as many antibodies of type c_1 are attached to as many epitopes of type x as possible;
      iii. as many antibodies of type c_2 are attached to as many epitopes of type y as possible;
      iv. as $\beta(x, c_1) > \beta(xy, a_1)$ and $\beta(y, c_2) > \beta(xy, a_1)$, antibodies of type a_1 are attached to all epitopes of type xy, to which no c_1 is attached at x or no c_2 is attached at y.
   (b) Unattached antibodies of type c_1, if $|S_1| > |R|$.
   (c) All antibodies of type a_2, unattached.
   (d) Unattached antibodies of type b, if $|B| > |R|$.
   (e) Unattached antibodies of type c_2, if $|S_2| > |R|$.
   (f) Unattached antibodies of type a_1, if $|A_1| > |R| - \max\{|S_1|, |S_2|\}$.
7. $P_5(Q_5) = +D$. In the resulting state Q_6, as $\beta(y, d) \not> \beta(xy, a_1)$ and $\beta(y, c_2) > \beta(y, d)$, antibodies of type d are attached to all epitopes of type y, where nothing is attached to the corresponding instance of xy.

Let $\chi(Q_i) = 0$ for $i < 6$ and let $\chi(Q_6) = 1$. The output decoding δ of Q_6 is defined by the following case distinction:

1. If there are no unattached antibodies of type c_2 and some attached antibodies of type d, then $\delta(Q_6) = $ "$|S_1| > |S_2|$".
2. If there are no unattached antibodies of type c_2 and no attached antibodies of type d, then $\delta(Q_6) = $ "$|S_1| = |S_2|$".
3. If there are unattached antibodies of type c_2, but not of type c_1, then $\delta(Q_6) = $ "$|S_1| < |S_2|$". Again R should have been larger.

4. If there are unattached antibodies of both, types c_1 and c_2, then $\delta(Q_6)$ is undefined, as R was not large enough.

Our presentation of Example 3 differs from the original in [30] in two ways: We have expressed it within our formal framework, thus treating it as a purely mathematical problem, and we have left out certain chemical features which are not required at this level of abstraction. Four cases are needed to define the output decoding $\delta(Q_6)$. This is a consequence of the fact that the multiset R may be too small. If $|R| > \max\{|S_1|, |S_2|\}$, then the third and fourth cases in the definition of δ cannot arise. The first and second cases can then be rephrased as follows:

1′. If there are some attached antibodies of type d, then $\delta(Q_6) = $ "$|S_1| > |S_2|$".
2′. If there are no attached antibodies of type d, then $\delta(Q_6) = $ "$|S_1| \leq |S_2|$".

These are the only cases needed to define δ, if R is large enough. In [30] this assumption is made implicitly.

The fact that the size of R influences the outcome of the computation is not surprising. To compute with inputs of unbounded size, one needs unbounded resources if the inputs are taken into account completely. A simple analogy with deterministic finite-state automata can illustrate the point:

Example 4. The task again is to compare the size of the sets S_1 and S_2. We consider two different methods to achieve this:

1. Encode the input in the form $c_1^{|S_1|} c_2^{|S_2|}$ and check whether the input word is in the language
$$L_> = \{c_1^n c_2^m \mid n, m \in \mathbb{N}_0, n > m\}.$$
   This encoding simulates the encoding by the multisets C_1 and C_2 in Example 3. As the language $L_>$ is not regular, a finite automaton will make mistakes on inputs which are too large. For instance, using a threshold counter[8] with threshold r, one can detect the four cases defining the output decoding in Example 3.

2. Encode the input over the alphabet $\{(c_1, c_2), (d, c_2), (c_1, d)\}$ of pairs in the form
$$\begin{cases} (c_1, c_2)^s (c_1, d)^{s_1 - s} & \text{if } s_1 > s_2, \\ (c_1, c_2)^s (d, c_2)^{s_2 - s} & \text{if } s_1 \leq s_2, \end{cases}$$
   where $s_1 = |S_1|$, $s_2 = |S_2|$ and $s = \min\{s_1, s_2\}$. Up to some details regarding the rôle of the symbol d and ignoring the size of R, these words represent the state Q_6 in Example 3. To determine whether $s_1 > s_2$, one checks whether the encoded input is in the language
$$L_>' = \{(c_1, c_2)^n (c_1, d)^m \mid n \in \mathbb{N}_0, m \in \mathbb{N}\}.$$
   This language is regular. It is accepted by a 2-state automaton. To make this work, the computation of the comparison has been shifted into the input encoding. The size of the input is not a relevant issue.

---

[8] A *threshold counter* is a finite automaton which can count up to a finite threshold and down again.

The examples establish the importance of specifying the input encoding and the output decoding and the relevance of the size (and nature) of the base of peptide configurations.

The computation in Example 3 is deterministic and takes 6 steps. The number of steps should not be confused with computation time. As in any model of computation, time costs accrue for several reasons:

1. Setting up the state Q_0.
2. Encoding the input.
3. Getting the next instruction.
4. Interpreting the instruction.
5. Executing the instruction.
6. Stabilizing the configuration.
7. Evaluating the state and deciding whether to stop.
8. Decoding the final state.

Only if the time cost for each of these items is uniformly bounded, can the time cost for a computation be measured in terms of the number of steps.

Continuing with Example 3, suppose one knows in advance that $|R| \geq r$ for some given large $r \in \mathbb{N}$ will be sufficient. One needs to prepare the multisets R, A_1, A_2, B and D of sizes no less than r, and also the multisets C_1 and C_2. We ignore the time to create Q_0. The input encoding could just result in terms of symbolic instructions, or it could involve creating the multisets involved. In the former case, the time cost for the encoding is a function depending on the number i of inputs and the program length[9] l; hence it is (roughly) in $O(i+l)$. In the latter case as many copies of the multisets as are used in the program need to be created; assuming that each multiset takes $O(g(r))$ time to create for some positive function[10] $g : \mathbb{N} \to \mathbb{R}$, then the time for the encoding is in $O((i+r)g(r))$.

In the former case, the cost for creating the multisets used by the program arises during the input encoding and is not part of the actual computation cost. In the latter case, the multisets are created during the computation. The cost of creating them is part of the computation cost.

The time for a configuration to stabilize depends strongly on the configuration, not just its size, as reactions cannot always occur in parallel. Similarly, the time needed to evaluate χ on the current configuration may depend on that configuration. In the example, the time for χ is a small constant. Decoding the final configuration may involve many complicated steps.

In summary, the concrete time cost is not sufficiently represented by the number of steps. In this respect, peptide computers are no better, but also no worse, than other computing models inspired by natural phenomena.

The paper by Hug and Schuler [30] contains informal programs for two further problems: Estimating the multiplicity of an element in a multiset; solving the satisfiability problem.

---

[9] The program length may depend on inputs, not just their number.

[10] The function $\log r$ could be a candidate for g; however, this depends on the nature of the multisets and the tools to create them.

In his thesis [3], Balan provides informal programs for the following problems: Hamiltonian path; exact cover by 3-sets; simulation of Turing machines; logical gates; simulation of Boolean circuits; addition and subtraction of binary numbers. Details and extensions of these results are also available as follows:

- Simulation of Turing machines and universality of peptide computers: [7, 8, 12]
- Logical gates and Boolean circuits: [2, 10]
- Arithmetic: [11]

A survey of these and related results as of 2006 is presented in [9]. Proposals for programming peptide computers are discussed in [5].

The applicability of the model of peptide computing relies strongly on the availability of enough types of antibodies and epitopes and of enough copies of these. Mathematically the former concerns the sizes of the components X, A and E of a peptide computer. The latter concerns the size and contents of configurations. In Example 3 above, only the latter is an issue. In several of the other simulations mentioned above, the sizes of at least the components A and E also depend on the problem at hand. For problems with bounded space requirements on a Turing machine, the sizes of A and E can be determined from the instance of the respective problem. For the simulation of arbitrary Turing machines there is no a priori bound on the sizes of A and E; this is a consequence of the fact that there is no a priori space bound for Turing machines. In [13] a peptide-computing-based method is proposed for creating the sets A and E incrementally as needed. This approach not only improves the universality result for peptide computing, but also leads to more economical solutions to the Hamilton-path and the exact-cover-by-3-sets problems.

One of the basic ideas of peptide computing is based on binding antibodies to epitopes and on inhibiting such bindings. In Example 3 this phenomenon appears on several occasions:

- Antibodies of type a_2 block attachments to epitopes y, z and yz.
- Antibodies of type b unblock such attachments, but block attachments to z.
- Antibodies of type a_1 block attachments to epitopes x, y and xy.

Several abstractions of the idea of using binding and blocking or permitting and inhibiting, based on naturally observed computation, lead to different models: As one extreme, on the theoretical end, one finds reaction systems [23]. Somewhere in the middle, one finds, for example, binding-blocking automata [3, 6]. At the other extreme end (ignoring all chemical details) one finds the present paper, where we attempt to model a rather detailed structure of the processes.

5   Non-determinism

For this section we rely on the article [4] in which non-determinism in peptide computers is classified. We review those results briefly. We postpone a more detailed discussion of non-determinism to a later paper.

Non-determinism in peptide computations arises at two levels:

1. At the micro-step level, reactions occur in an inherently non-deterministic way. Starting from an unstable configuration, successive steps may lead to further unstable configurations or to stable ones. A step is complete, when all successor configurations are stable. However, non-deterministically, the computation could continue much earlier as soon as one stable configuration has been reached even if other possible configurations are still unstable. In addition to the non-determinism regarding the sequence of reactions by which a stable configuration is reached, there is also non-determinism concerning the time it takes to reach a stable configuration and concerning the time it takes to perform a step.

2. At the step level, a configuration may lead to more than one stable configuration. At this level, non-determinism of peptide computers is similar to that of usual non-deterministic automata.

At the micro-step level, non-determinism can express massive parallelism, which could speed up computations. At the step level, non-determinism might not lead to a speed-up, but afford a natural model of non-deterministic computation.

Non-determinism of both kinds is found in Example 2. At the step level, $\mathsf{succ}^4(Q) = \mathsf{succ}^*(Q)$ consists of two stable configurations. At the micro-step level, already in $\mathsf{succ}^2(Q)$ does one find the two stable configurations, but also many unstable ones. If one knew that these are the only reachable stable configurations, then one could continue from these already two micro-steps earlier.

In the sequel we focus on the sources of non-determinism at the micro-step level. Consider a peptide computer $\mathcal{P}$ and an unstable configuration Q of $\mathcal{P}$. When reviewing the details in Section 3, one notices the following instance of non-determinism:

> According to Definition 7, a critical constellation of one of the forms (z, a, i, j) or $(z_1, z_2, i_2, j_2, i_1, j_1)$ is chosen non-deterministically. Here a, z, z_1 and z_2 are instances of symbols or words appearing in Q, not just symbols or words.

For the first type of critical constellations, within $w = \mathsf{strip}(z)$, the choice of $d \in \partial_E(w)$ and $\mathsf{dom}\,\tau$ is non-deterministic. For the second type of critical constellations a similar statement holds true.

In [4], three types of non-determinism are distinguished, and their sources are identified:

1. *Global non-determinism* is inherent non-determinism at the step level. This type of non-determinism can only arise, if there is an epitope to which two different antibodies can bind with the same affinity or if there are overlapping epitopes to which antibodies can bind with the same affinity, within the given configuration.

2. *Locally-global non-determinism* is non-determinism at the step level with respect to a specific configuration. According to Construction 1 this can also only arise, if there is an instance of an epitope to which two different antibodies can bind with the same affinity or if there are overlapping instances

of epitopes to which antibodies can bind with the same affinity, within the given configuration.
3. *Local non-determinism* is non-determinism at the micro-step level as explained above.

We leave the technical details to a later paper. In [4], some criteria are presented, by which the types of non-determinism of a peptide computer can be determined (Theorems 6–8) within the framework of that article.

6   Concluding Remarks

Peptide computers were first proposed by Hug and Schuler [30] in 2001. In subsequent work a semi-formal framework was used to demonstrate the theoretical feasibility, power and limitations of computing based on peptide-antibody interactions [2–13]. In [4] a formal model without all details was used to analyse the types of non-determinism occurring in peptide-based computations. In this paper we present this formal model in full detail. Micro-steps represent basic reactions, many of which may happen at unspecified moments in time and, thus, in non-deterministic order. We prove that reactions ultimately result in stable configurations. This leads to a well-defined notion of peptide computation with a rigorous formal foundation. Moreover, the sources of phenomena observed at the computational level can often be traced back to properties of the micro-step level. This issue, one may call it controllability, is likely to influence the design of programs using peptide-antibody interactions.

Our immediate plans include expanding the analysis of non-determinism, which was initiated in [4], describing the peptide programs for concrete problems in completely formal terms and relating the models of peptide computing and reaction systems.

Acknowledgments. This research was supported by the Natural Sciences and Engineering Council of Canada. We thank an anonymous referee for several very helpful suggestions.

References

1. Adamatzky, A.: Physarum Machines, Computers from Slime Mold, Nonlinear Science, Series A, vol. 74. World Scientific, Singapore (2010)
2. Balan, M.S., Krithivasan, K.: Modeling Boolean circuits using peptide-antibody interactions. In: Chandra, P., Rathish Kumar, B.V. (eds.) Mathematical Biology, pp. 187–193. Anamaya Publishers, New Delhi (2006), copublished by Anshan Ltd., Tunbridge Wells, UK
3. Balan, M.S.: Computational Models Using Peptide-Antibody Interactions. PhD Thesis, Indian Institute of Technology, Madras (2004)
4. Balan, M.S.: Non-determinism in peptide computer. In: Vaszil, G. (ed.) Proceedings of the International Workshop, Automata for Cellular and Molecular Computing, Budapest, Hungary, August 31, pp. 108–119. MTA, SZTAKI, Budapest (2007)

5. Balan, M.S.: A study on automation in peptide computing. In: Abraham, A., Carvalho, A., Herrera, F., Pai, V., Coelho, A., Menezes, R. (eds.) Proceedings of 2009 World Congress on Nature & and Biologically Inspired Computing, NABIC 2009, Coimbatore, India, December 9–11, pp. 128–133. IEEE (2009)
6. Balan, M.S.: Properties of binding-blocking automata: A study. In: Abdullah, R., Khader, A.T., Venkat, I., Wong, L.P., Subramanian, K.G. (eds.) Proceedings, Sixth International Conference on Bio-Inspired Computing: Theories and Applications, BIC-TA 2011, Penang, Malaysia, September 27–29, pp. 211–215. IEEE Computer Society, Los Alamitos (2011)
7. Balan, M.S., Jürgensen, H.: Peptide Computing – Universality and Theoretical Model. In: Calude, C.S., Dinneen, M.J., Păun, G., Rozenberg, G., Stepney, S. (eds.) UC 2006. LNCS, vol. 4135, pp. 57–71. Springer, Heidelberg (2006)
8. Balan, M.S., Jürgensen, H.: On the universality of peptide computing. Nat. Comput. 7, 71–94 (2008)
9. Balan, M.S., Jürgensen, H., Krithivasan, K.: Peptide computing: A survey. In: Krithivasan, K., Rama, R. (eds.) Formal Language Aspects of Natural Computing, Proceedings of Research Level Discussion Meeting on Natural Computing Held at Indian Institute of Technology Madras, Chennai. Ramanujan Mathematical Society Lecture Notes Series, vol. 3, pp. 63–76. Ramanujan Mathematical Society, India (2006)
10. Balan, M.S., Krithivasan, K.: Realizing Switching Functions Using Peptide-Antibody Interactions. In: Jonoska, N., Păun, G., Rozenberg, G. (eds.) Aspects of Molecular Computing. LNCS, vol. 2950, pp. 353–360. Springer, Heidelberg (2003)
11. Balan, M.S., Krithivasan, K.: Parallel computation of simple arithmetic using peptide-antibody interactions. BioSystems 76, 303–307 (2004)
12. Balan, M.S., Krithivasan, K., Sivasubramanyam, Y.: Peptide Computing - Universality and Complexity. In: Jonoska, N., Seeman, N.C. (eds.) DNA 2001. LNCS, vol. 2340, pp. 290–299. Springer, Heidelberg (2002)
13. Sakthi Balan, M., Seshan, P.: Incremental Building in Peptide Computing to Solve Hamiltonian Path Problem. In: Dediu, A.-H., Fernau, H., Martín-Vide, C. (eds.) LATA 2010. LNCS, vol. 6031, pp. 549–560. Springer, Heidelberg (2010)
14. Bartha, M., Krész, M.: Soliton circuits and network-based automata: Review and perspectives. In: Martín-Vide, C. (ed.) Scientific Applications of Language Methods, Mathematics, Computing, Language, and Life: Frontiers in Mathematical Linguistics and Language Theory, vol. 2, pp. 585–631. Imperial College Press, London (2011)
15. Blizard, W.D.: Multiset theory. Notre Dame J. Formal Logic 30, 36–66 (1989)
16. Calude, C.S., Păun, G.: Computing with Cells and Atoms: An Introduction to Quantum, DNA and Membrane Computing. Taylor & Francis, London (2001)
17. Cantor, C.R., Schimmel, P.R.: Biophysical Chemistry, vol. 3. W. H. Freeman, San Francisco (1980)
18. Czeizler, E., Czeizler, E.: On the power of parallel communicating Watson-Crick automata systems. Theoret. Comput. Sci. 358, 142–147 (2006)
19. Czeizler, E., Czeizler, E.: Parallel communicating Watson-Crick automata systems. Acta Cybernet 17, 685–700 (2006)
20. Czeizler, E., Czeizler, E.: A short survey on Watson-Crick automata. Bull. EATCS 88, 104–119 (2006)
21. Dasgupta, D., Niño, L.F.: Immunological Computation: Theory and Applications. CRC Press, Boca Raton (2009)

22. Dassow, J., Jürgensen, H.: Soliton automata. J. Comput. System Sci. 40, 158–181 (1990)
23. Ehrenfeucht, A., Rozenberg, G.: Reaction systems. Fund. Inform. 75, 263–280 (2007)
24. Gale, D., Shapley, L.S.: College admissions and the stability of marriage. Am. Math. Monthly 69, 9–15 (1962)
25. Gusfield, D., Irving, R.W.: The Stable Marriage Problem: Structure and Algorithms. MIT Press, Cambridge (1989)
26. Harada, K., Ishida, Y.: Antibody-based computing. Artif. Life Robotics 13, 180–183 (2008)
27. Head, T.: Biomolecular realizations of a parallel architecture for solving combinatorial problems. New Generation Computing 19, 301–312 (2001)
28. Heyting, A. (ed.): Constructivity in Mathematics, Proceedings of the Colloquium Held at Amsterdam 1957. Studies in Logic and the Foundations of Mathematics. North-Holland, Amsterdam (1959)
29. Hoeberechts, M.: On the Foundations of Computability Theory. PhD Thesis, The University of Western Ontario (2009)
30. Hug, H., Schuler, R.: Strategies for the development of a peptide computer. Bioinformatics 17, 364–368 (2001)
31. Ishida, Y.: Immunity-Based Systems: A Design Perspective. Springer, Berlin (2004)
32. Ishida, Y., Hayashi, T.: Asymmetric Phenomena of Segregation and Integration in Biological Systems: A Matching Automaton. In: Velásquez, J.D., Ríos, S.A., Howlett, R.J., Jain, L.C. (eds.) KES 2009, Part II. LNCS (LNAI), vol. 5712, pp. 789–796. Springer, Heidelberg (2009)
33. Ishida, Y., Sasaki, K.: Asymmetric Structure between Two Sets of Adaptive Agents: An Approach Using a Matching Automaton. In: König, A., Dengel, A., Hinkelmann, K., Kise, K., Howlett, R.J., Jain, L.C. (eds.) KES 2011, Part IV. LNCS (LNAI), vol. 6884, pp. 357–365. Springer, Heidelberg (2011)
34. Jürgensen, H.: Multisets, bags, families and crowds (2011) (manuscript)
35. Jürgensen, H., Konstantinidis, S.: Codes. In: Rozenberg, G., Salomaa, A. (eds.) Handbook of Formal Languages, vol. 1, pp. 511–607. Springer, Berlin (1997)
36. Kalmár, L.: An argument against the plausibility of Church's thesis. In: Heyting (ed.) [28], pp. 72–80
37. Knuth, D.E.: Mariages stables et leurs relations avec d'autres problèmes combinatoires: Introduction à l'analyse mathématique des algorithmes. Collection de la Chaire Aisenstadt, Les Presses de l'Université de Montréal, Montréal (1976), English translation: Stable Marriage and Its Relation to Other Combinatorial Problems: An Introduction to the Mathematical Analysis of Algorithms. Translated by M. Goldstein. CRM Proceedings & Lecture Notes, vol. 10. American Mathematical Society, Providence, Rhode Island (1997)
38. Păun, G., Rozenberg, G., Salomaa, A.: DNA Computing: New Computing Paradigms. Springer, Berlin (1998)
39. Păun, G.: Membrane Computing. Springer, Berlin (2002)
40. Péter, R.: Rekursivität und Konstruktivität. In: Heyting (ed.) [28], pp. 69–71
41. Rozenberg, G.: Computer science, informatics, and natural computing – personal reflections. In: Cooper, S.B., Löwe, B., Sorbi, A. (eds.) New Computational Paradigms. Changing Comceptions of What is Computable, pp. 373–379. Springer, New York (2008)
42. Surhone, L.M., Tennoe, M.T., Henssonow, S.F. (eds.): Peptide Computing. Betascript Publishing (August 2010), see footnote 1
43. Tarakanov, A.O., Skormin, V.A., Sokolova, S.P.: Immunocomputing, Principles and Applications. Springer, New York (2003)

On the Power of Randomness versus Advice in Online Computation*

Hans-Joachim Böckenhauer[1], Juraj Hromkovič[1], Dennis Komm[1],
Richard Královič[1], and Peter Rossmanith[2],**

[1] Department of Computer Science, ETH Zurich, Switzerland
{hjb,juraj.hromkovic,dennis.komm,richard.kralovic}@inf.ethz.ch
[2] Department of Computer Science, RWTH Aachen, Germany
rossmani@cs.rwth-aachen.de

Abstract. The recently introduced model of *advice complexity* of online problems tries to achieve a fine-grained analysis of the hardness of
online problems by asking how many bits of advice about the still unknown parts of the input an oracle has to provide for an online algorithm
to guarantee a specific competitive ratio. Until now, only deterministic
online algorithms with advice were considered in the literature. In this
paper, we consider, for the first time, online algorithms having access
to both random bits and advice bits. For this, we introduce the online
problem (n, k)-BOXES: Given a number of n closed boxes, an adversary
hides $k < \sqrt{n}$ items, each of unit value, within k consecutive boxes. The
goal is to open exactly k boxes and gain as many items as possible.

In the classical online setting without advice, we show that, if $k(k + 1) \leq n$, any deterministic algorithm is not competitive, because the adversary can ensure that not a single item is found. However, random-
ization drastically increases the gain in expectation. More precisely, we
prove that the expected gain is in the order of k^3/n and show that this
bound is tight up to some constant factor. A crucial result of our analysis
is the proof of the existence of two thresholds for the amount of random
bits used for solving (n, k)-BOXES. If the amount of random bits is below
the first threshold, randomization does not help at all. If, on the other
hand, the amount of randomness is above the second threshold of about
$\log n - 2 \log k$ random bits, then any additional random bit does not help
to improve the gain.

As our main result, we analyze the advice complexity of the boxes
problem both for deterministic and randomized online algorithms and
give a tight trade-off between the number of random bits and advice bits
needed for achieving a specific competitive ratio.

1 Introduction

In algorithmics, we seek for algorithms that produce high-quality output for
some specific problem within some given time or space bounds. In many real-
world applications, however, we are facing an additional hurdle when working in

---

* This work was partially supported by SNF grant 200021-141089.
** Part of this work was done while this author was staying at ETH Zurich.

H. Bordihn, M. Kutrib, and B. Truthe (Eds.): Dassow Festschrift 2012, LNCS 7300, pp. 30–43, 2012.

so-called *online environments*; here, the input arrives piecewise in consecutive discrete time steps while parts of the output have to be produced before the whole input is known (for an introduction and comprehensive discussion, we refer to the standard literature, e. g., [3, 7, 8, 10]). Formally, we are dealing with the following class of problems.

Definition 1. *A maximization online problem consists of a set $\mathcal{I}$ of inputs and a cost function. Every input $I \in \mathcal{I}$ is a sequence of requests $I = (x_1, \ldots, x_n)$. Furthermore, a set of feasible outputs (or solutions) is associated with every I; every output is a sequence of answers $O = (y_1, \ldots, y_n)$. The cost function assigns a positive real value $\mathrm{cost}(I, O)$ to every input I and any feasible output O. If the input is clear from the context, we omit I and denote the cost of O as $\mathrm{cost}(O)$. For every input I, we call any output O that is feasible for I and has highest possible cost an optimal solution of I, denoted by $\mathrm{Opt}(I)$.*

The established tool for measuring the performance of an online algorithm is the *competitive analysis* [3, 12] where one compares the cost of the solution computed by the online algorithms to the cost of an optimal offline solution computed by an algorithm knowing the whole input in advance.

Definition 2. *Consider an input $I = (x_1, \ldots, x_n)$ of a maximization online problem. An online algorithm A computes the output sequence $\mathrm{A}(I) = (y_1, \ldots, y_n)$ such that y_i is computed from $x_1, \ldots, x_i$. We denote the costs of the computed output by $\mathrm{cost}(\mathrm{A}(I))$.*

The online algorithm A is c-competitive if there exists a non-negative constant α such that, for every n and for each I of length at most n, $\mathrm{cost}(\mathrm{A}(I)) \geq 1/c \cdot \mathrm{cost}(\mathrm{Opt}(I)) - \alpha$. If $\alpha = 0$, A is called strictly c-competitive; A is called optimal if it is strictly 1-competitive.

Due to the definition of the problem we study in what follows, we only consider *strict* competitiveness. However, in many cases, competitive analysis does not seem very realistic because, by the nature of many real-world online scenarios, optimal results can never be reached. We want to get a better understanding of what online algorithms really lack. For a more fine-grained analysis of how much knowledge about the future parts of the input an online algorithm needs to compute a high-quality solution, we consider the *information content* of the given online problem as it was defined in [9]. This model can be viewed as a cooperation of an online algorithm A and an *oracle* O that has unlimited computational power, sees the whole input in advance, and writes binary information about it onto an *advice tape* before A reads any part of the input. Afterwards, A can access the bits from the advice tape in a sequential manner, just as a randomized algorithm would use its random tape. The *advice complexity* of the online problem is then defined as the minimum number of advice bits needed for achieving a good solution. The following definition formalizes this concept.

Definition 3. *Consider an input I of a maximization online problem. An online algorithm A with advice computes the output sequence $\mathrm{A}^\phi(I) = (y_1, \ldots, y_n)$ such* *that y_i is computed from $\phi, x_1, \ldots, x_i$, where ϕ is the content of the advice tape, i. e., an infinite binary sequence. We denote the costs of the computed output by* $\mathrm{cost}(\mathtt{A}^{\phi}(I))$. *$\mathtt{A}$ has advice complexity $s(n)$ if at most the first $s(n)$ bits of ϕ have been accessed during the computation of* $\mathtt{A}^{\phi}(I)$. *The c-competitiveness can be defined analogously to Definition 2.*

A first model of online algorithms with advice was introduced in [4] (see also [5]) and later refined in [2, 6]. For a survey of the concepts of advice complexity and the *information content* of online problems, see [9]. This model was applied to several classical online problems such as paging, job shop scheduling, the k-server problem, or metrical task systems in [1, 2, 5, 6, 11].

Another powerful tool for increasing the output quality of online algorithms is to allow randomized computation. One can easily extend Definition 2 for randomized algorithms as well. For the first time, we consider algorithms using both randomization *and* advice, and we give tight bounds on the trade-off between the amount of randomness and advice needed for achieving a specific competitive ratio.

We are now ready to introduce the online problem, denoted by (n, k)-BOXES, we deal with in the following.

Definition 4 ((n, k)-Boxes). *There are n boxes $b_1, \ldots, b_n$ standing in a row and we know that all are empty except for $k < \sqrt{n}$ boxes that are standing next to each other and which contain some expensive item each. An online algorithm $\mathtt{A}$ is allowed to open exactly k boxes of its choice aiming at opening as many full ones as possible. After $\mathtt{A}$ has opened k boxes, the (remaining) positions of the non-empty boxes are revealed and $\mathtt{A}$'s gain is the number of non-empty boxes it has opened.*

Note that the optimal solution of (n, k)-BOXES has always gain k. We call the position of the first full box the *starting position* and we note that, for any instance of size n, there are exactly $n - k + 1$ possible starting positions. Another way to look at the analysis of online problems is to view it as a game between the online algorithm $\mathtt{A}$ against an oblivious adversary $\mathtt{Adv}$. The adversary tries to hide the full boxes in such a way that $\mathtt{A}$'s gain is (in expectation and/or for any advice string) as small as possible.

At first, we can make the following straightforward observation about any deterministic online algorithm without advice for (n, k)-BOXES.

Theorem 1. *If $n \geq k(k + 1)$, $\mathtt{Adv}$ can ensure that no deterministic algorithm $\mathtt{A}$ has any gain.*

Proof. $\mathtt{Adv}$ knows $\mathtt{A}$'s deterministic strategy; however, for every box at position i that $\mathtt{A}$ opens, the number of starting positions to hide the k items is decreased by at most k. The removed positions are those between i to $i - k + 1$ where $i \geq k$ (see Fig. 1); not all of them exist if $i < k$. Hence, if $\mathtt{A}$ chooses boxes such that these intervals are disjoint, $\mathtt{Adv}$ is prevented to take at most k^2 starting positions. Accordingly, if $n \geq k^2 + k$, $\mathtt{Adv}$ may hide the items in such a way that $\mathtt{A}$ is not able to find any item at all.   □

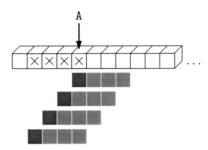

Fig. 1. A inspecting box b_5 and the "forbidden" starting positions for Adv

Thus, any deterministic algorithm is not competitive. In the next section, we show the power of randomization for (n, k)-BOXES.

2   Randomization

Since we are interested in values for k and n that are in the above order and the situation seems desperate for any deterministic online algorithm, we consider randomized algorithms, i. e., algorithms that use a source of randomness to make their decisions. Which box is chosen next can depend on which boxes that are already open are full and on the randomness that is available to the algorithm. Usually, the source of randomness is just a stream of random bits. It is then possible to measure the amount of randomness as the number of random bits used as a function of the input size. However, the input of the (n, k)-BOXES problem has always constant size, since n and k are fixed. As we are interested (without loss of generality) in randomized algorithms with a fixed upper bound on the running time, for any randomized algorithm R solving (n, k)-BOXES, there is an upper bound r on the number of random bits used by R.

Measuring the amount of randomness as the number of random bits is, however, rather coarse; in the sequel, we aim for a more fine-grained analysis. If R is allowed to use r random bits, then, equivalently, it can use a random number between 1 and 2^r instead. We generalize this model by considering algorithms that get one random number drawn uniformly from the set $\{1, \ldots, M\}$ where M is not restricted to be a power of two. Let $\mathcal{R}(M)$ denote the set of randomized algorithms with randomness restricted to the uniformly random choice of a number from $\{1, \ldots, M\}$.

If we fix an algorithm R, then the random variable $B_s(\mathtt{R})$ denotes the number of full boxes opened by R for starting position s. Then $\mathbb{E}[B_s(\mathtt{R})]$ is the expected number of full boxes opened if Adv chooses the starting position equal to s, and $\min_s \mathbb{E}[B_s(\mathtt{R})]$ is the worst-case expected gain of R. If R is clear from the context, we abbreviate $B_s(\mathtt{R})$ by B_s.

The following theorem gives a worst-case lower bound on the performance of any randomized algorithm $\mathtt{R} \in \mathcal{R}(M)$ solving (n, k)-BOXES with restricted randomness.

Theorem 2. *For any randomized online algorithm* R $\in \mathcal{R}(M)$ *for* (n, k)-BOXES, *there exists a starting position* s *(i. e., an input instance of* (n, k)-BOXES*), such that*

1. *if* $M < (n - k + 1)/k^2$, *then* $\mathbb{E}[B_s] = 0$,
2. *for any* M, $\mathbb{E}[B_s] \leq \frac{k^2(k+1)}{2(n-k+1)}$,
3. *if* $M = c(n - k + 1)/k^2$ *with* $c \in [1, 2k/(k + 3)]$, *then*

   (a) $\mathbb{E}[B_s] \leq \frac{2(k+\frac{3}{2})}{M} - \frac{2(n-k+1)}{M^2 k}$ *and also*

   (b) $\mathbb{E}[B_s] \leq \frac{2(c-1)}{c^2} \cdot \frac{k^3}{n-k+1} + \frac{3k^2}{c(n-k+1)}$.

Proof. R gets a random number from $\{1, \ldots, M\}$, so it can behave in M different ways and each behavior occurs with probability $1/M$. This is equivalent to choosing uniformly at random one deterministic algorithm from a set $\{A_1, \ldots, A_M\} = \text{Alg}(R)$.

Recall that the k full boxes start at some starting position, i. e., a position between 1 and $n - k + 1$.

1. Suppose that $M < (n - k + 1)/k^2$. Basically, we now extend the idea from the proof of Theorem 1. The number of starting positions that allow Adv to place the obstacles in a way that R does not find any item is $n - k + 1 - (Mk^2) > n - k + 1 - (n - k + 1) = 0$, which implies the claim.

2. At first, we can make a simple observation: Consider some specific (deterministic) algorithm $A_i \in \text{Alg}(R)$. There are at most k starting positions such that A_i opens a full box in the first step and gets at most k full boxes in total. Next, there are at most k starting positions such that A_i opens an empty box in the first step and a full box in the second step and gets at most $k - 1$ full boxes in total, and so on.

   Assume that the starting position is p. Let $\overline{O}_{i,p}$ denote the number of empty boxes opened by A_i until the first full box is found (or k if none) and let $O_{i,p} := k - \overline{O}_{i,p}$ and $O_p = O_{1,p} + \cdots + O_{M,p}$. Clearly, $O_{i,p}$ is an upper bound on the number of full boxes opened by A_i in total and O_p/M is an upper bound on the expected number of full boxes opened by R. Hence, taking s such that O_s is minimal, the expected number of full boxes opened by R for starting position s is

$$\mathbb{E}[B_s] = \frac{1}{M} O_s = \frac{1}{M} \min \{O_1, \ldots, O_{n-k+1}\}. \tag{1}$$

   From the observation above it follows that, for every A_i, we have

$$O_{i,1} + \cdots + O_{i,n-k+1} \leq k^2 + k(k - 1) + k(k - 2) + \cdots + k = \frac{k^2(k + 1)}{2}$$

and accordingly, using (1) and the fact that the minimum of a set of numbers can never be larger than its average, we get

$$\mathbb{E}[B_s] \leq \frac{1}{M} \cdot \frac{\sum_p O_p}{n-k+1} = \frac{1}{M(n-k+1)} \sum_p \sum_{i=1}^{M} O_{i,p}$$

$$= \frac{1}{M(n-k+1)} \sum_{i=1}^{M} \sum_p O_{i,p} = \frac{1}{M(n-k+1)} \sum_{i=1}^{M} \frac{k^2(k+1)}{2}$$

$$= \frac{1}{M} \cdot \frac{Mk^2(k+1)}{2(n-k+1)} = \frac{k^2(k+1)}{2(n-k+1)}. \tag{2}$$

3. Let us consider values of M such that $\frac{n-k+1}{k^2} \leq M \leq \frac{2(n-k+1)}{k(k+3)}$. Choose any value g such that $g \leq k/M$ and $g \geq \frac{k+1}{M} - \frac{n-k+1}{kM^2}$. We call a starting position p *big* if $O_{i,p} \geq gM$ for some i (i.e., there is a single algorithm A_i that makes a large contribution to O_p). Otherwise we call it *small*. Due to the assumptions on g, we have that $Mk(k - \lfloor gM \rfloor) \in [0, n-k+1]$.

For any A_i, p is made big by A_i if $O_{i,p}$ takes values between $\lceil gM \rceil$ and k, which leads to $k - \lceil gM \rceil + 1 \geq k - \lfloor gM \rfloor$ different values. For any such value x, there are at most k positions p such that $O_{i,p} = x$, and since there are M algorithms in total to consider, there are at most $Mk(k - \lfloor gM \rfloor)$ big positions and, accordingly, at least $n - k + 1 - Mk(k - \lfloor gM \rfloor) \geq 0$ small positions.

Let S be the set of all small positions. Since no algorithm A_j is allowed to contribute strictly more than $\lfloor gM \rfloor$ to these positions, we get, for any A_i,

$$\sum_{p \in S} O_{i,p} \leq k(1 + 2 + \cdots + \lfloor gM \rfloor) = \frac{k \lfloor gM \rfloor (\lfloor gM \rfloor + 1)}{2}$$

and consequently

$$\sum_{p \in S} O_p \leq \frac{Mk \lfloor gM \rfloor (\lfloor gM \rfloor + 1)}{2}.$$

In particular, there is some small position p with

$$O_p \leq \frac{Mk \lfloor gM \rfloor (\lfloor gM \rfloor + 1)}{2|S|} \leq \frac{Mk \lfloor gM \rfloor (\lfloor gM \rfloor + 1)}{2(n - k + 1 - Mk(k - \lfloor gM \rfloor))}$$

$$\leq \frac{MkgM(gM + 1)}{2(n - k + 1 - Mk(k - gM + 1))} \tag{3}$$

and, since $\mathbb{E}[B_p] = \frac{O_p}{M}$, there exists s such that

$$\mathbb{E}[B_s] \leq \frac{kgM(gM + 1)}{2(n - k + 1 - Mk(k - gM + 1))}. \tag{4}$$

We choose

$$g = \frac{2(k + \frac{3}{2})}{M} - \frac{2(n - k + 1)}{M^2 k}.$$

It is straightforward to verify that g satisfies our assumptions. Plugging this value of g into (4), we get $\mathbb{E}[B_s] \leq g$. Finally, by setting $M = c(n-k+1)/k^2$, we get

$$\mathbb{E}[B_s] \leq \frac{2(c-1)}{c^2} \cdot \frac{k^3}{n-k+1} + \frac{3k^2}{c(n-k+1)}.$$

This concludes the proof.   □

We now complement this lower bound by an upper bound that is tight up to a small constant factor.

Theorem 3. *Let M be an even number. There is a randomized algorithm* $\mathsf{R} \in \mathcal{R}(M)$ *for* (n,k)-Boxes *such that, for every starting position s,*

1. *if* $M \geq \frac{2n}{k(k-1)}$, *then* $\mathbb{E}[B_s] \geq \frac{8}{9}\frac{k^3}{2n} - 1$,
2. *if* $\frac{n-k+1}{k^2} < M \leq \frac{2n}{k(k-1)}$, *then* $\mathbb{E}[B_s] \geq \frac{2k-2}{M} - \frac{2n}{M^2 k}$.

Proof. As we mentioned, R is a probability distribution over $\mathrm{Alg}(\mathsf{R})$ where each deterministic algorithm $\mathsf{A} \in \mathrm{Alg}(\mathsf{R})$ gets chosen with the same probability $1/M$. Every algorithm A opens boxes within some interval of fixed length and performs a straightforward local search when a box is found, which enables A to find at least $k - i$ full boxes if a full box is discovered in step i. Moreover, consider an adversary Adv that tries to hide the boxes as good as possible from every algorithm in $\mathrm{Alg}(\mathsf{R})$ at once.

In the following, we focus on an equivalent problem to analyze the algorithm. We shrink the instance to an instance of size $\lfloor n/k \rfloor$, that is, we compress k boxes into one hyper-box, thereby neglecting the last $d = k\lfloor n/k \rfloor < k$ original boxes. There is exactly one non-empty hyper-box that has a value of $k-1$ in the beginning and whose value is decreasing by one with every unsuccessful opening of a hyper-box (except in the last step). The algorithm can open up to k hyper-boxes. When it opens the full hyper-box in the j-th trial, it achieves gain $k - j$ if $j < k$, and gain 1 if $j = k$.

Now we show that it is possible to reduce (n,k)-Boxes to its shrunk version. Indeed, assume that we have an algorithm A' for the shrunk version. We can construct an algorithm A for (n,k)-Boxes as follows. Whenever A' opens some hyper-box, A opens the last box corresponding to this hyper-box. As soon as A finds some full box, it continues with a local search.

Consider any input instance for A, which is specified by the starting position p. Then, A achieves at least the same gain as A' running on an instance where the hyper-box corresponding to p is full. Since p cannot be within the last $k - 1$ boxes, the hyper-box corresponding to p exists. If A' opens a full hyper-box, the starting position is within the distance k to the left of the box opened by A, therefore A opens a full box as well. If this happens in step j, the local search of A guarantees a gain of at least $k - j$ if $j < k$, and of 1 if $j = k$.

In the sequel, we provide a randomized algorithm R solving the shrunk version of (n,k)-Boxes, thereby proving that an equally good algorithm for the original problem exists. Consider some constant l such that $1 \leq l < 2k$. Starting with

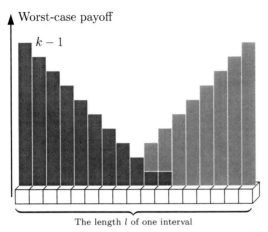

(a) An interval of size l and the gain of A and $\bar{A}$

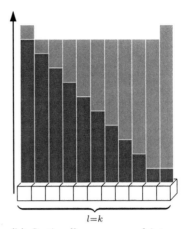

(b) Optimally compressed interval and the sums of the gains of A and $\bar{A}$

Fig. 2. The worst-case gain within one interval assigned to the $\lfloor n/k \rfloor$ hyper-boxes

some hyper-box q, a deterministic algorithm $A \in \text{Alg}(R)$ opens k consecutive hyper-boxes until k empty hyper-boxes are inspected or it finds a full hyper-box at the j-th trial, gaining $k - j$ if $j < k$, and 1 if $j = k$. Next, we define the symmetric algorithm to A denoted by $\bar{A} \in \text{Alg}(R)$: $\bar{A}$ initially opens hyper-box $q + l - 1$ and then continues to open $k - 1$ consecutive hyper-boxes left of $q + l - 1$ in reverse order until it arrives at $s + l - k$ or finds the item; A and $\bar{A}$ are called an *algorithm pair* because they work on the same interval.

Let us now bound the minimum of the total gain of the two algorithms within one interval of length l. Clearly, if $l = k$, the gain is at least $k - 1$ for every hyper-box, and, more general, if $l = k + i$ for $-k < i < k$, we get a gain of at least $k - 1 - i = 2k - l - 1$ (see Fig. 2). Starting at the first hyper-box, we assign every of the $M/2$ algorithm pairs to one interval of length l in a way such that, by allowing wrap-arounds, every hyper-box is covered by exactly c intervals.

It follows that

$$\mathbb{E}[B_s] \geq \frac{c(2k - l - 1)}{M}$$

if we can guarantee a number of c wrap-arounds (see Fig. 3); to do so, it clearly has to hold that

$$\frac{M}{2} \cdot l \geq \left\lfloor \frac{n}{k} \right\rfloor \cdot c$$

which can be guaranteed by satisfying

$$l \geq \frac{2cn}{Mk} \tag{5}$$

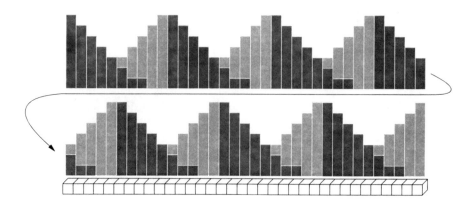

Fig. 3. The algorithm R with wrap-around $c = 2$

and we obviously aim at minimizing l while satisfying (5) which means that we may set $l := \lceil 2cn/Mk \rceil$, yielding

$$\mathbb{E}[B_s] \geq \frac{c\left(2k - \lceil \frac{2cn}{Mk} \rceil - 1\right)}{M} \geq \frac{c\left(2k - \frac{2cn}{Mk} - 2\right)}{M} = \frac{2ck - 2c}{M} - \frac{2c^2n}{M^2k}. \tag{6}$$

We now distinguish two cases according to the size of M.

1. Suppose $M \geq \frac{2n}{k(k-1)}$. Let

$$\delta := \frac{k^3}{2n} - \frac{k(16k - 8 + k^2)}{18n}.$$

In the following, we want to guarantee that

$$\frac{2ck - 2c}{M} - \frac{2c^2n}{M^2k} \geq \delta$$

and therefore

$$0 \geq k\delta M^2 - (2ck^2 - 2ck)M + 2c^2n \tag{7}$$

to prove the bound we claimed, that is, we have to show that there exists a number of wrap-arounds c for any $M \geq \frac{2n}{k(k-1)}$ such that (7) holds. A simple calculation gives that (7) is satisfied if and only if

$$\frac{3}{2} \cdot \frac{nc}{k(k-1)} \leq M \leq 3 \cdot \frac{nc}{k(k-1)}.$$

To show that we can cover all possible values of M for the right choice of c, note that, for $c = 1$, we have

$$\frac{3}{2} \cdot \frac{n}{k(k-1)} \leq 2 \cdot \frac{n}{k(k-1)} \leq M$$

and, for two consecutive values c' and c'' (that is, $c'' = c' + 1$), we have

$$\frac{3c'n}{k(k-1)} \geq \frac{3nc''}{2k(k-1)} = \frac{3n(c'+1)}{2k(k-1)} \iff 2 \geq \frac{c'+1}{c'}$$

which obviously holds for any $c' \geq 1$. From (6) and (7), we immediately conclude

$$\mathbb{E}[B_s] \geq \frac{2ck - 2c}{M} - \frac{2c^2 n}{M^2 k} \geq \frac{k^3}{2n} - \frac{k(16k - 8 + k^2)}{18n} \geq \left(1 - \frac{1}{9}\right)\frac{k^3}{2n} - 1.$$

2. Now suppose $\frac{n-k+1}{k^2} < M \leq \frac{2n}{k(k-1)}$. Here, we do not have enough randomness to do wrap-arounds. Thus, fixing $c := 1$ in (6), we immediately get

$$\mathbb{E}[B_s] \geq \frac{2k - l - 1}{M} \geq \frac{2k - 2}{M} - \frac{2n}{M^2 k}.$$

The claim follows immediately.   □

The intuitive idea behind the proof above is the following. With increasing M, as long as $M \leq \frac{2n}{k(k-1)}$, the gain of R increases, because we can choose from more and more deterministic strategies and shrink l. However, if M grows too much, the gain gets less and we start the wrap-around technique and thereby get a bound that does not depend on M anymore. If M increases over some threshold, another wrap-around is made and the intervals get decompressed a little. If M increases further, the intervals shrink until, for the some next threshold value, another wrap-around is made.

Up to this point, we restricted ourselves to even values for M. However, a simple observation resolves this issue.

Corollary 1. *For any M and any randomized algorithm $R \in \mathcal{R}(M)$, the bounds from Theorem 3 hold up to a multiplicative factor of $1 - 1/M$.*

Proof. Theorem 3 holds for any even M. Now if M is odd, R acts as above for any random choice from 1 to $M - 1$. In any of the cases, the expected gain X is the same as in Theorem 3. If M is chosen, R may open some arbitrary boxes. In this case, we assume that the gain is zero. We therefore get a total expected gain of at least

$$\left(1 - \frac{1}{M}\right) \cdot X + \frac{1}{M} \cdot 0.$$

□

Observe the two theorems above provide us with two sharp thresholds on the amount of randomness. If $M < (n+k-1)/k^2$, this small amount of randomness does not help at all. On the other hand, if $M > 2n/(k(k-1))$, which corresponds to roughly $\log n - 2\log k$ random bits, any further randomness does not help to improve the gain.

3   Randomized Algorithms with Advice

In this section, we present our main result by analyzing the trade-off between randomness and advice for (n, k)-Boxes. For this, we consider online algorithms that base their computation on both advice bits and randomness. In essence, we prove that, for the same amount of randomness, every additional advice bit allows us to find the same expected number of full boxes within a sequence of roughly double length. This implies that, for instance, the bound of $\frac{8}{9} \cdot \frac{k^3}{2n} - 1$ on the expected number of opened full boxes from the first claim of Theorem 3 can already be reached with a random number of roughly half the size.

To achieve this goal, we introduce the following notation. We denote by $F(n, k, M, b)$ the expected number of full boxes opened by the best algorithm that solves (n, k)-Boxes with randomness M and b bits of advice.

For $S \subseteq \{1, \ldots, n - k + 1\}$, we generalize (n, k)-Boxes to (S, n, k)-Boxes. In (S, n, k)-Boxes, the starting position Adv chooses for the full boxes, has to be chosen from the set S, otherwise it is identical to (n, k)-Boxes. An algorithm that solves (S, n, k)-Boxes is called *faithful* if it opens only boxes whose positions are in S until the first full box is encountered.

Lemma 1. *For every algorithm* A *that solves* (S, n, k)-Boxes *with randomness* M *and* b *advice bits, there exists a faithful algorithm* A' *that also solves* (S, n, k)-Boxes *with randomness* M *and* b *advice bits such that* $\mathbb{E}[B_s(\text{A}')] \geq \mathbb{E}[B_s(\text{A})]$, *for all* s.

Proof. Suppose we are given A as stated by the lemma. The following strategy is carried out by A' until it finds the first full box: If A opens box b_i, then A' opens b_j with $j = \max\{ s \in S \mid s \leq i \}$, i.e., the next smaller box that is in S, or it opens no box at all if this maximum does not exist. It is easy to see that A' opens its first full box not later than A: Assume A opens the first full box b_i, then $b_p, b_{p+1}, \ldots, b_i$ are all full, where p is the starting position of the sequence chosen by Adv. Of course, $p \in S$ and $p \leq j \leq i$. Afterwards, A' opens the same boxes as A, except for the case where A wants to open the box b_j which was already opened by A'. In this case, A' opens b_i.

In that way, A' opens at least the same number of full boxes as A.   □

Lemma 2. *Let* A *be an algorithm that solves* (S, n, k)-Boxes *with randomness* M *and no advice. Then*

$$\min_s \mathbb{E}[B_s(\text{A})] \leq F(|S| + k - 1, k, M, 0) + 1.$$

Proof. Due to Lemma 1, we can assume that A is faithful without loss of generality. We now convert A into an algorithm A' that solves $(|S| + k - 1, k)$-Boxes with randomness M and 0 advice bits such that

$$\min_s \mathbb{E}[B_s(\text{A}')] \geq \min_s \mathbb{E}[B_s(\text{A})] - 1.$$

Let $S = \{p_1, \ldots, p_{|S|}\}$ with $p_1 < p_2 < \cdots < p_{|S|}$. Let us further assume that A would open boxes at positions $p_{i_1}, \ldots, p_{i_k}$ if we would report them all as empty.

Then A' opens boxes at positions $i_1, \ldots, i_r$ until it finds the first full box b_{i_r}; afterwards A' continues with local search.

Let i be a worst-case starting position for A'. As we have already discussed, the last $k - 1$ boxes cannot be starting positions. We therefore have $i \leq |S| + k - 1 - (k - 1) = |S|$. We choose p_i as the starting position for A. If A' does not open a full box in the first r rounds, then neither does A, because A is faithful, and therefore only opens boxes from S. By construction, any full box it finds this way corresponds to a full box A' opens.   □

We are now ready to prove the main claims of this section.

Theorem 4. $F(n, k, M, b) \leq F(\lceil (n - k + 1)/2^b \rceil + k - 1, k, M, 0) + 1.$

Proof. If an algorithm A solves (n, k)-BOXES with b advice bits, then, by the pigeon-hole principle, there is at least one advice string that is used for at least $\lceil (n - k + 1)/2^b \rceil$ different starting positions. Let S be the set of those starting positions. Then the algorithm A can be used to solve (S, n, k)-BOXES without advice (but with randomness M). Hence, if A is optimal, then $F(n, k, M, b) \leq \min_s \mathbb{E}[B_s(A)] \leq F(|S| + k - 1, k, M, 0) + 1$ by Lemma 2. Clearly, $F(n, k, M, b)$ is anti-monotone in n (if k, M, and b are fixed, Adv obtains more positions to hide the boxes with growing n) and thus $F(|S| + k - 1, k, M, 0) \leq F(\lceil (n - k + 1)/2^b \rceil + k - 1, k, M, 0)$.   □

Theorem 5. $F(n, k, M, b) \geq F(\lceil (n - k + 1)/2^b \rceil + k - 1, k, M, 0).$

Proof. Consider an algorithm A for (n, k)-BOXES that uses b bits of advice. Again, there are $n - k + 1$ possible starting positions for Adv. On the other hand, we may subdivide these boxes into 2^b groups of size

$$\left\lceil \frac{n - k + 1}{2^b} \right\rceil,$$

and encode the position of the group that contains the starting position using b bits. Easily, we can extend this interval by $k - 1$ boxes (which cannot contain a starting position). Then, A simulates an optimal algorithm A' for an instance of this size. It directly follows that A gains at least as much as A'.   □

Combining the results from Section 2 with Theorems 4 and 5 immediately yields the following upper and lower bounds on the expected number of opened full boxes for randomized algorithms with advice.

Corollary 2

(a) If $M < \lceil \frac{n-k+1}{2^b} \rceil / k^2$, then $F(n, k, M, b) \leq 1$.
(b) For any M,

$$F(n, k, M, b) \leq \frac{k^2(k + 1)}{2\lceil \frac{n-k+1}{2^b} \rceil} + 1.$$

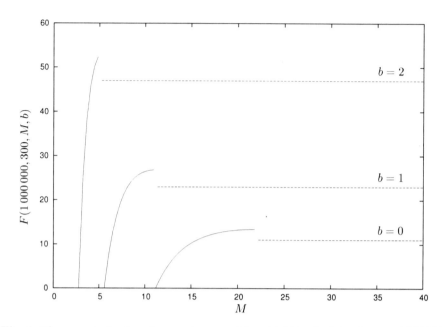

Fig. 4. The expected gain using randomness M and b bits of advice for $n = 1\,000\,000$, $k = 300$, different values of M and up to two advice bits

(c) If $M = c\lceil\frac{n-k+1}{2^b}\rceil/k^2$ with $c \in [1, 2k/(k+3)]$, then

$$F(n,k,M,b) \leq \frac{2(k+\frac{3}{2})}{M} - \frac{2\lceil\frac{n-k+1}{2^b}\rceil}{M^2k} + 1 \quad and$$

$$F(n,k,M,b) \leq \frac{2(c-1)}{c^2} \cdot \frac{k^3}{\lceil\frac{n-k+1}{2^b}\rceil} + \frac{3k^2}{c \cdot \lceil\frac{n-k+1}{2^b}\rceil} + 1.$$

Corollary 3. *Let M be an even number.*

(a) If $M \geq 2(\lceil\frac{n-k+1}{2^b}\rceil + k - 1)/(k(k-1))$, then

$$F(n,k,M,b) \geq \frac{8}{9} \cdot \frac{k^3}{2 \cdot \lceil\frac{n-k+1}{2^b}\rceil + k - 1} - 1.$$

(b) If $\lceil\frac{n-k+1}{2^b}\rceil \cdot \frac{1}{k^2} < M < 2(\lceil\frac{n-k+1}{2^b}\rceil + k - 1)/(k(k-1))$, then

$$F(n,k,M,b) \geq \frac{2k-2}{M} - \frac{2 \cdot \lceil\frac{n-k+1}{2^b}\rceil + 2k - 2}{M^2k}.$$

Note that these upper and lower bounds are almost tight. Obviously, Corollary 3 can easily be extended to the case of odd M using Corollary 1. A graphical illustration of the connection between the amount of randomness, the number of advice bits and the expected number of full boxes that are opened is given in Fig. 4.

4   Conclusion

In this paper, we have analyzed, for the first time, the trade-off between randomness and advice bits for online computation. We gave matching or almost matching upper and lower bounds on the competitive ratio for (n, k)-Boxes for any combination of randomness and advice bits. A goal for further research is to extend these results to a broader class of online problems.

References

1. Böckenhauer, H.-J., Komm, D., Královič, R., Královič, R.: On the Advice Complexity of the k-Server Problem. In: Aceto, L., Henzinger, M., Sgall, J. (eds.) ICALP 2011, Part I. LNCS, vol. 6755, pp. 207–218. Springer, Heidelberg (2011)
2. Böckenhauer, H.-J., Komm, D., Královič, R., Královič, R., Mömke, T.: On the Advice Complexity of Online Problems. In: Dong, Y., Du, D.-Z., Ibarra, O. (eds.) ISAAC 2009. LNCS, vol. 5878, pp. 331–340. Springer, Heidelberg (2009)
3. Borodin, A., El-Yaniv, R.: Online Computation and Competitive Analysis. Cambridge University Press (1998)
4. Dobrev, S., Královič, R., Pardubská, D.: How Much Information about the Future Is Needed? In: Geffert, V., Karhumäki, J., Bertoni, A., Preneel, B., Návrat, P., Bieliková, M. (eds.) SOFSEM 2008. LNCS, vol. 4910, pp. 247–258. Springer, Heidelberg (2008)
5. Dobrev, S., Královič, R., Pardubská, D.: Measuring the problem-relevant information in input. Theoretical Informatics and Applications (RAIRO) 43(3), 585–613 (2009)
6. Emek, Y., Fraigniaud, P., Korman, A., Rosén, A.: Online computation with advice. Theoretical Computer Science 412(24), 2642–2656 (2011)
7. Fiat, A., Woeginger, G.J. (eds.): Online Algorithms 1996. LNCS, vol. 1442. Springer, Heidelberg (1998)
8. Hromkovič, J.: Design and Analysis of Randomized Algorithms: Introduction to Design Paradigms. Springer (2005)
9. Hromkovič, J., Královič, R., Královič, R.: Information Complexity of Online Problems. In: Hliněný, P., Kučera, A. (eds.) MFCS 2010. LNCS, vol. 6281, pp. 24–36. Springer, Heidelberg (2010)
10. Irani, S., Karlin, A.R.: On online computation. In: Approximation Algorithms for NP-Hard Problems, ch. 13, pp. 521–564 (1997)
11. Komm, D., Královič, R.: Advice complexity and barely random algorithms. Theoretical Informatics and Applications (RAIRO) 45(2), 249–267 (2011)
12. Sleator, D.D., Tarjan, R.E.: Amortized efficiency of list update and paging rules. Communications of the ACM 28(2), 202–208 (1985)

Relevance of Entities in Reaction Systems

Andrzej Ehrenfeucht[1], Jetty Kleijn[2],
Maciej Koutny[3], and Grzegorz Rozenberg[1,2]

[1] Department of Computer Science, University of Colorado at Boulder,
430 UCB Boulder, CO 80309-0430, U.S.A.
andrzej@cs.colorado.edu
[2] LIACS, Leiden University, P.O. Box 9512, 2300 RA, The Netherlands
{kleijn,rozenber}@liacs.nl
[3] School of Computing Science, Newcastle University, NE1 7RU, UK
maciej.koutny@ncl.ac.uk

Abstract. Reaction systems are a model for the investigation of processes carried out by biochemical reactions in living cells. A reaction system consists of a set of reactions which transform a current system's state (a set of entities) into the successor state. In this paper we investigate which entities are actually relevant from the point of view of generating dynamic processes through such state transformations.

Keywords: reaction system, living cell, natural computing.

1 Introduction

The investigation of the computational nature of biochemical reactions is a research theme of Natural Computing. One of the goals of this research is to contribute to a computational understanding of the functioning of the living cell.

Reaction systems [1–7] are a formal framework for the investigation of processes carried out by biochemical reactions in living cells. The central idea of this framework is that the functioning of a living cell is based on interactions between (a large number of) individual reactions, and moreover these interactions are regulated by two main mechanisms: facilitation/acceleration and inhibition/retardation. These interactions determine the dynamic processes taking place in living cells, and reaction systems are an abstract model of these processes. This model is based on principles remarkably different from those underlying other *models of computation* in computer science. This is a consequence of the fact that on the one hand the model takes into account the basic bioenergetics of the living cell while on the other hand its (high) degree of abstraction allows it to be a qualitative rather than quantitative model.

In a nutshell, a reaction system consists of a finite set of reactions which can be applied to subsets (*states*) of a given set of entities, determining in this way the transformations of states. The specific question we address in this paper is which entities can be considered as relevant in the sense that state changes are "sensitive" to them.

H. Bordihn, M. Kutrib, and B. Truthe (Eds.): Dassow Festschrift 2012, LNCS 7300, pp. 44–55, 2012.

We provide a characterisation of relevant elements in terms of resources of reactions. In our considerations we use a specific "natural" notion of relevance, but we also discuss its relationship to other possible "natural" definitions of relevance.

The paper is organised in the following way. After setting up in Section 2 some mathematical notation used in the paper, we describe basic notions concerning reactions in Section 3, and basic notions concerning reaction systems in Section 4. In Section 5, we introduce the central notions of this paper: relevant/irrelevant sets and entities, and prove their basic properties. In Section 6, we demonstrate that for a reduced reaction system the set of relevant entities coincides with the resources used by the system's reactions. Then, in Section 7, we discuss two alternative formalisations of the notion of relevance. The last section contains a brief discussion of our results.

2   Preliminaries

Throughout the paper we use mostly standard mathematical notation. We use $X \div Y$ to denote the symmetric difference $(X \setminus Y) \cup (Y \setminus X)$ of two sets X and Y.

3   Reactions

In this section, we recall some key definitions concerning reactions and sets of reactions (see, e.g., [1, 5]).

Let Z be a finite nonempty set. A *reaction* over Z is a triplet of the form $a = (R, I, P)$, where $R, I, P \subseteq Z$ are nonempty sets such that $R \cap I = \varnothing$. The three component sets of reaction a are denoted by R_a, I_a and P_a, respectively, and called the *reactants*, *inhibitors* and *products* (of a). We denote by $rac(Z)$ the set of all possible reactions over Z.

Let $C \subseteq Z$. A reaction $a \in rac(Z)$ is *enabled* by C if $R_a \subseteq C$ and $I_a \cap C = \varnothing$. We denote this by $en_a(C)$. The *result* of a reaction $a \in rac(Z)$ on C is defined by

$$res_a(C) = \begin{cases} P_a \text{ if } a \text{ is enabled by } C \\ \varnothing \text{ otherwise .} \end{cases}$$

Moreover, the *result* of a set of reactions $B \subseteq rac(Z)$ on C, denoted by $res_B(C)$, is the union of the products of all the reactions from B, that is

$$res_B(C) = \bigcup_{b \in B} res_b(C) .$$

Note that $res_B(\varnothing) = \varnothing$ as the set of reactants of any reaction is nonempty and so no reaction is enabled by $C = \varnothing$. Also, $res_B(Z) = \varnothing$ as the set of inhibitors of any reaction is nonempty and so no reaction is enabled by Z.

Let $a, b \in rac(Z)$. Then b *covers* a if $res_b(C) = res_{\{a,b\}}(C)$, for all $C \subseteq Z$. We denote this by $b \geq a$; thus what a does (produces) is already covered (produced) by b. We also say that b *strictly covers* a if $b \geq a$ and $a \neq b$. Note that $\geq$ is a partial order.

As a matter of fact (see [5]), $b \geq a$ iff $R_b \subseteq R_a$, $I_b \subseteq I_a$ and $P_b \supseteq P_a$. Thus $b \geq a$ if b requires a subset of reactants of a and a subset of inhibitors of a but still produces at least all the products of a. Note that if $b \geq a$ then, for each $C \subseteq Z$, $en_a(C)$ implies $en_b(C)$.

4   Reaction Systems

A *reaction system* is a pair $\mathcal{A} = (S, A)$, where S is a finite nonempty *background* set comprising the *entities* of $\mathcal{A}$, and A is the set of reactions over S. To capture the *dynamic* behaviour of $\mathcal{A}$, we now describe all possible transitions between its states, where a *state* of $\mathcal{A}$ is any set C of its entities. Thus a reaction system with a background set S has exactly $2^{|S|}$ states.

Let $C \subseteq S$ be a state of a reaction system $\mathcal{A} = (S, A)$. Then $res_{\mathcal{A}}(C) = res_A(C)$ is the *result* of all the reactions of $\mathcal{A}$ enabled by C.

The state transformations captured by the above definition are deterministic. Thus, indeed, a reaction system $\mathcal{A} = (S, A)$ defines (specifies, implements) a function $res_{\mathcal{A}} : 2^S \rightarrow 2^S$, called the *result function* of $\mathcal{A}$. In the general model of reaction systems, processes of $\mathcal{A}$ are also influenced by the "environment" which reflects the fact that the living cell is an open system; it communicates and interacts with its environment. However, for the notions that we study in this paper it suffices to consider context-independent processes, i.e., processes determined by the system $\mathcal{A}$ only (without influence of its environment). In this way the successor state for a given state is determined solely by the result function $res_{\mathcal{A}}$.

Note that in this case, the successor $res_{\mathcal{A}}(C)$ of a current state C consists only of entities from the product sets of reactions of $\mathcal{A}$ enabled by C. This means that there is no *permanency* for entities $\mathcal{A}$: an entity from a current state will be present in (will carry over to) the successor state only if it is produced by at least one reaction enabled by the current state. This way of defining state transitions in reaction systems is motivated by the basic bioenergetics of the living cell, and it constitutes a fundamental difference with models of computations considered in computer science.

Since in this paper we are interested in state transitions in reaction systems, it is convenient to convey the subsequent discussion in terms of functions specified by reaction systems.

Proposition 1. *Let $\mathcal{A} = (S, A)$ be a reaction system. Then*

$$\bigcup_{C \in 2^S} res_{\mathcal{A}}(C) = \bigcup_{a \in A} P_a .$$

Proof. Follows from the fact that each reaction $a \in A$ is enabled by the state R_a. □

In other words, the entities occurring in the sets of the codomain of the result function of a reaction system are all the entities which occur in the products of the reactions of the system.

Let $\mathcal{A} = (S, A)$ be a reaction system and $b \in rac(S)$. Then b is *consistent* with $\mathcal{A}$ if $res_b(C) \subseteq res_{\mathcal{A}}(C)$, for all $C \subseteq S$; thus adding b to A yields a reaction system with the same result function.

A reaction system $\mathcal{A} = (S, A)$ is *reduced* if, for all $a \in A$,

(i) $res_{\mathcal{A}} \neq res_{A \setminus \{a\}}$.

(ii) there is no $b \in rac(S)$ which is consistent with $\mathcal{A}$ and strictly covers a.

Intuitively, (i) excludes reactions which do not add anything new to the results produced by other reactions in $\mathcal{A}$. As to the second condition, note that if b is consistent with $\mathcal{A}$ and b strictly covers a then b is (from the point of view of $\mathcal{A}$) a more 'efficient' version of a. Therefore, condition (ii) requires that all the reactions in $\mathcal{A}$ are in their most efficient version.

The two conditions in the definition of a reduced reaction system are independent. Consider, for example, the reaction system $\mathcal{A}_1 = (S, \{a, b\})$, where

$$S = \{1, 2\} \qquad a = (\{1\}, \{2\}, \{1\}) \qquad b = (\{1\}, \{2\}, \{2\}) \,.$$

Then both reactions are necessary to specify $res_{\mathcal{A}_1}$. On the other hand, a and b are covered by $c = (\{1\}, \{2\}, \{1, 2\})$ which is consistent with $res_{\mathcal{A}_1}$ and can be used to define a more efficient $\mathcal{A}_1' = (S, \{c\})$ specifying the same function as $\mathcal{A}_1$.

Conversely, let us consider the reaction system $\mathcal{A}_2 = (S, \{a, b, c\})$, where

$$S = \{1, 2, 3\} \quad a = (\{1, 2\}, \{3\}, \{1, 2\}) \quad b = (\{1\}, \{3\}, \{1\}) \quad c = (\{2\}, \{3\}, \{2\}) \,.$$

In this case, the first condition is not satisfied because reaction a is redundant (its enabledness implies enabledness of both b and c which together also produce $\{1, 2\}$). However, the second condition is satisfied as all reactions over $\{1, 2, 3\}$ strictly covering a or b or c are inconsistent with $res_{\mathcal{A}_2}$.

We close this section by demonstrating that considering only reduced reaction systems is not a restriction as far as result functions of reaction systems are concerned.

Theorem 1. *For every reaction system $\mathcal{A}$ there exists an* equivalent *reduced reaction system $\mathcal{A}'$, i.e., the two systems have the same background sets and the same result function.*

Proof. Let $\mathcal{A} = (S, A)$. Consider the set $con(\mathcal{A})$ of all the reactions from $rac(S)$ consistent with $\mathcal{A}$. Note that $(S, con(\mathcal{A}))$ is equivalent with $\mathcal{A}$ — as a matter of fact, it is the largest implementation of $res_{\mathcal{A}}$.

Let D be the set of all reactions in $con(\mathcal{A})$ which are $\geq$-maximal in $con(\mathcal{A})$.

Now we replace, in any order, each $a \in A$ which is not maximal in $con(\mathcal{A})$ by a reaction $b \in D$ such that $b \geq a$, Let A'' be the resulting set of reactions. Clearly, $\mathcal{A}'' = (S, A'')$ is equivalent with $\mathcal{A}$, and $\mathcal{A}''$ satisfies condition (ii) from the definition of a reduced system.

Next, in order to ensure that also (i) is satisfied, we inspect one by one all reactions, in any order, beginning with A'' and remove those reactions from the current set of reactions which can be removed without changing the result function. Let A' be the final outcome of this procedure. Clearly, $\mathcal{A}' = (S, A')$ still satisfies (ii), but it also satisfies (i). Thus $\mathcal{A}'$ is reduced, and moreover $\mathcal{A}'$ is equivalent to $\mathcal{A}$. Hence the theorem holds. □

5   Relevance in Reaction Systems

A central problem in the investigation of result functions of reaction systems is to understand when and why (for a given reaction system $\mathcal{A}$) $res_{\mathcal{A}}$ does not distinguish between two different states T and U, i.e., $res_{\mathcal{A}}(T) = res_{\mathcal{A}}(U)$. Intuitively, this means that the difference between T and U is irrelevant from the point of view of $res_{\mathcal{A}}$. In this paper, we define irrelevant sets of entities as the sets such that *whenever* two sets differ by an irrelevant set, then they will not be distinguishable by $res_{\mathcal{A}}$. Since the operation of symmetric difference is a mathematically natural way to define the difference between two sets, we use this operation in our definition of relevance. With this idea in mind, we say that:

- $X \subseteq S$ is *relevant* in $\mathcal{A}$ if

$$(\exists T, U \subseteq S) \ [T \div U = X \ \text{ and } res_{\mathcal{A}}(T) \neq res_{\mathcal{A}}(U)] \,. \tag{i}$$

- $x \in S$ is *relevant* in $\mathcal{A}$ if $\{x\}$ is relevant in $\mathcal{A}$, i.e.,

$$(\exists T \subseteq S) \ [\, res_{\mathcal{A}}(T \setminus \{x\}) \neq res_{\mathcal{A}}(T \cup \{x\})] \,. \tag{ii}$$

Intuitively, a set of entities X is *irrelevant* if any two sets of entities which 'differ' exactly by X are transformed to the same state, hence X is irrelevant from the $res_{\mathcal{A}}$ point of view. Thus, as expressed by (i), X is relevant if we can find two sets of entities which 'differ' exactly by X and for which $res_{\mathcal{A}}$ yields different results. What we are really interested in is whether *entities* are relevant or irrelevant, as expressed by part (ii) of the above definition. However, defining the relevance of sets through the relevance of their elements does not work, as shown in Section 6 (see the comments after Proposition 3). Thus we had to define (i) first.

Now, for a reaction system $\mathcal{A} = (S, A)$, we define:

- the *relevant domain* of $\mathcal{A}$ as $rdom(\mathcal{A}) = \{x \in S : x \text{ is relevant in } \mathcal{A}\}$.
- the *irrelevant domain* of $\mathcal{A}$ as $irdom(\mathcal{A}) = \{x \in S : x \text{ is irrelevant in } \mathcal{A}\}$.

Intuitively, $rdom(\mathcal{A})$ comprises those entities to which $res_{\mathcal{A}}$ is 'sensitive', and $irdom(\mathcal{A})$ those to which $res_{\mathcal{A}}$ is 'insensitive'.

It turns out that by combining irrelevant entities we never obtain a relevant set of entities. In other words, irrelevance is persistent, as shown next.

Proposition 2. *Let $\mathcal{A}$ be a reaction system. Then each $X \subseteq irdom(\mathcal{A})$ is irrelevant in $\mathcal{A}$.*

Proof. Let $\mathcal{A} = (S, A)$, and let X be a nonempty subset of $irdom(\mathcal{A})$. Let $T, U \subseteq S$ be such that $T \div U = X$. Let $T \setminus U = Y$ and $U \setminus T = Z$; thus $X = Y \cup Z$. Since $X \neq \emptyset$, at least one of Y, Z is nonempty.

Without loss of generality, assume that $Y \neq \varnothing$, thus $Y = \{y_1, y_2, \ldots, y_n\}$ for some $n \geq 1$. Let $T_0 = T$, $T_1 = T_0 \setminus \{y_1\}$, $T_2 = T_1 \setminus \{y_2\}$, $\ldots$, $T_n = T_{n-1} \setminus \{y_n\} = T \cap U$. Since, for each $i \in \{1, \ldots, n\}$, $y_i \in Y$ is irrelevant, we get

$$res_{\mathcal{A}}(T) = res_{\mathcal{A}}(T_0) = \ldots = res_{\mathcal{A}}(T_n) = res_{\mathcal{A}}(T \cap U) . \tag{$*$}$$

Similarly, one proves that

$$res_{\mathcal{A}}(T \cap U) = res_{\mathcal{A}}(U) . \tag{$**$}$$

It follows from $(*)$ and $(**)$ that

$$res_{\mathcal{A}}(T) = res_{\mathcal{A}}(T \cap U) = res_{\mathcal{A}}(U) .$$

This implies that, for all $T, U \subseteq S$ with $T \div U = X$, we have $res_{\mathcal{A}}(T) = res_{\mathcal{A}}(U)$. Therefore X is irrelevant. □

As a corollary of Proposition 2 we get the following property of the sets of reactants of reactions in a reaction system.

Lemma 1. *Let $\mathcal{A}$ be a reaction system. For each reaction $a \in A$, $R_a \nsubseteq irdom(\mathcal{A})$.*

Proof. Let $a \in A$. Assume to the contrary that $R_a \subseteq irdom(\mathcal{A})$. Then, by Proposition 2, R_a is irrelevant. Since $R_a \div \varnothing = R_a$ and $res_{\mathcal{A}}(\varnothing) = \varnothing$, this means that

$$res_{\mathcal{A}}(R_a) = \varnothing . \tag{$*$}$$

On the other hand, $en_a(R_a)$ and therefore

$$res_{\mathcal{A}}(R_a) = P_a . \tag{$**$}$$

But $(*)$ and $(**)$ imply that $P_a = \varnothing$, a contradiction with the definition of a reaction. Therefore $R_a \nsubseteq irdom(\mathcal{A})$. □

6 Characterising Relevant Domains

When it comes to sets of relevant entities, one should expect a relationship with resources used by the reaction system. Here by the *resources* of a single reaction a we mean $M_a = R_a \cup I_a$. The essence of the next result is that relevant entities must be resources.

Theorem 2. *Let $\mathcal{A} = (S, A)$ be a reaction system. Then*

$$rdom(\mathcal{A}) \subseteq \bigcup_{a \in A} M_a .$$

Proof. Let $x \in S$. If $x \notin \bigcup_{a \in A} M_a$, then it follows directly from the definition of $res_{\mathcal{A}}$ that, for each $T \subseteq S$, $res_{\mathcal{A}}(T \setminus \{x\}) = res_{\mathcal{A}}(T \cup \{x\})$. Hence x is irrelevant and so $x \notin rdom(\mathcal{A})$. □

The inclusion in the formulation of the above theorem can be replaced by equality in case of a reaction system with a single reaction.

Proposition 3. *Let* $\mathcal{A} = (S, \{a\})$ *be a reaction system. Then*

$$rdom(\mathcal{A}) = M_a \ .$$

Moreover, every nonempty set $X \subseteq R_a \cup I_a$ *is relevant.*

Proof. To show the second part of the statement of the theorem, let $X \subseteq R_a \cup I_a$ be such that $X \neq \varnothing$. Let $X' = X \cap R_a$ and $X'' = X \cap I_a$. To observe that X is relevant it then suffices (see (i) in Section 5) to take $T = R_a$ and $U = (R_a \setminus X') \cup X''$. We have then $T \div U = X$, but $res_a(T) \neq res_a(U)$. Hence all resources are relevant, and so from Theorem 2 it follows immediately that $rdom(\mathcal{A}) = M_a$.   □

Thus we also obtained a counterpart of Proposition 2 for sets of relevant entities in case of a system with a single reaction. However, any attempt to extend this to reaction systems with more reactions is bound to fail, as illustrated by the following example. Consider the reaction system $\mathcal{A}_3 = (S, \{a, b\})$, where

$$S = \{1, 2\} \qquad a = (\{1\}, \{2\}, \{1\}) \qquad b = (\{2\}, \{1\}, \{1\}) \ .$$

Then 1 is relevant because $\{1, 2\} \div \{2\} = \{1\}$ and $res_{\mathcal{A}_3}(\{1, 2\}) = \varnothing \neq \{1\} = res_{\mathcal{A}_3}(\{2\})$, and 2 is relevant because $\{1, 2\} \div \{1\} = \{2\}$ and $res_{\mathcal{A}_3}(\{1, 2\}) = \varnothing \neq \{1\} = res_{\mathcal{A}_3}(\{1\})$. However, $X = \{1, 2\}$ is not a relevant set of entities which is seen as follows. If $T, U \subseteq S$ are such that $T \div U = X$, then either $\{T, U\} = \{\{1\}, \{2\}\}$ or $\{T, U\} = \{\varnothing, S\}$. In the former case we obtain $res_{\mathcal{A}_3}(T) = \{1\} = res_{\mathcal{A}_3}(U)$, and in the latter $res_{\mathcal{A}_3}(T) = \varnothing = res_{\mathcal{A}_3}(U)$.

In general, not all resources are relevant. Consider, for example, the reaction system $\mathcal{A}_4 = (S, \{a, b\})$, where

$$S = \{1, 2, 3\} \qquad a = (\{1\}, \{2\}, \{1\}) \qquad b = (\{1, 3\}, \{2\}, \{1\}) \ .$$

Then entity 3 is not relevant since 3 is a resource only in the presence of entity 1 and then it has no additional influence on the result.

To strengthen the general results obtained so far, we turn our attention to reduced reaction systems which, intuitively, contain neither redundant nor inefficient reactions. Moreover, by Theorem 1, any reaction system is equivalent to a reduced reaction system, and so we still deal with all possible result functions of reaction systems.

It is easy to see that every reaction system with a single reaction is reduced. In the following main result of this paper which strengthens Theorem 2 we show that in the case of any reduced reaction system the relevant entities are precisely the resources used by the system.

Theorem 3. *Let* $\mathcal{A} = (S, A)$ *be a reduced reaction system. Then*

$$rdom(\mathcal{A}) = \bigcup_{a \in A} M_a \ .$$

Proof (Theorem 3). By Theorem 2 it suffices to prove that $\bigcup_{a \in A} M_a \subseteq rdom(\mathcal{A})$. We do this by showing that:

$$(\forall x \in S) \ [x \notin rdom(\mathcal{A}) \implies x \notin \bigcup_{a \in A} M_a] . \tag{\$}$$

To this aim we will now present two lemmas: the first demonstrates that all the reactants are relevant, and the second one demonstrates the same for inhibitors.

Lemma 2. *For each reaction* $a \in A$, $R_a \cap irdom(\mathcal{A}) = \varnothing$.

Proof (Lemma 2). Assume to the contrary that there exists $a \in A$ such that

$$R_a \cap irdom(\mathcal{A}) \neq \varnothing .$$

Let $b = (R_a \setminus irdom(\mathcal{A}), I_a, P_a)$. By Lemma 1, $R_b = R_a \setminus irdom(\mathcal{A}) \neq \varnothing$, and so $b \in rac(S)$. Clearly, b strictly covers a, and so, because $\mathcal{A}$ is reduced, b is *not* consistent with $res_{\mathcal{A}}$. Hence, there exists $T \subseteq S$ such that $en_b(T)$ and $res_b(T) = P_b \not\subseteq res_{\mathcal{A}}(T)$. Since $P_b = P_a$, we get

$$P_a \not\subseteq res_{\mathcal{A}}(T) . \tag{$*$}$$

Let $U = T \cup (R_a \cap irdom(\mathcal{A}))$. Since $en_b(T)$, we have (1) $R_b \subseteq T$ and (2) $I_b \cap T = \varnothing$. Since $R_a \setminus R_b = R_a \cap irdom(\mathcal{A})$, (1) implies that $R_a \subseteq U$. Since $I_b = I_a$ (and $I_a \cap R_a = \varnothing$), $I_a \cap U = \varnothing$. Therefore $en_a(U)$ and, consequently,

$$P_a \subseteq res_{\mathcal{A}}(U) . \tag{$**$}$$

Thus by $(*)$ and $(**)$ we get that

$$P_a \not\subseteq res_{\mathcal{A}}(T) \ \text{ and } \ P_a \subseteq res_{\mathcal{A}}(T \cup (R_a \cap irdom(\mathcal{A}))) .$$

This implies that the set $U \div T$ is relevant, which contradicts Proposition 2 (as $U \div T \subseteq R_a \cap irdom(\mathcal{A})$ and so, by Proposition 2, $U \div T$ must be irrelevant). Therefore Lemma 2 holds.   *(Lemma 2)* □

Lemma 3. *For each reaction* $a \in A$, $I_a \cap irdom(\mathcal{A}) = \varnothing$.

Proof (Lemma 3). Assume to the contrary that there exists $a \in A$ such that $I_a \cap irdom(\mathcal{A}) \neq \varnothing$. Clearly, for each $T \subseteq S$, $res_{A \setminus \{a\}}(T) \subseteq res_{\mathcal{A}}(T)$. Moreover, because $\mathcal{A}$ is reduced, there exists $T_a \subseteq S$ such that $res_{A \setminus \{a\}}(T_a) \neq res_{\mathcal{A}}(T_a)$. Thus

$$res_{A \setminus \{a\}}(T_a) \subset res_{\mathcal{A}}(T_a) . \tag{$*$}$$

Clearly, $en_a(T_a)$, as otherwise $res_{A \setminus \{a\}}(T_a) = res_{\mathcal{A}}(T_a)$ which contradicts $(*)$.

Let $U = T_a \cup irdom(\mathcal{A})$. By Lemma 2, for each $b \in A$, if $R_b \subseteq U$ then $R_b \subseteq T_a$. Consequently, if $b \in A$ is enabled by U, then it is also enabled by T_a, implying that

$$(\forall B \subseteq A) \ [res_B(U) \subseteq res_B(T_a)] . \tag{$**$}$$

Since we assumed that $I_a \cap irdom(\mathcal{A}) \neq \varnothing$, reaction a is not enabled by U and so $res_{\mathcal{A}}(U) \subseteq res_{A\setminus\{a\}}(U)$. Since, by $(**)$, $res_{A\setminus\{a\}}(U) \subseteq res_{A\setminus\{a\}}(T_a)$, we get that $res_{\mathcal{A}}(U) \subseteq res_{A\setminus\{a\}}(T_a)$. Consequently, by $(*)$, we obtain $res_{\mathcal{A}}(U) \subset res_{\mathcal{A}}(T_a)$. Since $U = T_a \cup irdom(\mathcal{A})$, this implies that the set $U \div T_a$ is relevant, which contradicts Proposition 2 (as $U \div T_a \subseteq irdom(\mathcal{A})$ and so, by Proposition 2, $U \div T_a$ must be irrelevant).

Hence it must be that $I_a \cap irdom(\mathcal{A}) = \varnothing$, and consequently Lemma 3 holds. *(Lemma 3)* $\square$

By Lemma 2 and Lemma 3, $irdom(\mathcal{A}) \cap \bigcup_{a \in A} M_a = \varnothing$, which implies that ($) holds and, consequently, the theorem holds.   *(Theorem 3)* $\square$

Our definition of a reduced reaction system $\mathcal{A}$ requires that $\mathcal{A}$ does not have *redundant* reactions, and moreover each reaction is in its most "efficient" form (as far as $\mathcal{A}$ is concerned). A redundant reaction is a reaction that can be removed without influencing the result function $res_{\mathcal{A}}$. Another sort of redundancy is the presence of resources which are not relevant: such entities influence the enabling of (some) reactions but do not influence state transitions! Theorem 3 says that also this kind of redundancy cannot happen in reduced reaction systems.

7   Alternative Notions of Relevance

In defining irrelevant/relevant sets of entities we relied on the operation of symmetric difference. In our view, this is just one of three natural choices to capture the notion of irrelevance/relevance. In this section, we analyse the relationships between them.

Let $X \subseteq S$ be a set of entities of a reaction system $\mathcal{A} = (S, A)$.

– X is *1-irrelevant* in $\mathcal{A}$ if:

$$(\forall T, U \subseteq S) \; [T \div U = X \implies res_{\mathcal{A}}(T) = res_{\mathcal{A}}(U)] \, .$$

– X is *2-irrelevant* in $\mathcal{A}$ if:

$$(\forall T, U \subseteq S) \; [U \subseteq T \text{ and } T \setminus U = X \implies res_{\mathcal{A}}(T) = res_{\mathcal{A}}(U)] \, .$$

– X is *3-irrelevant* in $\mathcal{A}$ if:

$$(\forall T \subseteq S) \; [res_{\mathcal{A}}(T \setminus X) = res_{\mathcal{A}}(T \cup X)] \, .$$

We will use the notations $irr1_{\mathcal{A}}(X)$, $irr2_{\mathcal{A}}(X)$ and $irr3_{\mathcal{A}}(X)$, respectively.

The first of the above three notions of irrelevance is the one investigated until now in this paper. The second considers X irrelevant if removing its elements from any set of entities does not change the result. The third notion of irrelevance considers X irrelevant if, as far as the result function is concerned, removing X from any set of entities has the same effect as adding X to this set of entities.

We now demonstrate relationships between these three notions of relevance.

Lemma 4. *For every $X \subseteq S$, $irr1_A(X)$ implies $irr2_A(X)$.*

Proof. Let $X \subseteq S$ and assume $irr1_A(X)$. Let $T, U \subseteq S$ with $U \subseteq T$ be such that $T \setminus U = X$. Then $T \div U = T \setminus U = X$, and since $irr1_A(X)$, we get $res_A(T) = res_A(U)$. Hence $irr2_A(X)$ and consequently the result holds. □

Lemma 5. *For every $X \subseteq S$, $irr2_A(X)$ implies $irr3_A(X)$.*

Proof. Let $X \subseteq S$ and assume $irr2_A(X)$, hence

$$(\forall T, U \subseteq S) \; [U \subseteq T \text{ and } T \setminus U = X \implies res_A(T) = res_A(U)] \, .$$

Consider arbitrary $T' \subseteq S$. Let $T' \setminus X = U$ and $T' \cup X = T$. Thus $T \setminus U = X$ and $U \subseteq T$. Hence, by $irr2_A(X)$, we get

$$res_A(T) = res_A(U) \, . \tag{$*$}$$

We note that

$$res_A(T' \cup X) = res_A(T) \text{ and } res_A(T' \setminus X) = res_A(U) \, . \tag{$**$}$$

By $(*)$ and $(**)$ we get $res_A(T' \cup X) = res_A(T' \setminus X)$. Therefore $irr3_A(X)$ and so the result holds. □

Lemma 6. *For every $X \subseteq S$, $irr3_A(X)$ implies $irr2_A(X)$.*

Proof. Let $X \subseteq S$ and assume $irr3_A(X)$, hence

$$(\forall T \subseteq S) \; [res_A(T \setminus X) = res_A(T \cup X)] \, .$$

Consider then arbitrary $T, U \subseteq S$ such that $U \subseteq T$ and $T \setminus U = X$. We note that, by $X \subseteq T$, we have

$$T \cup X = T \, . \tag{$\dagger$}$$

Moreover, by $irr3_A(X)$, we have

$$res_A(T \cup X) = res_A(T \setminus X) \, . \tag{$\ddagger$}$$

Hence, by $(\dagger)$ and $(\ddagger)$, $res_A(T) = res_A(T \setminus X)$. Since $U = T \setminus X$, we get $res_A(T) = res_A(U)$. Therefore $irr2_A(X)$ and so the result holds. □

We can therefore conclude that

Theorem 4. *1-irrelevance implies 2-irrelevance which in turn is equivalent to 3-irrelevance.*

Proof. The theorem follows directly from Lemma 4, Lemma 5 and Lemma 6. □

Hence the notion of relevant sets of entities as defined in Section 5 turns out to be the strongest among those discussed in this section, and therefore a reasonable choice for formalising the intuitive notion of relevance (from the point of view of result functions of reaction systems).

Finally, note that for singleton sets X the three notions of irrelevance coincide. This is no longer the case if X has two or more elements. Consider, for example, the reaction system $\mathcal{A}_5 = (S, \{a\})$, where

$$S = \{1, 2, 3\} \qquad a = (\{1, 2\}, \{3\}, \{1\}) \, .$$

Then the set $X = \{1, 3\}$ is not 1-irrelevant but it is 3-irrelevant. Hence the implication in the above theorem cannot be reversed.

8   Conclusions

In this paper, we presented an investigation of sets of entities of reaction systems which are relevant from the point of view of result functions. In particular, we proved that for the reduced reaction systems relevant entities are precisely those which are used as resources by the reactions. We have also discussed the relationship between the notion of relevance investigated in this paper and two alternative notions of relevance.

In our future work we intend to investigate derived notions of relevance where one is interested in establishing which entities become irrelevant 'sooner or later'. For example, one might say that a set of entities $X \subseteq S$ is *eventually irrelevant* in a reaction system $\mathcal{A}$ if

$$(\forall T, U \subseteq S)(\exists n \geq 1) \; [T \div U = X \implies res_{\mathcal{A}}^n(T) = res_{\mathcal{A}}^n(U)] \, ,$$

where $res_{\mathcal{A}}^n$ is the n-fold iteration of $res_{\mathcal{A}}$. In other words, eventual irrelevance implies that the initial distinction between states T and U will eventually disappear with the iteration of $res_{\mathcal{A}}$ whenever the two states differ by the set of entities X.

Acknowledgement. This research was supported by the Pascal Chair award from the Leiden Institute of Advanced Computer Science (LIACS) of Leiden University. The authors are indebted to the referees for useful comments; in particular, the detailed comments by one of the referees were very valuable in the production of the current version of this paper.

References

1. Brijder, R., Ehrenfeucht, A., Main, M.G., Rozenberg, G.: A tour of reaction systems. Int. J. Found. Comput. Sci. 22(7), 1499–1517 (2011)
2. Brijder, R., Ehrenfeucht, A., Rozenberg, G.: Reaction Systems with Duration. In: Kelemen, J., Kelemenová, A. (eds.) Computation, Cooperation, and Life. LNCS, vol. 6610, pp. 191–202. Springer, Heidelberg (2011)

3. Ehrenfeucht, A., Main, M.G., Rozenberg, G.: Combinatorics of life and death for reaction systems. Int. J. Found. Comput. Sci. 21(3), 345–356 (2010)
4. Ehrenfeucht, A., Rozenberg, G.: Events and modules in reaction systems. Theor. Comput. Sci. 376(1-2), 3–16 (2007)
5. Ehrenfeucht, A., Rozenberg, G.: Reaction systems. Fundam. Inform. 75(1-4), 263–280 (2007)
6. Ehrenfeucht, A., Rozenberg, G.: Introducing time in reaction systems. Theor. Comput. Sci. 410(4-5), 310–322 (2009)
7. Kleijn, J., Koutny, M., Rozenberg, G.: Modelling reaction systems with Petri nets. In: BioPPN-2011, International Workshop on Biological Processes & Petri Nets. CEUR-WS Workshop Proceedings, vol. 724, pp. 36–52 (2011)

Generalized Random Context Picture Grammars: The State of the Art

Sigrid Ewert and Max Rabkin

School of Computer Science, University of the Witwatersrand, Johannesburg
`sigrid.ewert@wits.ac.za`, `max.rabkin@students.wits.ac.za`

Abstract. Generalized random context picture grammars (grcpgs) are a method of syntactic picture generation. The terminals are subsets of the Euclidean plane and the replacement of variables involves the building of functions that will eventually be applied to terminals. Context is used to permit or forbid production rules.

Iterated function systems (IFSs) and their generalization, mutually recursive function systems (MRFSs), are among the best-known methods for constructing fractals. In earlier work it was shown that any picture sequence generated by an IFS or MRFS can be generated by a grcpg. Moreover, it was shown that grcpgs can generate a wider range of pictures than IFSs or MRFSs.

In this essay we give a summary of the above mentioned results. We then consider language-restricted iterated function systems (LRIFSs), a method of picture generation where a language controls which functions of an IFS are applied. We first show that LRIFSs are more powerful than IFSs. Then we show that any picture produced by an LRIFS where the restricting language is regular, can be approximated by a grcpg.

1   Introduction

A method of syntactic picture generation, using random context picture grammars (rcpgs), was described and studied elsewhere [6–9]. A summary of results can be found in [5]. In [10], Ewert and van der Walt introduced the notion of a generalized random context picture grammar (grcpg). These grammars use production rules to compose functions from some finite set of functions. These functions are then applied to terminals, which are subsets of the Euclidean plane, to create a picture. Context is used to permit or forbid production rules.

An iterated function system (IFS) is an iterative method for constructing fractals from a finite set of contractive maps defined on a complete metric space. The sequence of pictures generated by an IFS converges to a unique limit. The method was developed principally by Barnsley and co-workers, who obtained impressively life-like images both of nature scenes and the human face [1, 2]. Ewert and van der Walt [10] showed that any picture sequence generated by an IFS can also be generated by a grcpg that uses forbidding context only. Moreover, since grcpgs use context to control the sequence in which functions are applied, they can generate a wider range of fractals or, more generally, pictures than IFSs.

H. Bordihn, M. Kutrib, and B. Truthe (Eds.): Dassow Festschrift 2012, LNCS 7300, pp. 56–74, 2012.
© Springer-Verlag Berlin Heidelberg 2012

Mutually recursive function systems (MRFSs), called hierarchical iterated function systems by Peitgen and co-workers [13], are powerful methods of mathematical picture generation. MRFSs are a generalization of IFSs, and consist of networks or hierarchies of IFSs. Kruger and Ewert [12] generalized the above mentioned result for IFSs to show that for every MRFS, an equivalent grcpg can be constructed. They also showed that grcpgs are more general than MRFSs, in the sense that grcpgs can be constructed that generate sets of pictures that cannot be generated by any MRFS.

Language-restricted iterated function systems (LRIFSs) [15] are a generalization of IFSs, and consist of an IFS and a language that controls which functions of the IFS are applied. In this essay we first show that LRIFSs are more powerful than IFSs. Then we show that any picture produced by an LRIFS where the restricting language is regular, can be approximated by a grcpg.

The remainder of this paper is structured as follows. In Sect. 2, we review published results about the relationship between grcpgs and IFSs, and MRFSs, respectively. In Sect. 3 we focus on LRIFSs and in particular show that any picture produced by an LRIFS where the restricting language is regular, can be approximated by a grcpg. Future work is recommended in Sect. 4.

2   Previously Published Results

In this section we give the definitions of grcpgs, IFSs and MRFSs. Then we state the most important results about the relationship between grcpgs and IFSs, and grcpgs and MRFSs.

2.1   Generalized Random Context Picture Grammars

We define a generalized random context picture grammar and illustrate the main concepts with an example, the iteration sequence of the Sierpiński gasket.

Definition 1. *Let S be any set. Then $\wp(S)$ denotes the power set of S.*

Definition 2. *A generalized random context picture grammar $G = (V_N, V_T, V_F, P, (S, \epsilon))$ has a finite alphabet V of labels, consisting of disjoint subsets V_N of variables, V_T of terminals and V_F of function identifiers. The productions, finite in number, are of the form $A \rightarrow \{(A_1, \rho_1), (A_2, \rho_2), \ldots, (A_t, \rho_t)\}$ $(\mathfrak{P}; \mathfrak{F})$, where $A \in V_N$, $A_1, \ldots, A_t \in V_N \cup V_T$, $\rho_1, \ldots, \rho_t \in V_F^*$ and $\mathfrak{P}, \mathfrak{F} \subseteq V_N$. Finally, there is an initial configuration (S, ϵ), where $S \in V_N$ and ϵ denotes the empty string.*

Definition 3. *A pictorial form Π is a finite set $\{(B_1, \varphi_1), (B_2, \varphi_2), \ldots, (B_s, \varphi_s)\}$, where $B_1, \ldots, B_s \in V_N \cup V_T$ and $\varphi_1, \ldots, \varphi_s \in V_F^*$. We denote the set $\{B_1, \ldots, B_s\}$ by $l(\Pi)$.*

Definition 4. *For a grcpg G and pictorial forms Π and Γ we write $\Pi \Longrightarrow_G \Gamma$ if there is a production $A \rightarrow \{(A_1, \rho_1), (A_2, \rho_2), \ldots, (A_t, \rho_t)\}$ $(\mathfrak{P}; \mathfrak{F})$ in G, Π*

contains an element (A, φ), $l\left(\Pi \setminus \{(A, \varphi)\}\right) \supseteq \mathfrak{P}$ *and* $l\left(\Pi \setminus \{(A, \varphi)\}\right) \cap \mathfrak{F} = \emptyset$, *and* $\Gamma = \left(\Pi \setminus \{(A, \varphi)\}\right) \cup \{(A_1, \varphi \rho_1), (A_2, \varphi \rho_2), \ldots, (A_t, \varphi \rho_t)\}$. *As usual,* $\Longrightarrow_G^*$ *denotes the reflexive transitive closure of* $\Longrightarrow_G$.

Definition 5. *A picture is a pictorial form* Π *with* $l(\Pi) \subseteq V_T$.

Definition 6. *The gallery* $\mathcal{G}(G)$ *generated by a grcpg* G *is the set of pictures* Π *such that* $\{(S, \epsilon)\} \Longrightarrow_G^* \Pi$.

Definition 7. *The gallery of a grcpg* G *is rendered by specifying functions* $\Psi_G : V_T \to \wp\left(\mathbb{R}^2\right)$ *and* $\Upsilon_G : V_F \to F\left(\mathbb{R}^2\right)$, *where* $F\left(\mathbb{R}^2\right) = \{g \mid g : \mathbb{R}^2 \to \mathbb{R}^2\}$. *This yields a representation of a picture* $\Pi = \{(B_1, \varphi_1), (B_2, \varphi_2), \ldots, (B_s, \varphi_s)\}$ *in* $\mathbb{R}^2$ *by*

$$r(\Pi) = \bigcup_{i=1}^{s} \Upsilon_G(\varphi_i)(\Psi_G(B_i)) \quad,$$

where Υ_G *has been extended to* V_F^* *in the obvious manner,* $\Upsilon_G(\epsilon)$ *representing the identity function* id.

Definition 8. *If every production in* G *has* $\mathfrak{P} = \emptyset$, *we call* G *a generalized random forbidding context picture grammar (grFcpg).*

Note 1. It should be clear that we can also use (S, id) as initial configuration without that affecting the rendered gallery.

Note 2. For the sake of convenience, we write a production $A \to \{(A_1, \epsilon)\}$ $(\mathfrak{P}; \mathfrak{F})$ as $A \to A_1$ $(\mathfrak{P}; \mathfrak{F})$. Moreover, if $\mathfrak{P} = \mathfrak{F} = \emptyset$ in a production $A \to \{(A_1, \rho_1), (A_2, \rho_2), \ldots, (A_t, \rho_t)\}$ $(\mathfrak{P}; \mathfrak{F})$, then we write $A \to \{(A_1, \rho_1), (A_2, \rho_2), \ldots, (A_t, \rho_t)\}$.

We illustrate these concepts with an example.

Example 1. We generate the typical iteration sequence of the Sierpiński gasket with the grcpg $G_{\text{gasket}} = (\{S, T, U, F\}, \{b\}, \{g_{\text{lb}}, g_{\text{rb}}, g_{\text{t}}\}, P, (S, \epsilon))$, where P is the set:

$$S \to \{(T, g_{\text{lb}}), (T, g_{\text{rb}}), (T, g_{\text{t}})\} \ (\emptyset; \{U\}) \tag{1}$$

$$T \to U \ (\emptyset; \{S, F\}) \ | \tag{2}$$

$$F \ (\emptyset; \{S, U, F\}) \ | \tag{3}$$

$$b \ (\{F\}; \emptyset) \tag{4}$$

$$U \to S \ (\emptyset; \{T\}) \tag{5}$$

$$F \to b \ (\emptyset; \{T\}) \tag{6}$$

We give the derivation of a picture Π in $\mathcal{G}(G_{\text{gasket}})$ in detail.

$$\{(S,\epsilon)\}$$
$$\Longrightarrow_G \{(T,g_{lb}),(T,g_{rb}),(T,g_t)\} \qquad\qquad\text{(rule 1)}$$
$$\Longrightarrow_G^* \{(U,g_{lb}),(U,g_{rb}),(U,g_t)\} \qquad\qquad\text{(thrice rule 2)}$$
$$\Longrightarrow_G^* \{(S,g_{lb}),(S,g_{rb}),(S,g_t)\} \qquad\qquad\text{(thrice rule 5)}$$
$$\Longrightarrow_G^* \{(T,g_{lb}g_{lb}),(T,g_{lb}g_{rb}),(T,g_{lb}g_t)\} \cup$$
$$\{(T,g_{rb}g_{lb}),(T,g_{rb}g_{rb}),(T,g_{rb}g_t)\} \cup$$
$$\{(T,g_tg_{lb}),(T,g_tg_{rb}),(T,g_tg_t)\} \qquad\qquad\text{(thrice rule 1)}$$
$$\Longrightarrow_G \{(T,g_{lb}g_{lb}),(T,g_{lb}g_{rb}),(T,g_{lb}g_t)\} \cup$$
$$\{(T,g_{rb}g_{lb}),(F,g_{rb}g_{rb}),(T,g_{rb}g_t)\} \cup$$
$$\{(T,g_tg_{lb}),(T,g_tg_{rb}),(T,g_tg_t)\} \qquad\qquad\text{(rule 3)}$$
$$\Longrightarrow_G^* \{(b,g_{lb}g_{lb}),(b,g_{lb}g_{rb}),(b,g_{lb}g_t)\} \cup$$
$$\{(b,g_{rb}g_{lb}),(F,g_{rb}g_{rb}),(b,g_{rb}g_t)\} \cup$$
$$\{(b,g_tg_{lb}),(b,g_tg_{rb}),(b,g_tg_t)\} \qquad\text{(repeated application of rule 4)}$$
$$\Longrightarrow_G \{(b,g_{lb}g_{lb}),(b,g_{lb}g_{rb}),(b,g_{lb}g_t)\} \cup$$
$$\{(b,g_{rb}g_{lb}),(b,g_{rb}g_{rb}),(b,g_{rb}g_t)\} \cup$$
$$\{(b,g_tg_{lb}),(b,g_tg_{rb}),(b,g_tg_t)\} \qquad\qquad\text{(rule 6)}$$

Let $\Upsilon_G(g_{lb}) = (x,y) \to \left(\frac{x}{2},\frac{y}{2}\right)$, $\Upsilon_G(g_{rb}) = (x,y) \to \left(\frac{x}{2}+\frac{1}{2},\frac{y}{2}\right)$ and $\Upsilon_G(g_t) = (x,y) \to \left(\frac{x}{2}+\frac{1}{4},\frac{y}{2}+\frac{\sqrt{3}}{4}\right)$.

Then $r(\Pi) = \bigcup_{i=1}^9 \Upsilon_G(\varphi_i)(\Psi_G(b))$, where $\Upsilon_G(\varphi_1) = (x,y) \to \left(\frac{1}{2} \times \frac{x}{2}, \frac{1}{2} \times \frac{y}{2}\right)$, $\Upsilon_G(\varphi_2) = (x,y) \to \left(\frac{1}{2}\left(\frac{x}{2}+\frac{1}{2}\right),\frac{1}{2} \times \frac{y}{2}\right)$, $\Upsilon_G(\varphi_3) = (x,y) \to \left(\frac{1}{2}\left(\frac{x}{2}+\frac{1}{4}\right),\frac{1}{2}\left(\frac{y}{2}+\frac{\sqrt{3}}{4}\right)\right)$, ...

Let $\Psi_G(b)$ be the dark triangle with vertices $\left\{(0,0),(1,0),\left(\frac{1}{2},\frac{\sqrt{3}}{2}\right)\right\}$. Then $r(\Pi)$ represents the picture in Fig. 1(a). Alternatively, let $\Psi_G(b)$ be the dark square determined by the vertices $\{(0,0),(1,0),(1,1)\}$. Then $r(\Pi)$ represents Fig. 1(b).

2.2   Iterated Function Systems

Iterated function systems are among the best-known methods for constructing fractals. An extensive treatment of IFSs can be found in [11]. In this section we review results that show that grcpgs are more powerful than IFSs.

Definition 9. *An iterated function system* $\{X,\mathcal{F}\}$ *or* $\{X; f_1, f_2, \ldots, f_t\}$, $t > 0$, *is a pair consisting of a complete metric space* X *together with a finite set of contractive maps* $f_i : X \to X$, $1 \le i \le t$.

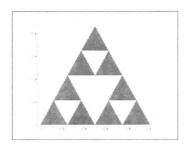
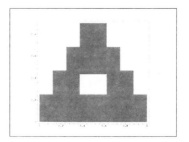

(a) $\Psi_G(\{b\})$ is a dark triangle   (b) $\Psi_G(\{b\})$ is a dark square

Fig. 1. Two pictures in the iteration sequence of the Sierpiński gasket

Let $\mathcal{H}(X)$ be the set of all nonempty compact subsets of X. For $E \in \mathcal{H}(X)$, let $F(E) = f_1(E) \cup f_2(E) \cup \ldots \cup f_t(E)$. By repeated application of F to E, we obtain a sequence in $\mathcal{H}(X)$, $E_0 = E, E_1 = F(E_0), E_2 = F(E_1), \ldots$.

The sequence $E_0, E_1, E_2, \ldots$ converges to a unique limit $\mathcal{E}$, called the attractor of the IFS, which is independent of the choice of starting set E_0, but completely determined by the choice of the maps f_i.

This sequence can be generated by a grFcpg, as was shown in [10]. We state the full result here—in Theorem 1—since the proof gives the translation from a given IFS to a grFcpg.

Theorem 1. *Let $\{X, \mathcal{F}\}$ be an IFS. Then there is a grFcpg G such that for every $l \geq 1$, G generates the set $\{(a, \varphi_1^l), (a, \varphi_2^l), \ldots, (a, \varphi_{t^l}^l)\}$, where the φ_i^l are all t^l possible sequences of length l of the $f_j \in \mathcal{F}$.*

Proof. Let $G = (\{S, I, T, U, F\}, \{a\}, \{f_1, f_2, \ldots, f_t\}, P, (S, \epsilon))$, where P is the set:

$$S \rightarrow \{(I, f_1), (I, f_2), \ldots, (I, f_t)\}$$

$$
\begin{aligned}
I \rightarrow \;& \{(T, f_1), (T, f_2), \ldots, (T, f_t)\} \; (\emptyset; \{F, U\}) \mid \\
& F \; (\emptyset; \{T, U\}) \\
T \rightarrow \;& U \; (\emptyset; \{I\}) \\
U \rightarrow \;& I \; (\emptyset; \{T\}) \\
\\
F \rightarrow \;& a \; (\emptyset; \{I\})
\end{aligned}
$$

$\square$

Example 2. We obtain the iteration sequence of the Sierpiński gasket with the IFS $\{\mathbb{R}^2; g_{\mathrm{lb}}, g_{\mathrm{rb}}, g_t\}$, where $g_{\mathrm{lb}} : (x, y) \rightarrow \left(\frac{x}{2}, \frac{y}{2}\right)$, $g_{\mathrm{rb}} : (x, y) \rightarrow \left(\frac{x}{2} + \frac{1}{2}, \frac{y}{2}\right)$ and $g_t : (x, y) \rightarrow \left(\frac{x}{2} + \frac{1}{4}, \frac{y}{2} + \frac{\sqrt{3}}{4}\right)$.

For any $E \in \mathcal{H}\left(\mathbb{R}^2\right)$, $F(E) = g_{\mathrm{lb}}(E) \cup g_{\mathrm{rb}}(E) \cup g_t(E)$. Let $E_0 = E$. Then $E_1 = F(E_0) = g_{\mathrm{lb}}(E_0) \cup g_{\mathrm{rb}}(E_0) \cup g_t(E_0)$, $E_2 = F(E_1) = g_{\mathrm{lb}}g_{\mathrm{lb}}(E_0) \cup g_{\mathrm{lb}}g_{\mathrm{rb}}(E_0) \cup g_{\mathrm{lb}}g_t(E_0) \cup g_{\mathrm{rb}}g_{\mathrm{lb}}(E_0) \cup g_{\mathrm{rb}}g_{\mathrm{rb}}(E_0) \cup g_{\mathrm{rb}}g_t(E_0) \cup g_t g_{\mathrm{lb}}(E_0) \cup g_t g_{\mathrm{rb}}(E_0) \cup g_t g_t(E_0), \ldots$. When we choose E_0 to be a dark triangle, respectively, a dark square, E_2 is represented by Fig. 1(a) and Fig. 1(b), respectively.

To this IFS corresponds the grFcpg $G = (\{S, I, T, U, F\}, \{a\}, \{g_{\mathrm{lb}}, g_{\mathrm{rb}}, g_t\}, P, (S, \epsilon))$, where P is the set:

$$S \to \{(I, g_{\mathrm{lb}}), (I, g_{\mathrm{rb}}), (I, g_t)\}$$

$$I \to \{(T, g_{\mathrm{lb}}), (T, g_{\mathrm{rb}}), (T, g_t)\} \ (\emptyset; \{F, U\}) \ |$$
$$\qquad F \ (\emptyset; \{T, U\})$$
$$T \to U \ (\emptyset; \{I\})$$
$$U \to I \ (\emptyset; \{T\})$$

$$F \to a \ (\emptyset; \{I\})$$

G generates the pictorial forms $\{(a, g_{\mathrm{lb}}), (a, g_{\mathrm{rb}}), (a, g_t)\}$, $\{(a, g_{\mathrm{lb}}g_{\mathrm{lb}}), (a, g_{\mathrm{lb}}g_{\mathrm{rb}}), (a, g_{\mathrm{lb}}g_t)\} \cup \{(a, g_{\mathrm{rb}}g_{\mathrm{lb}}), (a, g_{\mathrm{rb}}g_{\mathrm{rb}}), (a, g_{\mathrm{rb}}g_t)\} \cup \{(a, g_t g_{\mathrm{lb}}), (a, g_t g_{\mathrm{rb}}), (a, g_t g_t)\}, \ldots$.

In [10] it was also shown that there exists a set of pictures that can be generated by a grcpg, but that is not the sequence converging to the attractor of any IFS. Since grcpgs use context to control the sequence in which functions are applied, they can generate a wider range of pictures than IFSs. An example of such a picture set is $\mathcal{G}_{\mathrm{trail}}$, which is described below. $\mathcal{G}_{\mathrm{trail}}$ cannot be generated by a grFcpg, as becomes clear when inspecting the proof in [8], and therefore also not by an IFS.

$\mathcal{G}_{\mathrm{trail}} = \{\Theta_1, \Theta_2, \ldots\}$, where Θ_1, Θ_2 and Θ_3 are shown in Fig. 2(a), Fig. 2(b) and Fig. 2(c), respectively. For the sake of clarity, an enlargement of the lower lefthand ninth of Θ_3 is given in Fig. 2(d).

For $i = 2, 3, \ldots$, Θ_i is obtained by dividing each dark square in Θ_{i-1} into four and placing a copy of Θ_1, modified so that it has exactly $i + 2$ dark squares, all on the diagonal, into each quarter.

The modification of Θ_1 is effected in its middle dark square only and proceeds in detail as follows: The square is divided into four and the newly-created lower lefthand quarter coloured dark. The newly-created upper righthand quarter is again divided into four and its lower lefthand quarter coloured dark. This successive quartering of the upper righthand square is repeated until a total of $i - 1$ dark squares have been created, then the upper righthand square is also coloured dark. The new dark squares thus get successively smaller, except for the last two, which are of equal size.

2.3 Mutually Recursive Function Systems

Mutually recursive function systems, called hierarchical iterated function systems by Peitgen and co-workers [13], are a generalization of IFSs, and consist of

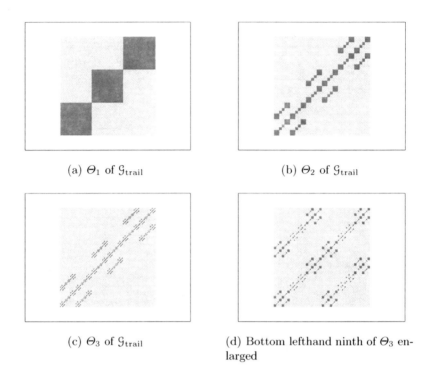

(a) Θ_1 of $\mathcal{G}_{\text{trail}}$ (b) Θ_2 of $\mathcal{G}_{\text{trail}}$

(c) Θ_3 of $\mathcal{G}_{\text{trail}}$ (d) Bottom lefthand ninth of Θ_3 enlarged

Fig. 2. Pictures of $\mathcal{G}_{\text{trail}}$

networks or hierarchies of IFSs. Mutually recursive function systems were developed to study wider ranges of fractal-like images that do not exhibit such high degrees of self-similarity as IFSs [13]. In this section we review results that show that grcpgs are more powerful than MRFSs.

There are a number of slight variations in the definitions of MRFSs that can be found in the literature. Here we use the definition used by Drewes [4].

Definition 10. *Let $n \in \mathbb{N}_+$. Then $I = (M, c)$ is an MRFS of rank n such that*

- *M is an $n \times n$ matrix $(m_{i,j})$ with*
  - *$m_{i,j} = f_{i,j}^1, \ldots, f_{i,j}^{t_{i,j}}$, $t_{i,j} \in \mathbb{N}$, and*
  - *$\forall i, j \in [n]$ and $k \in [t_{i,j}]$, $f_{i,j}^k : \mathbb{R}^2 \to \mathbb{R}^2$.*
- *$c = (c_1, \ldots, c_n)$ is a vector where each c_i is a possibly empty compact subset of $\mathbb{R}^2$. These sets are called condensation sets.*
- *For each i such that c_i is empty, $\exists j$ such that $t_{i,j} > 0$.*

Mutually recursive function systems generate pictures through application of the extended Hutchinson operator.

Definition 11. *Given an MRFS $I = (M, c)$, the Hutchinson operator $H_I : \left(\wp\left(\mathbb{R}^2\right)\right)^n \to \left(\wp\left(\mathbb{R}^2\right)\right)^n$ is defined as follows: for $v = (v_1, \ldots, v_n) \in \left(\wp\left(\mathbb{R}^2\right)\right)^n$, $H_I(v) = (v_1', \ldots, v_n')$, where $v_i' = c_i \cup \bigcup_{j \in [n]} H_{m_{i,j}}(v_j)$ for $i \in [n]$.*

Now, given an MRFS $I = (M, c)$ and a vector of initial pictures $u = (u_1, \ldots, u_n)$, with u_i a compact, possibly empty subset of $\mathbb{R}^2$, the sequence of pictures generated by I is $S_I(u, 1), S_I(u, 2), \ldots$, where $S_I(u, i) = H_I^i(u)$ [1] (the first component of the ith iteration of the Hutchinson operator). The picture language obtained from I is the set $L(I, u) = \{S_I(u, i) | i \in \mathbb{N}_+\}$.

Example 3. The MRFS I_S of rank 3 generates pictures which consist of a Sierpiński triangle with a "shadow" consisting of an inverse Sierpiński triangle. Figure 3 shows the first four pictures in $L(I_S, a)$ where $a = (a_1, a_2, a_3)$, a_1 is the empty set, a_2 is the filled-in triangle with vertices $(0, 0), (5, \sqrt{75})$ and $(-5, \sqrt{75})$ and a_3 is the filled-in triangle with vertices $(-5, 0), (5, 0)$ and $(0, \sqrt{75})$.

$$I_S = (M, c)$$
$$\text{where :}$$
$$M = \begin{pmatrix} \epsilon & f_7 & f_4, f_5, f_6 \\ \epsilon \text{ id}, f_1, f_2, f_3 & \epsilon \\ \epsilon & \epsilon & f_4, f_5, f_6 \end{pmatrix}$$
$$\text{and for } i \in [3], \ c_i = \emptyset,$$

with the functions defined as follows :

$$\text{id}(x, y) = (x, y)$$
$$f_1(x, y) = \left(\frac{x}{2} + 5, \frac{y}{2} \right)$$
$$f_2(x, y) = \left(\frac{x}{2} - 5, \frac{y}{2} \right)$$
$$f_3(x, y) = \left(\frac{x}{2}, \frac{y}{2} + \sqrt{75} \right)$$
$$f_4(x, y) = \left(\frac{x}{2}, \frac{y}{2} + \frac{\sqrt{75}}{2} \right)$$
$$f_5(x, y) = \left(\frac{x}{2} - 2.5, \frac{y}{2} \right)$$
$$f_6(x, y) = \left(\frac{x}{2} + 2.5, \frac{y}{2} \right)$$
$$f_7(x, y) = \left(\frac{x}{2} + y \tan \frac{\Pi}{8}, -\frac{y}{4} \right)$$

Kruger and Ewert [12] showed that for every MRFS, an equivalent grcpg can be constructed. We state the result here in full—in Theorem 2—since the proof gives the translation from a given MRFS to a grFcpg.

Theorem 2. *An MRFS $I = (M, c)$, of degree n with a vector of initial pictures $a = (a_1, \ldots, a_n)$, can be translated into a grcpg G_I.*

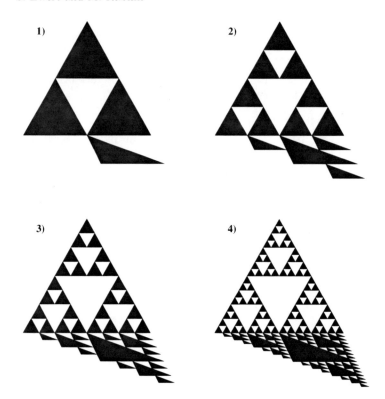

Fig. 3. Four pictures generated by the MRFS I_S

Proof.

$$G_I = (V_N, V_T, V_F, P, (S, \epsilon)) \text{ where}$$
$$V_N = \{S, I_1, \ldots, I_n, T_1, \ldots, T_n, U_1, \ldots, U_n, F_1, \ldots, F_n\}$$
$$V_T = \{a_1, \ldots, a_n, c_1, \ldots, c_n\}$$
$$V_F = \bigcup_{i,j \in [n]} \{f_{i,j}^1, \ldots, f_{i,j}^{t_{i,j}}\}$$

and P is the set of productions :

$$S \to \{(I_1, f_{1,1}^1), \ldots, (I_1, f_{1,1}^{t_{1,1}}), \ldots, (I_n, f_{1,n}^1), \ldots, (I_n, f_{1,n}^{t_{1,n}}), c_1\}$$
$$I_i \to \{(T_1, f_{i,1}^1), \ldots, (T_1, f_{i,1}^{t_{i,1}}), \ldots, (T_n, f_{i,n}^1), \ldots, (T_n, f_{i,n}^{t_{i,n}}), c_i\}$$
$$\qquad (\emptyset; \{I_1, \ldots, I_{i-1}, F_1, \ldots, F_n, U_1, \ldots, U_n\})$$
$$I_i \to F_i(\emptyset; \{I_1, \ldots, I_{i-1}, T_1, \ldots, T_n, U_1, \ldots, U_n\})$$
$$T_i \to U_i(\emptyset; \{I_1, \ldots, I_n, T_1, \ldots, T_{i-1}\})$$
$$U_i \to I_i(\emptyset; \{T_1, \ldots, T_n, U_1, \ldots, U_{i-1}\})$$
$$F_i \to a_i(\emptyset; \{I_1, \ldots, I_n, F_1, \ldots, F_{i-1}\}) \qquad\qquad \square$$

The language of G_I can be rendered in such a way that it is equal to the set of all approximations generated by I.

Example 4. The grcpg G_{shadow} was obtained by translating the MRFS I_S into a grcpg. With the terminals and functions defined as for I_S above, this grammar will generate exactly the same set of pictures as I_S.

$$G_{\text{shadow}} = (V_N, V_T, V_F, P, (S, \epsilon)) \text{ where}$$
$$V_N = \{S, I_1, I_2, I_3, T_1, T_2, T_3, U_1, U_2, U_3, F_1, F_2, F_3\}$$
$$V_T = \{a_1, a_2, a_3\}$$
$$V_F = \{\text{id}, f_1, f_2, f_3, f_4, f_5, f_6, f_7\}$$

and P is the set of productions :

$$S \to \{(I_2, f_7), (I_3, f_4), (I_3, f_5), (I_3, f_6)\}$$
$$I_1 \to \{(T_2, f_7), (T_3, f_4), (T_3, f_5), (T_3, f_6)\}(\emptyset; \{F_1, F_2, F_3, U_1, U_2, U_3\})$$
$$I_1 \to F_1(\emptyset; \{T_1, T_2, T_3, U_1, U_2, U_3\})$$
$$I_2 \to \{(T_2, \text{id}), (T_2, f_1), (T_2, f_2), (T_2, f_3)\}(\emptyset; \{I_1, F_1, F_2, F_3, U_1, U_2, U_3\})$$
$$I_2 \to F_2(\emptyset; \{I_1, T_1, T_2, T_3, U_1, U_2, U_3\})$$
$$I_3 \to \{(T_3, f_4), (T_3, f_5), (T_3, f_6)\}(\emptyset; \{I_1, I_2, F_1, F_2, F_3, U_1, U_2, U_3\})$$
$$I_3 \to F_3(\emptyset; \{I_1, I_2, T_1, T_2, T_3, U_1, U_2, U_3\})$$
$$T_1 \to U_1(\emptyset; \{I_1, I_2, I_3\})$$
$$T_2 \to U_2(\emptyset; \{I_1, I_2, I_3, T_1\})$$
$$T_3 \to U_3(\emptyset; \{I_1, I_2, I_3, T_1, T_2\})$$
$$U_1 \to I_1(\emptyset; \{T_1, T_2, T_3\})$$
$$U_2 \to I_2(\emptyset; \{T_1, T_2, T_3, U_1\})$$
$$U_3 \to I_3(\emptyset; \{T_1, T_2, T_3, U_1, U_2\})$$
$$F_1 \to a_1(\emptyset; \{I_1, I_2, I_3\})$$
$$F_2 \to a_2(\emptyset; \{I_1, I_2, I_3, F_1\})$$
$$F_3 \to a_3(\emptyset; \{I_1, I_2, I_3, F_1, F_2\})$$

In [12], Kruger and Ewert also showed that grcpgs can be constructed that generate sets of pictures that cannot be generated by any MRFS. Such a grcpg is easily obtained by simply modifying the context rules in a grcpg translated from some MRFS, to remove some (or all) of the restrictions that guarantee uniform refinement in the resulting pictures. Another easy way of obtaining such a grcpg is to simply add production rules to a grcpg translation of an MRFS.

Consider the set of all pictures that consist of a Sierpiński triangle with uniform refinement and a "shadow" made of an inverted Sierpiński triangle, also with uniform refinement, but the triangle and the "shadow" need not have the same level of refinement. Thus, this set contains all the pictures in $\mathcal{G}(G_{\text{shadow}})$ as well as pictures such as shown in Fig. 4. This set can be generated by a grFcpg, called G_{ext_1} in [12]. It should be clear that no MRFS can be constructed to generate all the pictures in this set.

Fig. 4. Two pictures from $\mathcal{G}(G_{\mathrm{ext}_1})$ that are not in $\mathcal{G}(G_{\mathrm{shadow}})$

3   Language-Restricted Iterated Function Systems

We can modify the picture produced by an IFS by using a language restriction, where a language controls which functions of the IFS are applied at different stages. This method of picture generation, introduced in [14], allows us to create pictures which are self-similar but not self-identical. For example, we can take an IFS which generates a picture of leaves on a stalk—Fig. 5(a)—and restrict it to get leaves on alternating sides—Fig. 5(b)—without changing the leaves themselves.

In this section we prove that the LRIFSs are strictly more powerful than the IFSs, and therefore investigate the relationship between LRIFSs and grcpgs. Although we do not investigate the relationship between LRIFSs and MRFSs, we use different types of approximation sequences for the two systems, so LRIFSs are of independent interest.

Definition 12. *A language-restricted iterated function system (LRIFS) is a tuple* $\mathcal{I}_L = \{X, \mathcal{F}, L\}$ *where* $\mathcal{I} = \{X, \mathcal{F}\}$ *is an IFS, called the underlying IFS of* $\mathcal{I}_L$, *and* $L \subseteq \mathcal{F}^*$.

Following [15], we interpret the words of $\mathcal{F}^*$ as functions by *reverse* composition; that is, if $f = f_1 f_2 \ldots f_{n-1} f_n$, where $f_1, f_2, \ldots, f_{n-1}, f_n \in \mathcal{F}$, then $f(\pi) = f_n(f_{n-1}(\ldots f_2(f_1(\pi))\ldots))$. Unlike [15], however, we give the symbol $\circ$ its usual meaning.

The definition of the attractor of an LRIFS is based on the fact that if $\{X, \mathcal{F}\}$ is an IFS and π is a point in its attractor, then the attractor is equal to $\overline{\{f(\pi) \mid f \in \mathcal{F}^*\}}$, where $\overline{}$ denotes the topological closure.

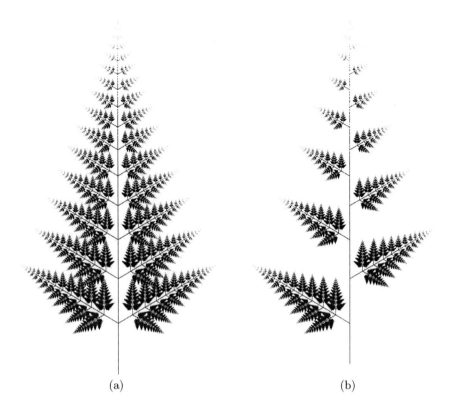

(a)   (b)

Fig. 5. An IFS and a language-restricted variation

Definition 13. *If $\mathfrak{I}_L = (X, \mathfrak{F}, L)$ is an LRIFS and $\pi \in X$ then the attractor of $\mathfrak{I}_L$ at π is*

$$\mathcal{A}_\pi(\mathfrak{I}_L) = \overline{\{f(\pi) \mid f \in L\}} \ .$$

Thus, every word of L contributes a single point to the picture. In [15], π is required to be in the attractor of the underlying IFS, but Lemma 1 shows that that is unnecessary.

In all examples in this work and in [15], L is a regular language and $\mathfrak{F}$ a set of affine functions (an affine function is a translation composed with linear function). We call such LRIFSs regular and affine, respectively. Affine regular LRIFSs can generate a wide variety of pictures, even with a single underlying IFS. For example, see Fig. 5, already mentioned, and Fig. 6, which shows two fractals described in [15] and a version of the Sierpiński triangle (restricted with the language $(F_1 + F_3 + F_4)^*$ in the notation of that paper). The functions which generate Fig. 5 are

$$f_1 = t\left(0, \frac{1}{8}\right) \circ s\left(\frac{7}{8}\right)$$

$$f_2 = t\left(0, \frac{1}{8}\right) \circ r\,(60) \circ s\left(\frac{1}{3}\right)$$

$$f_3 = t\left(0, \frac{1}{8}\right) \circ r\,(-60) \circ s\left(\frac{1}{3}\right)$$

$$f_4 = \mathrm{proj}_y \circ s\left(\frac{1}{4}\right)$$

where t denotes translation, s scaling, r rotation (in degrees) and proj_y projection onto the y axis. Fig. 5(b) is restricted with the language $(f_1 + f_2 + f_3 + f_4)^*(f_2 + f_3 f_1)(f_1 f_1)^*(f_4 f_1^* + \varepsilon)$.

Fig. 6. Attractors of a single IFS restricted by three different languages

Every attractor of an IFS is also an attractor of a regular LRIFS, which can be seen by using the language $L = \mathcal{F}^*$. On the other hand, there are pictures which are the attractor of an LRIFS but not of any IFS (at least when we restrict ourselves to the affine functions). To prove this, we will need some basic facts about the closure properties of LRIFS attractors.

Lemma 1. *If A is the attractor of an (affine, regular) LRIFS at a point π, then there is another (affine, regular) LRIFS whose attractor is A at every point.*

Proof. Suppose $\mathcal{I}_L = \{X, \mathcal{F}, L\}$ and $A = A_\pi(\mathcal{I}_L)$. Let g be the function which is constantly π, and $\mathcal{I}_{gL} = \{X, \mathcal{F} \cup \{g\}, gL\}$. Then

$$\begin{aligned}
A &= A_\pi(\mathcal{I}_L) \\
&= \overline{\{f(\pi) \mid f \in L\}} \\
&= \overline{\{(f \circ g)(\rho) \mid f \in L\}} \\
&= \overline{\{f(\rho) \mid f \in gL\}} \\
&= A_\rho(\mathcal{I}_{gL})
\end{aligned}$$

for any ρ. Thus $\mathcal{I}_{gL}$ is the desired LRIFS; furthermore it is regular (resp. affine) if $\mathcal{I}_L$ is.   □

Thus the starting point π is essentially arbitrary: if we want to generate a single picture, we can find an LRIFS which generates it from any starting point.

Lemma 2. *If $\mathcal{A}$ is the attractor of an (affine, regular) LRIFS at a point π and $0 < a < 1$, then there is another (affine, regular) LRIFS whose attractor at π is $s(a)(\mathcal{A})$.*

Proof. Suppose $\mathfrak{I}_L = \{X, \mathcal{F}, L\}$ and $\mathcal{A} = \mathcal{A}_\pi(\mathfrak{I}_L)$. Let $g = s(a)$. Then $\mathfrak{I}_{Lg} = \{X, \mathcal{F} \cup \{g\}, Lg\}$ has the desired attractor. $\qquad\square$

Thus the class of LRIFS attractors is closed under downscaling. Furthermore, we will now show that they are closed under union.

Lemma 3. *If $\mathcal{A}$ and $\mathcal{A}'$ are attractors of (affine, regular) LRIFSs, then so is $\mathcal{A} \cup \mathcal{A}'$.*

Proof. Let $\mathfrak{I}_L = \{X, \mathcal{F}, L\}$ and $\mathfrak{I}'_{L'} = \{X, \mathcal{F}', L'\}$ be LRIFSs whose attractors are $\mathcal{A}$ and $\mathcal{A}'$ respectively. By Lemma 1 we can assume, without loss of generality, that they can be generated from the same starting point, π. Then

$$
\begin{aligned}
\mathcal{A} \cup \mathcal{A}' &= \mathcal{A}_\pi(\mathfrak{I}_L) \cup \mathcal{A}_\pi(\mathfrak{I}'_{L'}) \\
&= \overline{\{f(\pi) \mid f \in L\}} \cup \overline{\{f(\pi) \mid f \in L'\}} \\
&= \overline{\{f(\pi) \mid f \in L\} \cup \{f(\pi) \mid f \in L'\}} \\
&= \overline{\{f(\pi) \mid f \in (L \cup L')\}} \\
&= \mathcal{A}_\pi(\mathfrak{J}_{L \cup L'})
\end{aligned}
$$

where $\mathfrak{J} = \{X, \mathcal{F} \cup \mathcal{F}'\}$. $\qquad\square$

The previous three lemmas allow us to generate an LRIFS attractor by overlaying the (possibly downscaled) attractors of other LRIFSs (with all three operations preserving affineness and regularity). This contrasts with IFSs, as the following theorem shows by an example.

Theorem 3. *There is an LRIFS $\mathfrak{I}_L = \{X, \mathcal{F}, L\}$ with $\mathcal{F}$ a set of affine functions $X \to X$ whose attractor is not the attractor of any IFS $\mathfrak{J} = \{X, \mathcal{F}'\}$ with $\mathcal{F}'$ a set of affine functions.*

Proof. Let $\mathcal{A}$ be the Cantor square, suitably scaled down, surrounded by a square, as depicted in Fig. 7. Since the Cantor square and the square are both attractors of IFSs, $\mathcal{A}$ is the attractor of an LRIFS by the above theorems.

Suppose $\mathcal{A}$ is the attractor of an IFS $\{X, \mathcal{F}\}$ where $\mathcal{F}$ is a set of affine functions.

Let $S \subseteq \mathcal{A}$ be the square, and $C = \mathcal{A} \setminus S$ be the Cantor square. Let $f \in \mathcal{F}$. Since S is connected, either $f(S) \subseteq S$ or $f(S) \subseteq C$.

If $f(S) \subseteq S$, then $f(S)$ must be a point or a line segment, since the image of S under an affine map is either a quadrilateral (but $f(S) \neq S$ since f is a contraction), a triangle (but no triangle is a subset of S), a line segment or a point.

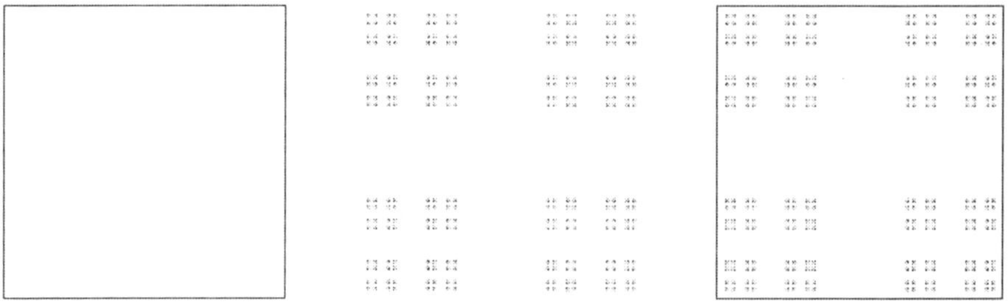

Fig. 7. A square, S, and the Cantor square, C, which are attractors of IFSs, and their union, $\mathcal{A}$

On the other hand, if $f(S) \subseteq C$ then $f(S)$ is a singleton, because C is totally disconnected.

Thus each $f \in \mathcal{F}$ maps S to either a line segment or a singleton. Since C is inside S and each f is affine, $f(\mathcal{A})$ is either a line segment or a singleton for each f. However, $\mathcal{A} = f_1(\mathcal{A}) \cup \ldots \cup f_n(\mathcal{A})$ but $\mathcal{A}$ is not a finite union of line segments and singletons, which is a contradiction. Thus $\mathcal{A}$ is not the attractor of any IFS.   □

Since we have proven the LRIFSs are strictly more powerful than IFSs, we wish to extend our main result for IFSs—that they can be generated by a grFcpg—to LRIFSs. We use a different notion of approximation than for IFSs, because we wish to retain the information provided by all the strings in the language, rather than discarding them in better approximations.

Definition 14. If L is a language, let $L_{\leq n} = \{x \in L : |x| \leq n\}$.

We use the concept of $L_{\leq n}$ to generate approximations to the attractors of an LRIFS which are uniform, in that they are not closer approximations in one part than another. This is illustrated by Fig. 8, which shows three approximations based on $L_{\leq n}$ for different n, and one based on an arbitrary subset of L.

Theorem 4. Let $\mathcal{I}_L = \{X, \mathcal{F}, L\}$ be a regular LRIFS, and π a point in X. Then there is a grFcpg G that can be rendered as

$$\{\mathcal{A}_\pi(\mathcal{I}_{L_{\leq n}}) : n \in \mathbb{N}\},$$

and the functions used in rendering G are exactly those in $\mathcal{F}$ along with the identity.

Proof. Let $M = (Q, \mathcal{F}, q_0, A, \delta)$ be a deterministic finite automaton which recognizes L^r, the reverse language of L. The use of L^r is due to the fact that reverse composition is used in LRIFSs. We simulate all the paths of M by a grcpg G, and control G so that all paths are truncated at the same length.

Let $G = ((Q \times \{0, 1\}) \cup \{S, C_0, C_1, C_2\}, \{e, p\}, \mathcal{F} \cup \{\mathrm{id}\}, P, (S, \mathrm{id}))$, where S, C_0, C_1, C_2, e and p are fresh symbols, id is the identity function on X, and P is constructed from N as follows:

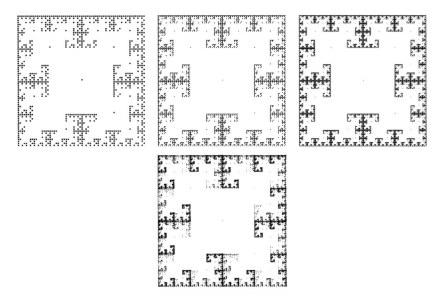

Fig. 8. Three uniform approximations of an LRIFS and a non-uniform approximation

1. $S \to (C_0, \mathrm{id}) \, ((q_0, 0), \mathrm{id})$
2. For each accepting state $q \in A$, with edges to $q'_1, \ldots, q'_k \in Q$ labelled with $f_1, \ldots, f_k \in \mathcal{F}$ respectively, add a production

$$(q, 0) \to ((q'_1, 1), f_1) \; \cdots \; ((q'_k, 1), f_k) \, (p, \mathrm{id}) \; (\emptyset; \{C_1\})$$

3. For each non-accepting state $q \in Q \backslash A$, with edges to $q'_1, \ldots, q'_k \in Q$ labelled with $f_1, \ldots, f_k \in \mathcal{F}$ respectively, add a production

$$(q, 0) \to ((q'_1, 1), f_1) \; \cdots \; ((q'_k, 1), f_k) \; (\emptyset; \{C_1\})$$

4. $C_0 \to (C_1, \mathrm{id}) \; (\emptyset; Q \times \{0\})$
5. For each state $q \in Q$, add a production $(q, 1) \to ((q, 0), \mathrm{id}) \; (\emptyset; \{C_0, C_2\})$
6. $C_1 \to (C_0, \mathrm{id}) \; (\emptyset; Q \times \{1\})$
7. $C_0 \to (C_2, \mathrm{id}) \; (\emptyset; Q \times \{0\})$
8. For each state $q \in Q$, add a production $(q, 1) \to (e, \mathrm{id}) \; (\emptyset; \{C_0, C_1\})$
9. $C_2 \to (e, \mathrm{id}) \; (\emptyset; Q \times \{1\})$

Any derivation in this grammar proceeds in phases. First the start symbol is rewritten by production 1. At any point where C_0 and $(q, 0)$ appear in the string, we can only apply productions from 2 and 3, since all other productions are forbidden or cannot be applied, and therefore these productions will be applied to all non-terminals of the form $(q, 0)$. When this is done, there is a single C_0 and all other non-terminals are of the form $(q, 1)$, and there is a choice of productions: 4 or 7.

If we apply production 4, we rewrite every $(q, 1)$ into $(q, 0)$ by production 5. When this is done we rewrite C_1 into C_0 by production 6 so that another iteration can be applied.

If, instead, we apply production 7, we proceed to delete the $(q, 1)$ non-terminals by rewriting them to a symbol (production 8) which will be rendered as the empty set. Once they are all deleted, we delete C_2 by production 9.

Thus all branches are extended in tandem until they terminate. The branches, besides those containing the control symbols C_0, C_1 and C_2, correspond to paths through M up to a certain length, and are labelled by a composition of the symbols along the paths (interspersed with id). Thus if we render the generated gallery by interpreting each function in $\mathcal{F} \cup \{\text{id}\}$ as itself and rendering p by $\{\pi\}$ and e by $\emptyset$, then we obtain $\mathcal{A}_\pi(\mathcal{I}_{L_{\leq n}})$ for each $n \in \mathbb{N}$ (since $L_{\leq n}$ is finite, the closure operation in the definition of the attractor makes no difference).   □

An example illustrating the method used in this proof is given in Fig. 9.

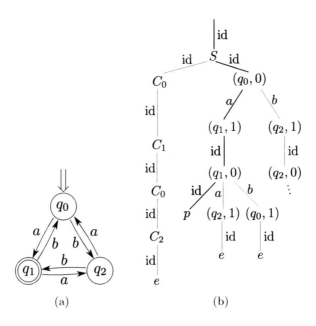

Fig. 9. (a) An automaton for a language L and (b) the derivation tree corresponding to $L_{\leq 1}$, with the highlighted path corresponding to the word $a \in L_{\leq 1}$

4   Future Work

Culik and Dube [3] showed that any uniformly growing, deterministic, context-free Lindenmayer system (D0L-system) can be simulated by an MRFS. As mentioned above, Kruger and Ewert [12] showed that for any MRFS, an equivalent grcpg can be constructed, and that the grcpg can be modified to generate sequences of pictures that cannot be generated by the basis MRFS. Therefore it would be interesting to simulate uniformly growing D0L-systems by grcpgs and then modify the grcpg to generate pictures that cannot be generated by the basis D0L-system, as we have done in this paper for IFSs and MRFS.

Our notion of a uniform approximation to an LRIFS is based exclusively on the length of the strings in the language; it would be interesting to formulate a notion which depends on the area, to obtain an approximation which looks uniform rather than having uniform depth, and determine whether this approximation can also be generated by a grcpg.

We showed that LRIFSs are more powerful than IFSs, but did not investigate their relationship to other extensions of IFSs (in particular MRFSs), and this topic is worthy of investigation.

Acknowledgements. We would like to thank the referees for their helpful comments.

References

1. Barnsley, M.F.: Fractals Everywhere. Academic Press, Boston (1988)
2. Barnsley, M.F., Hurd, L.P.: Fractal Image Compression. Peters, Wellesley (1993)
3. Culik II, K., Dube, S.: L-systems and mutually recursive function systems. Acta Informatica 30, 279–302 (1993)
4. Drewes, F.: Tree-based picture generation. Theoretical Computer Science 246, 1–51 (2000)
5. Ewert, S.: Random context picture grammars: The state of the art. In: Drewes, F., Habel, A., Hoffmann, B., Plump, D. (eds.) Manipulation of Graphs, Algebras and Pictures, pp. 135–147. Hohnholt, Bremen (2009)
6. Ewert, S., van der Walt, A.: Generating pictures using random forbidding context. International Journal of Pattern Recognition and Artificial Intelligence 12(7), 939–950 (1998)
7. Ewert, S., van der Walt, A.: Generating pictures using random permitting context. International Journal of Pattern Recognition and Artificial Intelligence 13(3), 339–355 (1999)
8. Ewert, S., van der Walt, A.: A hierarchy result for random forbidding context picture grammars. International Journal of Pattern Recognition and Artificial Intelligence 13(7), 997–1007 (1999)
9. Ewert, S., van der Walt, A.: Random context picture grammars. Publicationes Mathematicae (Debrecen) 54 (supp.), 763–786 (1999)
10. Ewert, S., van der Walt, A.: Shrink indecomposable fractals. Journal of Universal Computer Science 5(9), 521–531 (1999), http://www.jucs.org/jucs_5_9
11. Hoggar, S.G.: Mathematics for Computer Graphics. Cambridge University Press, Cambridge (1992)
12. Kruger, H., Ewert, S.: Translating mutually recursive function systems into generalised random context picture grammars. South African Computer Journal (36), 99–109 (2006)

13. Peitgen, H.O., Jürgens, H., Saupe, D.: Chaos and Fractals. New Frontiers of Science. Springer, New York (1992)
14. Prusinkiewicz, P., Hammel, M.: Automata, languages, and iterated function systems. In: Hart, J.C., Musgrave, F.K. (eds.) Fractal Modeling in 3D Computer Graphics and Imagery, pp. 115–143. ACM SIGGRAPH (1991)
15. Prusinkiewicz, P., Hammel, M.: Escape-time visualization method for language-restricted iterated function systems. In: Proceedings of Graphics Interface 1992, Vancouver, British Columbia, Canada, pp. 213–223 (May 1992)

Cooperating Distributed Tree Automata

Henning Fernau

Universität Trier, FB IV—Abteilung Informatik,
54286 Trier, Germany
fernau@uni-trier.de

Abstract. We propose a study on cooperating distributed tree automata, proving in particular characterizations of the yields of such automata systems.

1 Introduction

Jürgen Dassow is one of the pioneers of cooperating distributed grammar systems (CDGS), as testified by one of the first papers on this topic that introduced the very name of this area, as well as by the fact that he co-authored the (first) textbook on this subject; see [10,11]. Although the subject comes close to its silver jubilee soon, there are quite a number of questions still open ever since they were introduced. For instance, it is still open whether (context-free) matrix languages can be characterized by CDGS working in $= k$-mode. Can results from tree automata be helpful here?

A related issue is that of cooperating distributed automata. As pioneering papers, we refer to [13,15,16].

There is a third line of research that we will try to continue with this paper, namely that of CDGS with regular components, as studied in [11,17]. Let us recall that possibly surprising fact that in many cases, regular (meaning: right-linear) components are just as powerful as context-free grammar components. This is by large due to the fact that chain rules are permitted. This explains why results using finite automata instead might yield different results, as presented in this paper.

Finally, we continue our studies on accepting grammar systems; see [23,25], as we exhibit tight relations between the questions of generating versus accepting grammar systems on the one hand and of top-down versus bottom-up tree automata systems on the other hand.

2 Definitions

2.1 Classical CDGS

A *CD grammar system* of degree n, with $n \geq 1$, is a $(n+3)$-tuple

$$G = (N, T, S, P_1, \ldots, P_n),$$

H. Bordihn, M. Kutrib, and B. Truthe (Eds.): Dassow Festschrift 2012, LNCS 7300, pp. 75–85, 2012.
© Springer-Verlag Berlin Heidelberg 2012

where N, T are disjoint alphabets of nonterminal and terminal symbols, respectively, $S \in N$ is the axiom, and $P_1, \ldots, P_n$ are finite sets of rewriting rules over $N \cup T$.

Throughout this paper, we consider only context-free rewriting rules. Since we are interested in generating and accepting systems, we further distinguish between so-called *generating rules*, which have the form $A \to z$, with $A \in N$ and $z \in (N \cup T)^*$, and *accepting rules*, which are of the form $z \to A$, with $A \in N$ and $z \in (N \cup T)^*$.

Let $G = (N, T, S, P_1, \ldots, P_n)$ be a CD grammar system with only generating context-free rules. For $x, y \in (N \cup T)^*$ and $1 \le i \le n$, we write $x \Rightarrow_i y$ iff $x = x_1 A x_2$, $y = x_1 z x_2$ for some $A \to z \in P_i$. Hence, subscript i refers to the production set to be used. In addition, we denote by $\Rightarrow_i^{=k}$ ($\Rightarrow_i^{\le k}$, $\Rightarrow_i^{\ge k}$, $\Rightarrow_i^*$, respectively) a derivation consisting of exactly k steps (at most k steps, at least k steps, an arbitrary number of steps, respectively) as above. Moreover, we write

$$x \Rightarrow_i^t y \quad \text{iff} \quad x \Rightarrow_i^* y \text{ and there is no } z \text{ such that } y \Rightarrow_i z.$$

For a CD grammar system $G = (N, T, S, P_1, \ldots, P_n)$ with only accepting context-free rules, we define the above relations $x \Rightarrow_i y$, $\Rightarrow_i^{=k}$, $\Rightarrow_i^{\le k}$, $\Rightarrow_i^{\ge k}$, $\Rightarrow_i^*$, and $\Rightarrow_i^t$ appropriately. In particular, for $x, y \in (N \cup T)^*$ and $1 \le i \le n$, we write $x \Rightarrow_i y$ iff $x = x_1 z x_2$, $y = x_1 A x_2$ for some $z \to A \in P_i$.

Let $D := \{*, t\} \cup \{\le k, = k, \ge k \mid k \in \mathbb{N}\}$. The language *generated* in the f-mode, $f \in D$, by a CD grammar system G with only generating rules is defined as:

$$L_f^{gen}(G) := \{ w \in T^* \mid S \Rightarrow_{i_1}^f \alpha_1 \Rightarrow_{i_2}^f \ldots \Rightarrow_{i_{m-1}}^f \alpha_{m-1} \Rightarrow_{i_m}^f \alpha_m = w \text{ with}$$
$$m \ge 1, 1 \le i_j \le n, \text{ and } 1 \le j \le m \}.$$

Similarly, one can define the language *accepted* in f-mode, $f \in D$, by a CD grammar system G with only accepting rules:

$$L_f^{acc}(G) := \{ w \in T^* \mid w \Rightarrow_{i_1}^f \alpha_1 \Rightarrow_{i_2}^f \ldots \Rightarrow_{i_{m-1}}^f \alpha_{m-1} \Rightarrow_{i_m}^f \alpha_m = S \text{ with}$$
$$m \ge 1, 1 \le i_j \le n, \text{ and } 1 \le j \le m \}$$

If $f \in D$, then the families of languages generated (accepted, respectively) in f-mode by CD grammar systems with at most n components using context-free [λ-free] rules are denoted by $\mathcal{L}^{gen}(\mathrm{CD}_n, \mathrm{CF}[-\lambda], f)$ ($\mathcal{L}^{acc}(\mathrm{CD}_n, \mathrm{CF}[-\lambda], f)$, respectively). If the number of components is not restricted, then we write $\mathcal{L}^{gen}(\mathrm{CD}_\infty, \mathrm{CF}[-\lambda], f)$, $\mathcal{L}^{acc}(\mathrm{CD}_\infty, \mathrm{CF}[-\lambda], f)$, respectively. Notice that in order to save space, we have used (and will employ later) a bracket notation; when used within a sentence or statement, this means that the reader is asked to either consistently neglect the symbols in brackets or to consistently read all the contents in the brackets. So, in a sense, two statements are presented at the same time, mostly differentiating between the cases that allow or forbid erasing productions. One more comment is in order here: As it is usually the case in the theory of Regulated Rewriting, we consider two languages to be equal if they differ at most by the empty word. With this convention, the result $\mathcal{L}^{gen}(\text{CD}_\infty, \text{CF} - \lambda, t) = \mathcal{L}(\text{ET0L})$ (contained in the theorem below) is true, although clearly ET0L systems can generate languages that contain the empty word and that can, hence, not be generated by non-erasing CDGS.

Finally, the reader should be familiar with the extended Chomsky hierarchy:

$$\mathcal{L}(\text{REG}) \subsetneq \mathcal{L}(\text{CF}) \subsetneq \mathcal{L}(\text{ET0L}) \subsetneq \mathcal{L}(\text{CS}) \subsetneq \mathcal{L}(\text{RE}).$$

From [25], we recall the following result:

Theorem 1. *(1) If $n \in \mathbb{N} \cup \{\infty\}$ and $f \in D \setminus \{t\}$, then*

$$\mathcal{L}^{gen}(\text{CD}_n, \text{CF}[-\lambda], f) = \mathcal{L}^{acc}(\text{CD}_n, \text{CF}[-\lambda], f).$$

(2) $\mathcal{L}^{gen}(\text{CD}_\infty, \text{CF}[-\lambda], t) = \mathcal{L}(\text{ET0L}) \subsetneq \mathcal{L}^{acc}(\text{CD}_\infty, \text{CF}[-\lambda], t) = \mathcal{L}(\text{CS})$.

Later on, we will need a normal form result that relies on the following notion: A CDGS $G = (N, T, S, P_1, \ldots, P_n)$ is called *arity-deterministic* if for each nonterminal $A \in N$, there exists a unique number $\alpha(A)$ (called the *arity* of A) such that any rule $A \to w \in \bigcup_{i=1}^n P_i$ (in the generating case) or $w \to A \in \bigcup_{i=1}^n P_i$ (in the accepting case) obeys $|w| = \alpha(A)$.

Theorem 2 (Arity-deterministic normal form). *For any $f \in D$, $n \in \mathbb{N}$, and $L \in \mathcal{L}^{acc}(\text{CD}_n, \text{CF}[-\lambda], f)$, there exists an arity-deterministic context-free [λ-free] CDGS $G = (N, T, S, P_1, \ldots, P_n)$ with $L_f^{acc}(G) = L$. An analogous statement is true for generating systems.*

Proof. If $L \in \mathcal{L}^{acc}(\text{CD}_n, \text{CF}[-\lambda], f)$, then there is a context-free [λ-free] CDGS $G_0 = (N_0, T, S_0, P_{0,1}, \ldots, P_{0,n})$ with $L_f^{acc}(G_0) = L$. We will construct a sequence of grammar systems $G_k = (N_k, T, S_k, P_{k,1}, \ldots, P_{k,n})$ with $L_f^{acc}(G_k) = L$. We will show by an inductive argument that $G_{|N_0|}$ is arity-deterministic. To this end, we need an auxiliary notion that allows us to quantify the distance of a CDGS from being arity-deterministic. We say that G_k *fails arity-determinism by r* if there are r nonterminals A for which we can find rules $w \to A$ and $w' \to A$ in G_k with $|w| \neq |w'|$. Obviously, G_0 fails arity-determinism by at most $|N_0|$. Assume (by induction hypothesis) that G_k fails arity-determinism by at most $|N_0| - k$ and that $L^{acc}(G_k) = L$. If $k = |N_0|$, then G_k is arity-deterministic as claimed. Otherwise, there is a nonterminal A such that

$$\beta(A) = \{|w| \mid \exists j \in \{1, \ldots, n\} \exists w \in (T \cup N_k)^* : w \to A \in P_{k,j}\}$$

contains at least two elements. Replacing A, we introduce nonterminals (A, n) with $n \in \beta(A)$.

(1) Left-hand sides w of rules that contain A are replaced by any (A, n) for $n \in \beta(A)$ (so that this may yield quite an increase on the number of rules).
(2) Right-hand sides A of rules $w \to A$ (where w does not contain any A after step 1)) are replaced by $(A, |w|)$.

This describes a new CDGS G_{k+1} that is obviously equivalent to G_k and that fails arity-determinism by at most $|N_0| - k - 1$ by construction.

Clearly, there is some $k \leq |N_0|$ such that $L^{acc}(G_k) = L$ and G_k is arity-deterministic by induction. $\square$

2.2   Tree Automata

Let $\mathbb{N}$ be the set of nonnegative integers and let $(\mathbb{N}^*, \cdot, \lambda)$ (or simply $\mathbb{N}^*$) be the free monoid generated by $\mathbb{N}$. For $y, x \in \mathbb{N}^*$, we write $y \leq x$ iff there is a $z \in \mathbb{N}^*$ with $x = y \cdot z$. "$y < x$" abbreviates: $y \leq x$ and $y \neq x$. As usual, $|x|$ denotes the length of the word x.

We are now giving the necessary definitions for trees and tree automata. More details can be found, e.g., in the chapter written by Gécseg and Steinby in [36].

A *ranked alphabet* V is a finite set of symbols together with a finite relation called rank relation $r_V \subset V \times \mathbb{N}$. Define $V_n := \{f \in V \mid (f, n) \in r_V\}$. Since elements in V_n are often considered as *function symbols* (standing for functions of *arity* n), elements in V_0 are also called *constant symbols*. A *tree over* V is a mapping $t : \Delta_t \to V$, where the domain Δ_t is a finite subset of $\mathbb{N}^*$ such that (1) if $x \in \Delta_t$ and $y < x$, then $y \in \Delta_t$; (2) if $y \cdot i \in \Delta_t$, $i \in \mathbb{N}$, then $y \cdot j \in \Delta_t$ for $1 \leq j \leq i$. An element of Δ_t is also called a *node* of t, where the node λ is the *root* of the tree. Then $t(x) \in V_n$ whenever, for $i \in \mathbb{N}$, $x \cdot i \in \Delta_t$ iff $1 \leq i \leq n$. If $t(x) = A$, A is the *label* of x. Let V^t denote the set of all finite trees over V. By this definition, trees are rooted, directed, acyclic graphs in which every node except the root has one predecessor and the direct successors of any node are linearly ordered from left to right. Interpreting V as a set of function symbols, V^t can be identified with the well-formed terms over V. A *frontier node* in t is a node $y \in \Delta_t$ such there is no $x \in \Delta_t$ with $y < x$. If $y \in \Delta_t$ is not a frontier node, it is called *interior node*. The *depth* of a tree t is defined as $\text{depth}(t) = \max\{|x| \mid x \in \Delta_t\}$, whereas the *size* of t is given by $|\Delta_t|$. Letters will be viewed as trees of size one and depth zero.

We are now going to define a catenation on trees. Let \$ be a new symbol, i.e., $\$ \notin V$, of rank 0. Let $V_\$^t$ denote the set of all trees over $V \cup \{\$\}$ which contain exactly one occurrence of label \$. By definition, only frontier nodes can carry the label \$. For trees $u \in V_\$^t$ and $t \in (V^t \cup V_\$^t)$, we define an operation $\#$ to replace the frontier node labelled with \$ of u by t according to

$$u\#t(x) = \begin{cases} u(x), & \text{if } x \in \Delta_u \wedge u(x) \neq \$, \\ t(y), & \text{if } x = z \cdot y \wedge u(z) = \$ \wedge y \in \Delta_t. \end{cases}$$

If $U \subseteq V_\$^t$ and $T \subseteq (V^t \cup V_\$^t)$, then $U\#T := \{u\#t \mid u \in U \wedge t \in T\}$. For $t \in V^t$ and $x \in \Delta_t$, the *subtree* of t at x, denoted by t/x, is defined by $t/x(y) = t(x \cdot y)$ for any $y \in \Delta_{t/x}$, where $\Delta_{t/x} := \{y \mid x \cdot y \in \Delta_t\}$. $\text{ST}(T) := \{t/x \mid t \in T \wedge x \in \Delta_t\}$ is the set of subtrees of trees from $T \subseteq V^t$. Furthermore, for any $t \in V^t$ and any tree language $T \subseteq V^t$, the *quotient* of T and t is defined as:

$$U_T(t) := \begin{cases} \{u \in V_\$^t \mid u\#t \in T\}, & \text{if } t \in V^t \setminus V_0, \\ t, & \text{if } t \in V_0. \end{cases}$$

Let V be a ranked alphabet and m be the maximum rank of the symbols in V. A *(bottom-up) (finite-state) tree automaton* over V is a quadruple $A = (Q, V, \delta, F)$ such that Q is a finite state alphabet (disjoint with V_0), $F \subseteq Q$ is a set of final states, and $\delta = (\delta_0, \ldots, \delta_m)$ is an $m+1$-tuple of state transition functions, where $\delta_0(a) = \{a\}$ for $a \in V_0$ and $\delta_k : V_k \times (Q \cup V_0)^k \to 2^Q$ for $k = 1, \ldots, m$. In this definition, the constant symbols at the frontier nodes are taken as sort of initial states. Now, a transition relation (also denoted by δ) can be recursively defined on V^{t} by letting

$$\delta(f(t_1, \ldots, t_k)) := \begin{cases} \{f\}, & \text{if } k = 0, \\ \bigcup_{q_i \in \delta(t_i), i=1,\ldots,k} \delta_k(f, q_1, \ldots, q_k), & \text{if } k > 0. \end{cases}$$

A tree t is accepted by A iff $\delta(t) \cap F \neq \emptyset$. The tree language accepted by A is denoted by $T(A)$. A is *deterministic* if each of the functions δ_k maps each possible argument to a set of cardinality of at most one. Deterministic tree automata can be viewed as algorithms for labelling the nodes of a tree with states. Analogously to the case of string automata, it can be shown that nondeterministic and deterministic bottom-up finite-state tree automata accept the same class of tree languages, namely the *regular tree languages*, at the expense of a possibly exponential state explosion.

An alternative view on the work of tree automata is that of labelling a tree with states. To this end, we will formally view all symbols from Q as having rank zero, so that they may serve as labels of frontier nodes. Now, A (or more specifically, its transition function δ) defines a derivation relation $\vdash_\delta$ on $(V \cup Q)^{\mathrm{t}}$ by $s \vdash_\delta t$ if $s \neq t$ and there are trees $u \in (V \cup Q)^{\mathrm{t}}_\$$, $s' = f(q_1, \ldots, q_k)$, $t' = q$ with $s = u\#s'$, $t = u\#t'$, $f \in V_k$, $q \in \delta_k(f, q_1, \ldots, q_k)$. Clearly, $s \in V^{\mathrm{t}}$ is accepted by a tree automaton $A = (Q, V, \delta, F)$ if $s \vdash_\delta^* q_f$ for some $q_f \in F$.

We are now defining the central new notion of this paper. A *cooperating distributed (bottom-up finite-state) tree automata system*, or CDTAS for short, of degree n, with $n \geq 1$, is a $(n + 3)$-tuple $A = (Q, V, \delta_1, \ldots, \delta_n, F)$, where $A_i = (Q, V, \delta_i, F)$ are bottom-up finite-state tree automata. As with CDGS, $\vdash_i$ denotes the derivation relation $\vdash_{\delta_i}$, and we can introduce $\vdash_i^f$ for $f \in D := \{*, t\} \cup \{\leq k, = k, \geq k \mid k \in \mathbb{N}\}$ as before, leading to the tree language

$$L_f(A) := \{ t \in V^{\mathrm{t}} \mid t \Rightarrow_{i_1}^f t_1 \Rightarrow_{i_2}^f \cdots \Rightarrow_{i_{m-1}}^f t_{m-1} \Rightarrow_{i_m}^f t_m \in F \text{ with}$$

$$m \geq 1, 1 \leq i_j \leq n, \text{ and } 1 \leq j \leq m \}$$

and corresponding tree language families $\mathcal{L}^{\mathrm{t}}(\mathrm{CD}_n, f)$.

Notice that we can assume $|F| = 1$ as a normal form, as otherwise we might introduce a new (unique) final state q_f and (to each transition component δ_i) rules that might lead into q_f whenever a transition into some $q \in F$ has been possible in δ_i before. This construction will possibly turn a deterministic δ_i into a nondeterministic one, but this is not of any concern in this paper.

3   Basic Results

The main result shows that the famous theorem usually attributed to Doner, Thatcher (and also to Wright) [18,37] transfers and generalizes to the case of cooperating distributed systems. This theorem characterizes context-free languages via yields of regular tree languages. We therefore provide the necessary key notion: For $t \in V^{\mathbf{t}}$, we define the *yield-operator* $\mathcal{Y}$ as follows:

$$\mathcal{Y}(t) = \begin{cases} t(\lambda), & \text{if } t(\lambda) \in V_0 \\ \mathcal{Y}(t/1) \cdots \mathcal{Y}(t/k), & \text{if } t(\lambda) \in V_k, k > 0 \end{cases}$$

The operator naturally extends to tree languages and tree language families.

Lemma 1

$$\forall n \in \mathbb{N} \cup \{\infty\} \forall f \in D : \mathcal{Y}(\mathcal{L}^{\mathbf{t}}(\mathrm{CD}_n, f)) \subseteq \mathcal{L}^{acc}(\mathrm{CD}_n, \mathrm{CF} - \lambda, f).$$

Proof. As it is usually the case with nondeterministic automata, we can assume (without loss of generality) that CDTAS have only one final (accepting) state. So, consider a CDTAS $A = (Q, V, \delta_1, \ldots, \delta_n, F)$ with $|F| = 1$, i.e., $F = \{q_f\}$. Let m be the maximum rank of symbols from V. We construct a simulating accepting CDGS $G_A = (N, T, q_f, P_1, \ldots, P_n)$, as follows: $N = Q$, $T = V_0$, $w \to q \in P_i$ if $q \in Q$, $w = w_1 \cdots w_k$, $w_j \in V_0 \cup Q$ whenever $q \in \delta_k(g, w_1, \ldots, w_k)$ for some $g \in V_k$ (*).

Now, each derivation of A producing a certain yield can be simulated by G_A, where the correct labels of inner nodes are guessed during the derivation due to (*). Conversely, these guesses according to (*) label the inner nodes of a derivation tree in a way corresponding to a tree that can be accepted by A.

A formal reasoning would be a relatively tedious exercise, based on the ideas originating from Doner, Thatcher and Wright in the late sixties, which can be also found in any textbook on tree languages. Therefore, we only sketch the basic idea of the inductive step of the proof in the following. Recall that the definition of $\vdash_i$ transforms trees with leaf labels from $(V_0 \cup Q)$ into trees with leaf labels from $(V_0 \cup Q)$; extending the definition of the yield operator $\mathcal{Y}$ accordingly, this means that sentential forms of G_A are transformed, so that the basic claim can be written as:

$$\forall 1 \leq i \leq n : (s \vdash_i t) \implies (\mathcal{Y}(s) \Rightarrow_i \mathcal{Y}(t)), \tag{1}$$

where $\Rightarrow_i$ refers to an application of a rule from P_i. The converse of this claim would contain the mentioned guessing step of labels of inner nodes, i.e., whenever $u \Rightarrow_i v$ according to G_A, then there are trees s, t such that $\mathcal{Y}(s) = u$ and $\mathcal{Y}(t) = v$, and also $s \vdash_i t$.

Moreover, notice that the derivation modes translate 1-1 between the tree automaton model and the accepting grammar model. For instance, Equation (1) would translate into:

$$\forall f \in D \forall 1 \leq i \leq n : (s \vdash_i^f t) \implies (\mathcal{Y}(s) \Rightarrow_i^f \mathcal{Y}(t)),$$

This claim follows easily (again) by induction apart from the t-mode. Here, we need the following additional argument: There is no tree t such that $s \vdash_i t$ if and only if there is no sentential form v such that $u \Rightarrow_i v$ for $u = \mathcal{Y}(s)$. But this claim follows immediately from the definition of the rules in G_A.

For the converse, we would have to argue for all modes $f \in D$ that, whenever $u \Rightarrow_i^f v$ according to G_A, then there are trees s, t such that $\mathcal{Y}(s) = u$ and $\mathcal{Y}(t) = v$, and also $s \vdash_i^f t$. $\qquad \square$

Lemma 2

$$\forall n \in \mathbb{N} \cup \{\infty\} \forall f \in D : \mathcal{Y}(\mathcal{L}^{\mathrm{t}}(\mathrm{CD}_n, f)) \supseteq \mathcal{L}^{acc}(\mathrm{CD}_n, \mathrm{CF} - \lambda, f).$$

Proof. Consider an accepting CDGS $G = (N, T, S, P_1, \ldots, P_n)$. As we have shown above, we may assume that G is arity-deterministic. The rank of $A \in N$ is hence given by an arity function α. The idea is to construct tree automata that accept the derivation trees of the component grammars.

As G contains no erasing rules, this means that we can consider $N \cup T$ as a ranked alphabet V, with $V_0 = T$. We construct an equivalent CDTAS $A_G = (Q, V, \delta_1, \ldots, \delta_n, F)$ as follows. $Q = \{A' \mid A \in N\}$. Let $h : Q \cup V \to V$ be defined as being constant on V and removing the prime from A' to yield A; h can be easily extended to a morphism $h : (Q \cup V)^+ \to V^*$. We build δ_i as follows: For any rule $w \to A \in P_i$, we introduce a rule $\delta_{i,k}(A, w_1, \ldots, w_k) = \{A'\}$, where $|w| = \alpha(A) = k$ and $h(w_1) \cdots h(w_k) = w$. Finally, set $F = \{S'\}$.

A_G accepts derivation trees of G. Conversely, the yield of any tree that derives S' (in A_G) corresponds to a sentential form that derives S (in G). The straightforward induction proof showing the correctness of the construction is left to the reader. $\qquad \square$

This allows us to state:

Theorem 3

$$\forall n \in \mathbb{N} \cup \{\infty\} \forall f \in D : \mathcal{Y}(\mathcal{L}^{\mathrm{t}}(\mathrm{CD}_n, f)) = \mathcal{L}^{acc}(\mathrm{CD}_n, \mathrm{CF} - \lambda, f).$$

Without going into further formal details, let us remark that, apart from the notion of bottom-up tree automata as introduced above, there also exists a notion of top-down tree automata. The according formalizations can be found (again) in any textbook on tree languages and are suppressed here in the interest of space. The yields of of top-down tree automata naturally correspond to word languages generated by grammars. This can be generalized towards grammar systems completely analogous as before. In our notations, we will write for instance $\mathcal{L}^{\mathrm{t}}(\mathrm{TD\text{-}CD}_n, f)$ to refer to tree languages that are accepted by cooperating distributed tree automata systems that work top-down.

We only summarize the corresponding statements below.

Corollary 1

$$\forall n \in \mathbb{N} \cup \{\infty\} \forall f \in D : \mathcal{Y}(\mathcal{L}^{\mathrm{t}}(\mathrm{TD\text{-}CD}_n, f)) = \mathcal{L}^{gen}(\mathrm{CD}_n, \mathrm{CF} - \lambda, f).$$

From Theorem 1, we can hence deduce:

Corollary 2. *(1) If $n \in \mathbb{N} \cup \{\infty\}$ and $f \in D \setminus \{t\}$, then*

$$\mathcal{L}^{\mathrm{t}}(\mathrm{TD\text{-}CD}_n, f) = \mathcal{L}^{\mathrm{t}}(\mathrm{CD}_n, f).$$

(2) $\mathcal{L}^{\mathrm{t}}(\mathrm{TD\text{-}CD}_n, t) \subsetneq \mathcal{L}^{\mathrm{t}}(\mathrm{CD}_n, t)$.

4   Comments and Remarks

Remark 1. There is another line of interpretation of our results, namely purely on the level of tree grammars. It is well-known that regular tree languages can be (also) characterized by regular tree grammars. The usual proof shows that (generating) regular tree grammars can be seen as a re-interpretation of top-down finite tree automata. Basically the same reasoning (which is hence not repeated here) shows that accepting regular tree grammars (a notion that was not fomerly studied in the literature to the best of the knowledge of the author) are in one-to-one correspondance with bottom-up finite tree automata. Of course, one can (now) define generating and accepting CDGS whose components are regular tree grammars. Then, we may state that for any mode but the *t*-mode, generating and accepting CDGS having regular tree grammar components describe the same tree language families, while accepting CDGS with regular tree grammar components are strictly more powerful than their generating counterparts. Such results are perfectly in line with earlier findings in the case of word languages; see [7,23,25].

Remark 2. The relations between CDGS and systems of tree automata should easily extend beyond what has been stated so far. We only discuss four extensions here:

1. Mitrana and Păun introduced in [32,34] hybrid CDGS, i.e., systems where each component runs according to an individual mode.
2. There habe been also many other modes introduced for CDGS whose definitions easily transfer to CDTAS, for instance, those derived by internal hybridization [26,22] or those based on competence notions [3,6,12,14].
3. Occasionally, CDGS have been considered whose components work accepting or generating (bidirectional); see [24]. This finds natural analogues in CDTAS whose components work top-down or bottom-up.
4. The notion of multi-dimensional trees allows to have, for instance, trees as yields of higher-dimensional trees, whose (string) yields finally characterize important classes of mildly context-sensitive languages, one of the core topics of (computer-)linguistics; see [28,35]. For those mainly interested in Regulated Rewriting, the connections to the Weir hierarchy of control languages described in [35,38] should be an intersting aspect in itself.

It looks as if in most cases (at least), the equivalences between yields of CDTAS and the power of CDGS hold for these variants, as well.

Remark 3. While the standard notion of derivation in grammars is that of a grammar as a *generative* device, standard tree automata work bottom-up. Hence, these standard notions define the same word language classes (in the case of tree automata, via the yield operation) only because in most cases, the power of generating and accepting grammars coincide [7,8,25,23,31]. This is not the case with various types of Regulated Rewriting, so that investigating accordingly defined tree automta might yield interesting results.

5   Conclusions and Future Research

Tree automata have become an increasingly important topic in formal language theory, due to their versatility and applicability. Can aspects of cooperation and distribution stir even more interest, i.e., can automata systems as proposed in this note be useful in applications? Or will techniques from tree automata be useful for solving old open questions in CDGS or other areas within Regulated Rewriting, see Remark 3? In this context, we like to draw the reader's attention to the papers [20,21,19] which are the only ones we could recall that is somehow interconnecting regulated rewriting and tree automata theory. Further relations to linguistics have been described in the preceding Remark 2; in particular, we mention again the tight relation between the Weir hierarchy (within Regulated Rewriting) and the formalism of multi-dimensional trees.

Furthermore, there is some hope that the viewpoint of tree automata might bring CDGS closer to being applied in areas like natural language processing. For instance, the way these automata interface may be seen as a special case of the interaction possible within so-called millstream systems that were introduced from linguistic motivations; see [4,5]. It might be nice to continue this line of research in more depth. Another aspect might be that bidirectional grammars (also discussed in the preceding Remark) have their foundings in linguistics, as well; see [1,2,27]. In this context, also research relating context-free languages to classes of graphs more general than trees might revive; see [9].

In the relations we obtained in this note, we considered tree automata basically working step-by-step as transformators of trees corresponding to derivation trees. More generally, it might be interesting to explore cooperating distributed systems of tree transducers (as defined in textbooks on tree automata), which might also give interesting applications in the world of XML document processing; see [29,30,33].

Acknowledgements. We are grateful to Anna Kasprzik for some discussions and to the anonymous referees for their insightful comments.

References

1. Appelt, D.E.: Bidirectional grammars and the design of natural language generation systems. In: Proceedings of Third Conference on Theoretical Issues in Natural Language Processing (TINLAP-3), January 7-9, pp. 185–191. New Mexico State University, Las Cruces (1987)
2. Asveld, P.R.J., Hogendorp, J.A.: On the generating power of regularly controlled bidirectional grammars. International Journal of Computer Mathematics 40 (1991)
3. ter Beek, M.H., Csuhaj-Varjú, E., Holzer, M., Vaszil, G.: Cooperating Distributed Grammar Systems: Components with Nonincreasing Competence. In: Kelemen, J., Kelemenová, A. (eds.) Computation, Cooperation, and Life. LNCS, vol. 6610, pp. 70–89. Springer, Heidelberg (2011)
4. Bensch, S., Björklund, H., Drewes, F.: Algorithmic Properties of Millstream Systems. In: Gao, Y., Lu, H., Seki, S., Yu, S. (eds.) DLT 2010. LNCS, vol. 6224, pp. 54–65. Springer, Heidelberg (2010)

5. Bensch, S., Drewes, F.: Millstream systems — a formal model for linking language modules by interfaces. In: Workshop on Applications of Tree Automata in Natural Language Systems, pp. 28–36. Association for Computational Linguistics ACL (2010)

6. Bordihn, H., Csuhaj-Varjú, E.: On competence and completeness in CD grammar systems. Acta Cybernetica 12, 347–360 (1996)

7. Bordihn, H., Fernau, H.: Accepting grammars with regulation. International Journal of Computer Mathematics 53, 1–18 (1994)

8. Bordihn, H., Fernau, H.: Accepting grammars and systems via context condition grammars. Journal of Automata, Languages and Combinatorics 1(2), 97–112 (1996)

9. Courcelle, B., Lapoire, D.: Facial Circuits of Planar Graphs and Context-Free Languages. In: Brim, L., Gruska, J., Zlatuška, J. (eds.) MFCS 1998. LNCS, vol. 1450, pp. 616–624. Springer, Heidelberg (1998)

10. Csuhaj-Varjú, E., Dassow, J.: On cooperating/distributed grammar systems. J. Inf. Process. Cybern. EIK (formerly Elektron. Inf.verarb. Kybern.) 26(1/2), 49–63 (1990)

11. Csuhaj-Varjú, E., Dassow, J., Kelemen, J., Păun, G.: Grammar Systems: A Grammatical Approach to Distribution and Cooperation. Gordon and Breach, London (1994)

12. Csuhaj-Varjú, E., Dassow, J., Vaszil, G.: Variants of competence-based derivations in cd grammar systems. International Journal of Foundations of Computer Science 21(4), 549–569 (2010)

13. Csuhaj-Varjú, E., Mitrana, V., Vaszil, G.: Distributed Pushdown Automata Systems: Computational Power. In: Ésik, Z., Fülöp, Z. (eds.) DLT 2003. LNCS, vol. 2710, pp. 218–229. Springer, Heidelberg (2003)

14. Dassow, J.: On cooperating distributed grammar systems with competence based start and stop conditions. Fundamenta Informaticae 76(3), 293–304 (2007)

15. Dassow, J., Mitrana, V.: Cooperating distributed push-down automata. Tech. rep., Universität Magdeburg, Fakultät für Informatik (1996)

16. Dassow, J., Mitrana, V.: Stack cooperation in multistack pushdown automata. Journal of Computer and System Sciences 58(3), 611–621 (1999)

17. Dassow, J., Păun, G., Vicolov, S.: On the power of cooperating/distributed grammar systems with regular components. Foundations of Computing and Decision Sciences 18(2), 83–108 (1993)

18. Doner, J.: Tree acceptors and some of their applications. Journal of Computer and System Sciences 4(5), 406–451 (1970)

19. Drewes, F., van der Merwe, B.: Path languages of random permitting context tree grammars are regular. Fundamenta Informaticae 82(1-2), 47–60 (2008)

20. Drewes, F., du Toit, C., Ewert, S., van der Merwe, B., van der Walt, A.P.J.: Random context tree grammars and tree transducers. South African Computer Journal 34, 11–25 (2005)

21. Drewes, F., du Toit, C., Ewert, S., van der Merwe, B., van der Walt, A.P.J.: Bag context tree grammars. Fundamenta Informaticae 86(4), 459–480 (2008)

22. Fernau, H., Freund, R., Holzer, M.: Hybrid modes in cooperating distributed grammar systems: combining the t-mode with the modes $\leq k$ and $= k$. Theoretical Computer Science 299, 633–662 (2003)

23. Fernau, H., Holzer, M.: Accepting multi-agent systems II. Acta Cybernetica 12, 361–379 (1996)

24. Fernau, H., Holzer, M.: Bidirectional cooperating distributed grammar systems. Publicationes Mathematicae, Debrecen 54(supplement), 787–806 (1999)

25. Fernau, H., Holzer, M., Bordihn, H.: Accepting multi-agent systems: the case of cooperating distributed grammar systems. Computers and Artificial Intelligence 15(2-3), 123–139 (1996)
26. Fernau, H., Holzer, M., Freund, R.: Hybrid modes in cooperating distributed grammar systems: internal versus external hybridization. Theoretical Computer Science 259(1-2), 405–426 (2001)
27. Hogendorp, J.A.: Controlled bidirectional grammars. International Journal of Computer Mathematics 27, 159–180 (1989)
28. Kasprzik, A.: Making finite-state methods applicable to languages beyond context-freeness via multi-dimensional trees. In: Piskorski, J., Watson, B., Yli-Jyrä, A. (eds.) Post-proceedings of the 7th International Workshop on Finite-State Methods and Natural Language Processing, pp. 98–109. IOS Press, Amsterdam (2009)
29. Martens, W., Neven, F.: Frontiers of tractability for typechecking simple XML transformations. Journal of Computer and System Sciences 73(3), 362–390 (2007)
30. Martens, W., Neven, F., Gyssens, M.: Typechecking top-down XML transformations: Fixed input or output schemas. Information and Computation 206(7), 806–827 (2008)
31. Mihalache, V.: Accepting cooperating distributed grammar systems with terminal derivation. EATCS Bulletin 61, 80–84 (1997)
32. Mitrana, V.: Hybrid cooperating/distributed grammar systems. Computers and Artificial Intelligence 12(1), 83–88 (1993)
33. Neven, F.: Automata, Logic, and XML. In: Bradfield, J. (ed.) CSL 2002. LNCS, vol. 2471, pp. 2–26. Springer, Heidelberg (2002)
34. Păun, G.: On the generative capacity of hybrid CD grammar systems. J. Inf. Process. Cybern. EIK (formerly Elektron. Inf.verarb. Kybern.) 30(4), 231–244 (1994)
35. Rogers, J.: Syntactic structures as multi-dimensional trees. Research on Language & Computation 1, 265–305 (2003)
36. Rozenberg, G., Salomaa, A. (eds.): Handbook of Formal Languages, vol. III. Springer, Berlin (1997)
37. Thatcher, J.W.: Characterizing derivation trees of context-free grammars through a generalization of finite automata theory. Journal of Computer and System Sciences 1, 317–322 (1967)
38. Weir, D.J.: A geometric hierarchy beyond context-free languages. Theoretical Computer Science 104, 235–261 (1992)

A Note on Combined Derivation Modes
for Cooperating Distributed Grammar Systems

Markus Holzer

Institut für Informatik, Universität Giessen,
Arndtstraße 2, 35392 Giessen, Germany
holzer@informatik.uni-giessen.de

Abstract. We investigate the generative power of cooperating distributed grammar systems (CDGS) with context-free rules working in cut-f-mode of derivation, when f is a full-competence mode in combination with another derivation mode, combined sf-mode, for short. Cut-modes were introduced in [BORDIHN, HOLZER: Cooperating distributed grammar systems as models of distributed problem solving, *Fund. Inform.* 76, 2007] in order to model distributed problem solving by CDGSs, and combined sf-modes were recently investigated in [BORDIHN, HOLZER: A note on cooperating distributed grammar systems working in combined modes. *Inform. Process. Lett.* 108, 2008]. While cut-f-modes were investigated for the classical CDGS derivation modes and moreover also for combined t-modes, the generative capacity for cut-modes based on combined sf-modes was left open in the literature. This paper closes this gap.

1 Introduction

The theory of cooperating distributed grammar systems—for an overview we refer to [6]—has become a well established field in formal language theory since its origin in [3], with the forerunner paper [10]. A cooperating distributed grammar system (CDGS, for short) consists of a finite set of (context-free) grammars, called components, performing derivation steps on a common sentential form in turns, according to some cooperation protocol. Standard cooperation protocols are the so-called $*$-mode, $\leq k$-mode, $= k$-mode, or $\geq k$-mode, where a component, once started, has to perform an arbitrary number, at most k, exactly k, or at least k derivation steps, respectively. Moreover, there are two basic cooperation protocols which are based on the feature of competence of the problem solving agents on the current state of the problem solving: (1) In the t-mode, a component can start and has to remain deriving unless and until there is no nonterminal left in the sentential form to which one of its productions is applicable (that is, the component is not able to contribute to the problem solving any more), and (2) in the sf-mode, a component is allowed to become and has to remain active unless and until there is some nonterminal present in the sentential form which cannot be rewritten by this component (that is, the component does not possess the full competence on the current state of the problem

H. Bordihn, M. Kutrib, and B. Truthe (Eds.): Dassow Festschrift 2012, LNCS 7300, pp. 86–98, 2012.

solving). Forcing a derivation to fulfill two or multiple requirements simultaneously leads to the so called combined modes from [7]. In particular, combined t- and combined sf-modes were investigated in the literature [2,7]. We summarize the results for CDGSs working in the above mentioned derivation modes in Table 1. Here $\mathcal{L}(\mathrm{LIN})$ denotes the family of linear context-free languages,

Table 1. Generative capacity of CD grammar systems working in the classical modes, the combined t-modes, and the combined sf-modes of derivation compared

Derivation mode f	$\mathcal{L}(\mathrm{CD}, \mathrm{CF}[-\lambda], f)$	$\mathcal{L}(\mathrm{CD}, \mathrm{CF}[-\lambda], (t \wedge f))$	$\mathcal{L}(\mathrm{CD}, \mathrm{CF}[-\lambda], (sf \wedge f))$
$*$		$\mathcal{L}(\mathrm{ET0L})$	$\mathcal{L}(\mathrm{P}, \mathrm{CF}[-\lambda], \mathrm{ac})$
≤ 1		$\mathcal{L}_{fin}(\mathrm{P}, \mathrm{CF}[-\lambda])$	$\mathcal{L}(\mathrm{LIN})$
$= 1$	$\mathcal{L}(\mathrm{CF})$		
≥ 1		$\mathcal{L}(\mathrm{ET0L})$	
$\leq k$ with $k \geq 2$		$\mathcal{L}_{fin}(\mathrm{P}, \mathrm{CF}[-\lambda])$	
$= k$ with $k \geq 2$	$\mathcal{L}(\mathrm{CF}) \subset \cdot \subseteq \mathcal{L}(\mathrm{P}, \mathrm{CF}[-\lambda])$		$\mathcal{L}(\mathrm{P}, \mathrm{CF}[-\lambda], \mathrm{ac})$
$\geq k$ with $k \geq 2$		$\mathcal{L}(\mathrm{RP}, \mathrm{CF}[-\lambda], \mathrm{ac})$	
t	$\mathcal{L}(\mathrm{ET0L})$		$\mathcal{L}(\mathrm{CF})$
sf	$\mathcal{L}(\mathrm{P}, \mathrm{CF}[-\lambda], \mathrm{ac})$	$\mathcal{L}(\mathrm{CF})$	$\mathcal{L}(\mathrm{P}, \mathrm{CF}[-\lambda], \mathrm{ac})$

$\mathcal{L}(\mathrm{CF})$ the family of context-free languages, and $\mathcal{L}(\mathrm{ET0L})$ the family of languages generated by ET0L systems. Moreover, $\mathcal{L}(\mathrm{P}, \mathrm{CF}[-\lambda], \mathrm{ac})$ ($\mathcal{L}(\mathrm{P}, \mathrm{CF}[-\lambda])$, respectively) refers to the family of all languages generated by programmed context-free grammars with (without, respectively) appearance checking. The definition of a programmed context-free grammar is briefly recalled in one of the next sections. We note that $\mathcal{L}(\mathrm{P}, \mathrm{CF}, \mathrm{ac})$ equals the family of recursively enumerable languages, while $\mathcal{L}(\mathrm{P}, \mathrm{CF} - \lambda, \mathrm{ac})$ is a proper subset of the family of context-sensitive languages. The family of languages generated by programmed context-free grammars without appearance checking of finite index is denoted by $\mathcal{L}_{fin}(\mathrm{P}, \mathrm{CF}[-\lambda])$. Loosely speaking, the index of a grammar is the maximal number of nonterminals simultaneously appearing in a sentential form during a terminating derivation, considering the most economical derivation for each string. Finally, $\mathcal{L}(\mathrm{RP}, \mathrm{CF}[-\lambda], \mathrm{ac})$ refers to the family of languages generated by recurrent programmed grammars introduced in [11].

Originally CDGSs were introduced to formally model communities of cooperating autonomous problem solving agents which use the blackboard model of problem solving [3,4]. Recently in [1] this original motivation was used to revisit the known cooperation protocols for CDGSs under the effect that agents contribute to the solution by solving sub-tasks of the whole problem. In terms of CD grammar systems, solving sub-tasks corresponds to working on a substring of the current sentential form. This led to the concept of CDGSs working in the *cut-f-mode* of derivation, where f is one of the aforementioned cooperation protocols. In these *cut-f-modes*, in any derivation step the sentential form is partitioned (cut) into several substrings which can be associated to different components. In order to avoid any rigidity, this association is done via a partial mapping, such that both some substrings of the sentential form and some components may be disregarded. Then, each component to which a substring has been associated works in (one and the same) derivation mode, more precisely, in the f-mode of derivation if the CDGS as a whole is driven in the *cut-f-mode*. Finally, the new sentential form is obtained after finishing this procedure by concatenating all the subwords, regarded or disregarded by the partial mapping, in their original order. Clearly from the AI motivation this approach seems to be more adequate. The results presented in [1] only deal with the cut-mode versions of the classical derivation modes, as well as the cut-mode versions of the combined t-modes. To our knowledge, surprisingly, the cut-mode versions of the combined sf-modes were not investigated so far. In this paper we close this gap and complete the picture of CDGSs working in these sub-task oriented cooperation protocols. We summarize the results on CDGS working in cut-mode derivations for the above mentioned derivation modes in Table 2, which can be found in Section 4. The next section contains preliminaries, where we provide the basic definition of cut-f-mode derivations, in particular cut-mode versions of combined sf-modes. Then in Section 3 we investigate the generative power of CD grammar systems working in these newly defined derivations modes and, finally, we summarize our results and highlight the remaining open questions in Section 4.

2   Definitions

We assume the reader to be familiar with the standard notions of formal language theory as contained in [5]. In particular, for some alphabet V, let $v \in V^*$ be a word over V, and if $W \subseteq V$, then $|v|_W$ denotes the number of occurrences of symbols from W in the word v. In what follows, we consider two languages to be equal if they differ at most by the empty word λ.

A context-free cooperating distributed grammar system (CDGS) with n components, $n \geq 1$, is a construct $G = (N, T, P_1, P_2, \ldots, P_n, S)$, where each $G_i = (N, T, P_i, S)$ is a context-free grammar and P_i is called a *component* of G. For $1 \leq i \leq n$, let

$$\text{dom}(P_i) = \{\, A \in N \mid \text{there is a word } v \text{ such that } A \to v \in P_i \,\}$$

denote the set of all nonterminals which can be rewritten by the component P_i. For $x, y \in (N \cup T)^*$ and $1 \leq i \leq n$, we write $x \Rightarrow_i y$ if and only if $x = x_1 A x_2$ and $y = x_1 z x_2$ for some $A \to z \in P_i$. Hence, subscript i refers to the component to be used. By $\Rightarrow_i^{\leq k}, \Rightarrow_i^{=k}, \Rightarrow_i^{\geq k}, \Rightarrow_i^*$, for $k \geq 1$, we denote a derivation consisting of at most k steps, exactly k steps, at least k steps, an arbitrary number of steps, respectively, executed by component P_i. Furthermore, we write $x \Rightarrow_i^t y$ if and only if $x \Rightarrow_i^* y$ and there is no z such that $y \Rightarrow_i z$. Moreover, $x \Rightarrow_i^{sf} y$ if and only if $x \Rightarrow_i^* x'$, $x' \Rightarrow_i y$, and P_i is sf-competent on x' but is *not* sf-competent on y, where a component P_i is said to be sf-*competent* on a word x if and only if (1) $x = u_0 A_1 u_1 A_2 u_2 \ldots u_{m-1} A_m u_m$ with $m \geq 1$, $u_j \in T^*$, for $0 \leq j \leq m$, and

$A_j \in N$, for $1 \leq j \leq m$, and (2) for each j, for $1 \leq j \leq m$, there is a production $A_j \to w_j$ in P_i. Note that the definition of the derivation relation implies that component P_i is *sf*-competent on x and on all intermediate sentential forms in the derivation $x \Rightarrow_i^* x'$, too.

Combining the former modes with the requirement concerning the *t*-mode we obtain the modes $(t \wedge *)$, $(t \wedge \leq k)$, $(t \wedge = k)$, and $(t \wedge \geq k)$ which are defined as follows—see, e.g., [7,8]: There exists a derivation which satisfies both properties at once, e.g., $x \Rightarrow_i^{(t \wedge \leq k)} y$ if and only if there exists an m-step derivation from x to y using P_i such that $m \leq k$ and there is no z such that $y \Rightarrow_i z$. These *combined t-modes* were investigated in [7,8] in more detail. Moreover, in [2] the authors generalized combined *t*-modes to *combined sf-modes*, which are analogously defined as their corresponding *t*-mode counterparts. In this way one obtains the derivation modes $(sf \wedge *)$, $(sf \wedge \leq k)$, $(sf \wedge = k)$, $(sf \wedge \geq k)$, and even $(sf \wedge t)$. For results on these modes we refer to [2].

Applying the idea on distributed problem solving from [1] to one of the aforementioned modes f leads to the *cut-f-mode*, f_c-mode for short, which is defined as follows: $x \Rightarrow^{f_c} y$ if and only if

1. $x = x_0 x_1 \ldots x_m$ with $m \geq 0$, $x_i \in (N \cup T)^*$, for $0 \leq i \leq m$,
2. there is a partial injective mapping $\rho : \{0, 1, \ldots m\} \hookrightarrow \{1, 2, \ldots, n\}$ such that
   $y_i = x_i$, if $i \notin \mathrm{dom}(\rho)$, and $y_i = z_i$, if $x_i \Rightarrow_{\rho(i)}^f z_i$, and
3. $y = y_0 y_1 \ldots y_m$.

Here $\mathrm{dom}(\rho) = \{ i \mid \rho(i)$ is defined $\}$ denotes the set of indices in the decomposition $x = x_0 x_1 \ldots x_m$ to which some component is associated by ρ.

Let f be some of the aforementioned ordinary derivation modes or a corresponding cut-mode derivation, then the language generated by G working in the f-mode is the set

$$L(G) = \{ w \in T^* \mid S = w_0 \Rightarrow^f w_1 \Rightarrow^f \ldots \Rightarrow^f w_m = w, \text{ for } m \geq 0 \}$$

where $\Rightarrow^f$ denotes the f-mode derivation relation induced by the CDGS G. For any derivation mode f, the family of languages generated by CDGSs with context-free components working in the f-mode of derivation is denoted by $\mathcal{L}(\mathrm{CD}, \mathrm{CF}, f)$. If λ-rules are forbidden, then we use the notation $\mathcal{L}(\mathrm{CD}, \mathrm{CF} - \lambda, f)$.

In order to clarify our notation we give an example, which we literally take from [2], extended by the discussion on the cut-mode derivation.

Example 1. Let $G = (\{S, S', A, A', B, B'\}, \{a, b, c\}, P_1, P_2, P_3, P_4, S)$ be a CD grammar system with the production sets $P_1 = \{S \to S', S' \to AB\}$, $P_2 = \{A \to A', A' \to A', B \to cB'\}$, $P_3 = \{B' \to B, B \to B, A' \to aAb\}$, and $P_4 = \{A \to ab, B \to c\}$.

Then it is not difficult to see that the CDGS G generates the non-context-free language $\{ a^n b^n c^n \mid n \geq 1 \}$ in $(sf \wedge = 2)$-mode. To this end we argue as follows: The derivation has to start with component P_1 resulting in the sentential form AB. Assume the sentential form to be $a^n A b^n c^n B$, for $n \geq 0$. Then either P_2 or P_4 is competent on the sentential from. In the latter case the $(sf \wedge = 2)$-derivation terminates with the terminal word $a^{n+1} b^{n+1} c^{n+1}$. In the former case, i.e., when P_2 is active, the only possible way to successfully continue the $(sf \wedge = 2)$-derivation is to apply $A \to A'$ followed by $B \to cB'$, which makes P_2 non-competent on $a^n A' b^n c^{n+1} B'$. Otherwise, if $B \to cB'$ is applied first, then P_2 is non-competent on the sentential form, thus the $= 2$-requirement is not satisfied; if the application of $A \to A'$ is followed by $A' \to A'$, then P_2 remains competent on the sentential form after two steps, thus the $(sf \wedge = 2)$-condition is not fulfilled. Next, the derivation can only be continued by P_3. By a similar reasoning as above we find that the only successful sequence of rules is $B' \to B$ followed by $A' \to aAb$. This leads to the sentential form $a^{n+1} Ab^{n+1} c^{n+1} B$. Hence, the stated claim on $(sf \wedge = 2)$-mode of derivation follows.

When considering the $(sf \wedge = 2)_c$-mode a similar reasoning applies, since no production set is successfully applicable to a sentential form containing *one* non-terminal only, except from the application of P_1 to the axiom S. Thus, a production set P_i with $2 \leq i \leq 4$ is successful only, if two appropriate nonterminals in the sentential form are present. Therefore, the only possible derivations in $(sf \wedge = 2)_c$-mode are the derivations shown above. Thus, we obtain the non-context-free language mentioned above.   □

Since in the previous example the ordinary and the cut-mode derivation coincides, we give another example were these modes differ.

Example 2. Consider the CDGS $G = (\{S, A, A', B, B'\}, \{a, b\}, P_1, P_2, S)$ with the production sets $P_1 = \{S \to AB\}$ and $P_2 = \{A \to a, B \to b\}$.

The language generated by the CDGS in $(sf \wedge = 1)$-mode is the empty set $\emptyset$. This is seen as follows: The derivation starts with P_1 resulting in the sentential form AB. Then the only component that is competent on this sentential form is P_2. It is easy to see that P_2 cannot perform a successful derivation in $(sf \wedge = 1)$-mode. Thus, no terminal word can be derived from the axiom. This shows the stated claim.

When considering the $(sf \wedge = 1)_c$-mode of derivation, nonterminals can be hidden from a particular production set by the cutting of the sentential form and an appropriate assignment of the values to the partial injective mapping. As above, the derivation starts by $S \Rightarrow^{(sf \wedge = 1)_c} AB$, but now it can be continued by $AB \Rightarrow^{(sf \wedge = 1)_c} aB \Rightarrow^{(sf \wedge = 1)_c} ab$, $AB \Rightarrow^{(sf \wedge = 1)_c} Ab \Rightarrow^{(sf \wedge = 1)_c} ab$, or $AB \Rightarrow^{(sf \wedge = 1)_c} ab$, which are also the only possible derivation sequences at all. Therefore, the language generated by G in $(sf \wedge = 1)$-mode of derivation is the finite set $\{ab\}$.   □

3   Simulation Results

We consider context-free CDGSs working together according to the cut-mode derivation of a combined *sf*-mode protocol. First, we recall some known facts about the generative capacity of CDGSs working in combined *sf*-modes. We start with the cut-mode versions of the $(sf \wedge \leq 1)$- and $(sf \wedge = 1)$-mode of derivation. It was shown in [2] that

$$\mathcal{L}(\text{CD}, \text{CF}[-\lambda], f) = \mathcal{L}(\text{LIN}),$$

for $f \in \{(sf \wedge \leq 1), (sf \wedge = 1)\}$. For the cut-mode version of these two modes we find a characterization in terms of context-free languages (cf. [2]).

Theorem 1. *If $f \in \{(sf \wedge \leq 1), (sf \wedge = 1)\}$, then $\mathcal{L}(\mathrm{CD}, \mathrm{CF}[-\lambda], f_c) = \mathcal{L}(\mathrm{CF})$.*

Proof. Let $G = (N, T, P, S)$ be a context-free grammar. We construct an equivalent CDGS $G' = (N', T, P_1, P_2, S)$ working in the $(sf \wedge = 1)_c$- or $(sf \wedge \leq 1)_c$-mode as follows. Let $N' = N \cup \{A' \mid A \in N\}$, the union being disjoint, and define $P_1 = \{A \rightarrow h(w) \mid A \rightarrow w \in P\}$ and $P_2 = \{h(A) \rightarrow A \mid A \in N\}$, where $h : (N \cup T)^* \rightarrow (N' \cup T)^*$ is a homomorphism given by $h(a) = a$ for $a \in T$ and $h(A) = A'$ for $A \in N$. It is easy to see that $L(G') = L(G)$, when G' works in one of the aforementioned cut-derivation modes.

Conversely, let $G = (N, T, P_1, P_2, \ldots, P_n, S)$ be a context-free CDGS working in the $(sf \wedge = 1)_c$- or $(sf \wedge \leq 1)_c$-modes. First observe, that any f_c-mode of derivation, for any derivation mode f, can be sequentialized, i.e., we may assume that the domain of the partial injective function ρ used in an f_c-derivation step is a singleton set—the straightforward proof is left to the reader. Next we consider a derivation in the $(sf \wedge = 1)_c$- or $(sf \wedge \leq 1)_c$-mode in more detail. First, let $x \Rightarrow^{(sf \wedge = 1)_c} y$ be an arbitrary derivation step performed by the CDGS G from above. By definition and the above reasoning on sequentialization we may assume that (1) $x = x_0 x_1 x_2$ with $x_i \in (N \cup T)^*$, for $0 \leq i \leq 2$, (2) there is a partial injective mapping $\rho : \{0, 1, 2\} \hookrightarrow \{1, 2, \ldots n\}$ satisfying $\mathrm{dom}(\rho) = \{1\}$, and (3) $y = x_0 z_1 x_2$, for $x_1 \Rightarrow^{(sf \wedge = 1)}_{\rho(i)} z_1$. Since $x_1 \Rightarrow^{(sf \wedge = 1)}_{\rho(i)} z_1$ there is a derivation that satisfies both, the sf- and the $= 1$-mode property. Hence, component $P_{\rho(i)}$ is fully competent on x_1 and after applying one single rule from $P_{\rho(i)}$ this component is either not fully competent anymore or has derived a word from T^*. Therefore, we can write $x_1 = x_0' A x_2'$ and $z_1 = x_0' z x_2'$ for some rule $A \rightarrow z$ from $P_{\rho(i)}$, such that $z \in (N \cup T)^* (N \setminus \mathrm{dom}(P_{\rho(i)}))(N \cup T)^*$ or $z \in T^*$. In the latter case both x_0' and x_2' are from T^*, too. But then we can alter the decomposition of x chosen in (1) such that the middle word, for which ρ assigns a value, is the nonterminal A instead of x_1, still leading to y under the $(sf \wedge = 1)_c$ derivation, when choosing the partial injective mapping ρ from (2) and the rule $A \rightarrow z$ from $P_{\rho(i)}$ for the derivation in $(sf \wedge = 1)$-mode on the nonterminal A. Obviously, a similar reasoning applies for the $(sf \wedge \leq 1)_c$-mode of derivation.

Thus, in $(sf \wedge = 1)_c$- or $(sf \wedge \leq 1)_c$-derivations which eventually terminate, only those rules of the CDGS are relevant, which introduce a nonterminal not in $\mathrm{dom}(P_i)$ or terminate. We set

$$Q_i = \{A \rightarrow z \mid A \rightarrow z \in P_i, z \in T^* \cup (N \cup T)^*(N \setminus \mathrm{dom}(P_i))(N \cup T)^*\},$$

for each i with $1 \leq i \leq n$, and define the context-free grammar $G' = (N, T, Q, S)$, where $Q = \bigcup_{1 \leq i \leq n} Q_i$. By our previous investigation we find $L(G') = L(G)$. Thus, the stated claim follows. □

As shown in [10] context-free [λ-free] CDGSs working in the sf-mode precisely characterize the family family $\mathcal{L}(\mathrm{P}, \mathrm{CF}[-\lambda], \mathrm{ac})$ of languages generated by programmed context-free [λ-free] grammars with appearance checking, that is

$$\mathcal{L}(\mathrm{CD}, \mathrm{CF}[-\lambda], sf) = \mathcal{L}(\mathrm{P}, \mathrm{CF}[-\lambda], \mathrm{ac}).$$

Let us briefly recall the definition of programmed context-free grammars. A *programmed context-free grammar* (see, e.g., [5]) is a tuple $G = (N, T, P, S, \Lambda, \sigma, \phi)$, where N, T, P, and S, for $S \in N$, are as in the definition of context-free grammars. Let Λ be a finite set of labels (for the productions in P), such that Λ can be interpreted as a function which outputs a production when being given a label; σ and ϕ are functions from Λ into the set of subsets of Λ. Usually, the productions are written in the form $(r : A \to \alpha, \sigma(r), \phi(r))$, where r is the label of $A \to \alpha$. For (x, r_1) and (y, r_2) in $(N \cup T)^* \times \Lambda$ and $\Lambda(r_1) = A \to \alpha$, we write $(x, r_1) \Rightarrow (y, r_2)$ if and only if either $x = x_1 A x_2$, $y = x_1 \alpha x_2$ and $r_2 \in \sigma(r_1)$, or $x = y$ and rule $A \to \alpha$ is not applicable to x, i.e., nonterminal A is not present in x, and $r_2 \in \phi(r_1)$. In the latter case, the derivation step is done in *appearance checking* mode. The set $\sigma(r_1)$ is called success field and the set $\phi(r_1)$ failure field of r_1. As usual, the reflexive transitive closure of $\Rightarrow$ is denoted by $\Rightarrow^*$. The language generated by G is defined as $L(G) = \{ w \in T^* \mid (S, r_1) \Rightarrow^* (w, r_2) \text{ for some } r_1, r_2 \in \Lambda \}$. A programmed context-free grammar G is a *recurrent programmed context-free grammar* if for every $p \in \Lambda$ of G, if $\phi(p) = \emptyset$, then $p \in \sigma(p)$, and if $\phi(p) \neq \emptyset$, then $p \in \sigma(p) = \phi(p)$.

Now we are ready to continue our considerations on the *sf*- and *sf$_c$*-mode. For the context-free [λ-free] CDGS working in *sf$_c$*-mode of derivation the following upper and lower bounds

$$\mathcal{L}(\mathrm{RP}, \mathrm{CF}, \mathrm{ac}) \subseteq \mathcal{L}(\mathrm{CD}, \mathrm{CF}, \mathit{sf}_c) \subseteq \mathcal{L}(\mathrm{P}, \mathrm{CF}, \mathrm{ac}) = \mathcal{L}(\mathrm{RE})$$

and

$$\mathcal{L}(\mathrm{RP}, \mathrm{CF} - \lambda, \mathrm{ac}) \subseteq \mathcal{L}(\mathrm{CD}, \mathrm{CF} - \lambda, \mathit{sf}_c) \subseteq \mathcal{L}(\mathrm{CS})$$

were reported in [1]. Since by definition of the combined *sf*-modes both $(\mathit{sf} \wedge *)$- and $(\mathit{sf} \wedge \geq 1)$-mode obviously coincide with the ordinary *sf*-mode, the following theorem is immediate.

Theorem 2. *If* $f \in \{(\mathit{sf} \wedge *)\} \cup \{(\mathit{sf} \wedge \geq 1)\}$, *then*

$$\mathcal{L}(\mathrm{CD}, \mathrm{CF}[-\lambda], f_c) = \mathcal{L}(\mathrm{CD}, \mathrm{CF}[-\lambda], \mathit{sf}_c). \qquad \square$$

Thus, the previously mentioned upper and lower bounds on the generative capacity of CDGSs working in *sf$_c$*-mode carry over to the cut-modes under consideration. Hence, we obtain the following corollary.

Corollary 1. *If* $f \in \{(\mathit{sf} \wedge *)\} \cup \{(\mathit{sf} \wedge \geq 1)\}$, *then*

1. $\mathcal{L}(\mathrm{RP}, \mathrm{CF}, \mathrm{ac}) \subseteq \mathcal{L}(\mathrm{CD}, \mathrm{CF}, f_c) \subseteq \mathcal{L}(\mathrm{P}, \mathrm{CF}, \mathrm{ac}) = \mathcal{L}(\mathrm{RE})$ *and*
2. $\mathcal{L}(\mathrm{RP}, \mathrm{CF} - \lambda, \mathrm{ac}) \subseteq \mathcal{L}(\mathrm{CD}, \mathrm{CF} - \lambda, f_c) \subseteq \mathcal{L}(\mathrm{CS})$. $\qquad \square$

It is left open which of the inclusions in this corollary are strict. Next we turn our attention to the remaining combined *sf*-modes.

Regardless, whether the [λ-free] context-free CDGS works in one of the combined *sf*-modes that count the derivation steps in a non-trivial manner in [2] a characterization of $\mathcal{L}(\mathrm{P}, \mathrm{CF}[-\lambda], \mathrm{ac})$ was obtained, i.e.,

$$\mathcal{L}(\mathrm{CD}, \mathrm{CF}[-\lambda], f) = \mathcal{L}(\mathrm{P}, \mathrm{CF}[-\lambda], \mathrm{ac}),$$

for $f \in \{ (sf \wedge \leq k) \mid k \geq 2 \} \cup \{ (sf \wedge = k) \mid k \geq 2 \} \cup \{ (sf \wedge \geq k) \mid k \geq 2 \}$.
For the cut-mode versions of these derivation modes the situation is similar, but
the proofs are much more involved. First we show that for [λ-free] context-free
CDGS running in $(sf \wedge \leq k)_c$- or $(sf \wedge = k)_c$-mode, for $k \geq 2$, a characterization
in terms of [λ-free] context-free programmed grammar with appearance checking
is obtained. The proof of the following theorem parallels the proof given in [2]
for the ordinary sf-combined modes.

Theorem 3. *If* $f \in \{ (sf \wedge \leq k) \mid k \geq 2 \} \cup \{ (sf \wedge = k) \mid k \geq 2 \}$, *then*

$$\mathcal{L}(P, CF[-\lambda], ac) \subseteq \mathcal{L}(CD, CF[-\lambda], f_c).$$

Proof. Obviously the language families $\mathcal{L}(CD, CF[-\lambda], (sf \wedge = k)_c)$, for $k \geq 2$, are
closed under union and embrace the finite languages. Let $L \subseteq T^*$ be a language
in $\mathcal{L}(P, CF[-\lambda], ac)$, then

$$L = \bigcup_{a,b \in T} (a \cdot \delta_{a,b}(L) \cdot b) \cup (L \cap T^2) \cup (L \cap T) \cup (L \cap \{\lambda\}),$$

where $\delta_{a,b}(L) = \{ w \in T^+ \mid awb \in L \}$. Since L is in $\mathcal{L}(P, CF[-\lambda], ac)$, the
language $\delta_{a,b}(L)$ is in $\mathcal{L}(P, CF[-\lambda], ac)$ due to the closure of these families under
derivatives. Thus, for the proof of the present assertion, it is sufficient to show
that $a \cdot \delta_{a,b}(L) \cdot b$ is in $\mathcal{L}(CD, CF[-\lambda], (sf \wedge = k)_c)$, provided that $\delta_{a,b}(L)$ is in
$\mathcal{L}(P, CF[-\lambda], ac)$. Let $G = (N, T, P, S, \Lambda, \sigma, \phi)$ be a programmed context-free
grammar generating $\delta_{a,b}(L)$.

Now, we design a CDGS G' working in $(sf \wedge = 2)_c$-mode generating the language $a \cdot \delta_{a,b}(L) \cdot b$, where

$$N' = N \cup \Lambda \cup \{ p', p'', \tilde{p} \mid p \in \Lambda \} \cup \{ A_p, A_p' \mid A \in N, p \in \Lambda \}$$
$$\cup \{ R_p \mid p \in \Lambda \} \cup \{ R, S', S'' \}$$

is the set of nonterminals, the unions being pairwise disjoint, T is the set of
terminals, S' is the axiom, and G' has the production sets given below.

To start the derivation one uses

$$P_{init} = \{ S' \to S'' \} \cup \{ S'' \to pSR \mid p \in \Lambda \}.$$

Then for every production $A \to w$ with label p we consider the following two
cases:

1. To apply the rule under consideration we define

$$P_{p,\sigma,1} = \{ p \to p' \} \cup \{ A \to A_p, A_p \to A_p \} \cup \{ B \to B \mid B \in N \},$$
$$P_{p,\sigma,2} = \{ p' \to p'', p'' \to p'' \} \cup \{ A_p \to A_p' \} \cup \{ B \to B \mid B \in N \},$$
and
$$P_{p,\sigma,3} = \{ p'' \to q \mid q \in \sigma(p) \} \cup \{ A_p' \to A_p', A_p' \to w \} \cup \{ B \to B \mid B \in N \}.$$

2. To verify that the rule is not applicable one uses

$$P_{p,\phi,1} = \{p \to \tilde{p}\} \cup \{R \to R_p, R_p \to R_p\} \cup \{B \to B \mid B \in N \setminus \{A\}\}$$

and

$$P_{p,\phi,2} = \{\tilde{p} \to q, q \to q \mid q \in \phi(p)\} \cup \{R_p \to R\} \cup \{B \to B \mid B \in N \setminus \{A\}\}.$$

Finally,

$$P_{term} = \{p \to a \mid p \in \Lambda\} \cup \{R \to b\}$$

is used to terminate the derivation.

The only component applicable to the axiom S' is P_{init}. Now, in every non-terminal sentential form, either R or R_p appears as right marker and one label symbol $p \in \Lambda$ or its primed, double primed, or tilded version appears as left marker. Given a sentential form paR with $\alpha \in (N \cup T)^*$, the components $P_{p,\sigma,1}$, $P_{p,\sigma,2}$, and $P_{p,\sigma,3}$ can be used in order to simulate the successful application of the production with label p of G, yielding a string $q\beta R$, for $q \in \sigma(p)$ and string $\beta \in (N \cup T)^*$, and $P_{p,\phi,1}$ and $P_{p,\phi,2}$ can simulate its application in appearance checking mode, yielding $q\alpha R$, for $q \in \phi(p)$. For this, the complete sentential form is given to the components. When a string of the form pwR with $w \in T^*$ is obtained, the component P_{term} can be applied yielding w. This proves every word of $L(G)$ can be derived by the CDGS G' in $(sf \wedge = 2)_c$-mode of derivation.

For the converse inclusion observe the following. A terminal word can only be obtained by applying P_{term} to a sentential form in $\Lambda T^*\{R\}$. Starting off with a sentential form $\alpha = p\beta R$, for $p \in \Lambda$ and $\beta \in (N \cup T)^*$, neither the components $P_{q,\sigma,2}$ and $P_{q,\sigma,3}$ nor $P_{q,\phi,2}$, for $q \in \Lambda$, are applicable to any substring of α, since the presence of some symbol A_q, q'', or $\tilde{q}$ is needed, respectively, in order to stop deriving. Therefore, we have to distinguish the following two cases:

1. Some $P_{q,\phi,1}$ becomes active first. If a component $P_{q,\phi,1}$ is applied to a substring α' of α, this component can become inactive only after the production $q \to \tilde{q}$ has been used. Therefore, $q = p$ has to hold. Since $\tilde{p}$ can be rewritten only with the help of $P_{p,\phi,2}$ and $P_{p,\phi,2}$ can stop deriving only by application of $R_p \to R$ the symbol R_p must have been introduced when $P_{q,\phi,1}$ was active. Thus, $\alpha' = \alpha$ has to hold. In conclusion, $|\alpha|_A = 0$ since $P_{p,\phi,1}$ is not fully competent on α otherwise. Furthermore, $\tilde{p}$ must be replaced together with R_p during one and the same application of $P_{p,\phi,2}$. Therefore,

$$\alpha = p\beta R \Rightarrow_{p,\phi,1}^{(sf \wedge = 2)_c} \tilde{p}\beta R_p \Rightarrow_{p,\phi,2}^{(sf \wedge = 2)_c} q\beta R,$$

for $q \in \phi(p)$, is the only successful continuation of such derivation, simulating the application of the production with label p in appearance checking mode. Note that some $P_{q,\sigma,1}$, for $q \in \Lambda$, cannot be applied to any substring of $\tilde{p}\beta R_p$ since it would need a symbol q to terminate its derivation in $(sf \wedge = 2)_c$-mode. Similarly both $P_{q,\sigma,2}$ and $P_{q,\sigma,3}$ cannot be applied to any substring since they require A_p and p'' (or A'_p if the simulated production is terminating).

2. Some $P_{q,\sigma,1}$ becomes active first. Similarly as above, if a component $P_{q,\sigma,1}$ is applied to a substring α' of α this component can become inactive only after the production $q \to q'$, which implies $q = p$. Since p' can be rewritten only with the help of $P_{p,\sigma,2}$ and this component can stop deriving only by applying the rule $A_p \to A'_p$ the symbol A_p must have been introduced during the application of $P_{q,\sigma,1}$ in $(sf \wedge = 2)_c$-mode of derivation. Since the component works in $(sf \wedge = 2)_c$-mode only one nonterminal A_p can be introduced in a successful derivation. Hence, we have found the derivation

$$\alpha = p\beta A\gamma R \Rightarrow^{(sf \wedge = 2)_c}_{p,\sigma,1} p'\beta A_p\gamma R.$$

Then with similar argumentation as above, one can show that these derivation can be continued with an application of component $P_{p,\sigma,2}$ and finally with $P_{p,\sigma,3}$, i.e.,

$$p'\beta A_p\gamma \Rightarrow^{(sf \wedge = 2)_c}_{p,\sigma,2} p''\beta A'_p\gamma R \Rightarrow^{(sf \wedge = 2)_c}_{p,\sigma,3} q\beta w\gamma R,$$

where $q \in \sigma(p)$ and $A \to w$ is the rule with label p. Hence, this simulates a successful application of the rule $A \to w$ with label p of the programmed grammar G. Finally, it is easy to see that by construction, no other sequence of components is applicable to α and its intermediate sentential forms. The details are left to the reader.

The result for arbitrary k, where $k \geq 3$, follows by a similar construction, where the productions in P_{init} and P_{term} have to be replaced by sets containing chains of productions of appropriate length, and in each other component the rules that force the component to quit its derivation must be also replaced by chains of productions accordingly. For instance the component $P_{p,\sigma,1}$ must be replaced by

$$\{p \to [p,1], [p,1] \to [p,2], \ldots, [p, k-2] \to p'\}$$
$$\cup \{A \to A_p, A_p \to A_p\} \cup \{B \to B \mid B \in N\},$$

where all $[p, i]$, for $1 \leq i \leq k-2$, are new nonterminals, in order to work properly in the $(sf \wedge = k)_c$-mode of derivation. This technique was used in [7,8] under the name *prolongation technique*.

A careful analysis of the designed CDGS G' in fact reveals that it still works properly, even in the $(sf \wedge \leq 2)_c$-mode, which also generalizes to the $(sf \wedge \leq k)_c$-mode, for $k \geq 2$. Hence, also in these cases the [λ-free] context-free programmed grammar is simulated by a [λ-free] context-free CDGS. This proves the stated claim.   □

What about the remaining combined *sf*-mode that counts? Recall, that if the CDGSs are driven in the ordinary $(sf \wedge \geq k)$-mode then we have

$$\mathcal{L}(\text{CD}, \text{CF}[-\lambda], (sf \wedge \geq k)) = \mathcal{L}(\text{P}, \text{CF}[-\lambda], \text{ac}),$$

if $k \geq 2$. For the cut-mode version of the $(sf \wedge \geq k)$-mode, for $k \geq 2$, we find the following situation. As already observed, the $(sf \wedge \geq 1)$-mode is equal to the *sf*-mode, which also holds for their cut-mode versions. For the $(sf \wedge \geq k)$-mode in general, it is easy to see that one can do a simulation of the ordinary *sf*-mode by the prolongation technique, as elaborated by [7,8]. Whether the converse simulation is also possible is left open. Thus, by the argumentation above and Corollary 1 we have shown the following theorem.

Theorem 4. *If $f \in \{ (sf \wedge \geq k) \mid k \geq 2 \}$, then*

$$\mathcal{L}(\mathrm{RP}, \mathrm{CF}[-\lambda], ac) \subseteq \mathcal{L}(\mathrm{CD}, \mathrm{CF}[-\lambda], f_c).$$ □

For the upper bound we state the following theorem.

Theorem 5. *If $f \in \{ (sf \wedge \leq k) \mid k \geq 2 \} \cup \{ (sf \wedge = k) \mid k \geq 2 \} \cup \{ (sf \wedge \geq k) \mid k \geq 2 \}$ then*

$$\mathcal{L}(\mathrm{CD}, \mathrm{CF}[-\lambda], f_c) \subseteq \mathcal{L}(\mathrm{P}, \mathrm{CF}[-\lambda], ac).$$

Proof. We only sketch the construction. In [2] the simulation of a [λ-free] context-free CDGS working in *f*-mode of derivation by a [λ-free] context-free programmed grammar with appearance checking was shown. During this simulation the sentential form is considered as a whole, since *f* is a non-cut-mode. In order to design a simulation of a CDGS for a cut-*f*-mode of derivation by a programmed grammar in a step by step way one has to (1) mark a substring of the sentential form that is *consecutive*, then (2) to do the simulation of the *f*-mode derivation on the marked substring, and finally (3) to unmark the substring. In principle the tasks of marking symbols, simulation of the *f*-mode, and unmarking symbols can be done by the programmed grammar. But how to ensure that the marked symbols are consecutive using context-free rules only? To this end recall the simulation of a phrase structure grammar in Kuroda normal form[1] [9] by a context-free programmed grammar with appearance checking, see, e.g., [5]. By a coding and decoding mechanism of the sentential form that can be implemented by a [λ-free] context-free programmed grammar, one is able to verify whether two symbols A and B are consecutive (and in the right order AB or BA) in the underlying sentential form. It is easy to see that this mechanism can also be used to verify whether more than two symbols are consecutive. Now combining this with the idea of the step by step simulation results in a [λ-free] context-free programmed grammar with appearance checking that is able to simulate the CDGS under consideration that runs its derivation in cut-*f*-mode. The tedious details are left to the reader.   □

Combining the aforementioned statements results in the following corollary.

Corollary 2.   *1. If $f \in \{ (sf \wedge \leq k) \mid k \geq 2 \} \cup \{ (sf \wedge = k) \mid k \geq 2 \}$, then*

$$\mathcal{L}(\mathrm{CD}, \mathrm{CF}[-\lambda, f_c) = \mathcal{L}(\mathrm{P}, \mathrm{CF}[-\lambda], ac).$$

---

[1] A Chomsky grammar with set of nonterminals N and set of terminals T is in *Kuroda normal form*, if every production is either of the form $A \rightarrow BC$, $A \rightarrow B$, $A \rightarrow a$, or $AB \rightarrow CD$, where $A, B, C, D \in N$ and $a \in T$.

2. If $f \in \{ (sf \wedge \geq k) \mid k \geq 2 \}$ then

$$\mathcal{L}(\mathrm{RP}, \mathrm{CF}[-\lambda], \mathrm{ac}) \subseteq \mathcal{L}(\mathrm{CD}, \mathrm{CF}[-\lambda], f_c) \subseteq \mathcal{L}(\mathrm{P}, \mathrm{CF}[-\lambda], \mathrm{ac}). \qquad \square$$

Although the previous corollary looks very similar to Corollary 1, there is a subtle difference, because by similar techniques as in [7,8], one can show that

$$\mathcal{L}(\mathrm{CD}, \mathrm{CF}[-\lambda], (sf \wedge \geq k)_c) \subseteq \mathcal{L}(\mathrm{CD}, \mathrm{CF}[-\lambda], (sf \wedge \geq k \cdot \ell)_c),$$

for $k, \ell \geq 1$. In particular, this means that

$$\mathcal{L}(\mathrm{CD}, \mathrm{CF}[-\lambda], sf_c) = \mathcal{L}(\mathrm{CD}, \mathrm{CF}[-\lambda], (sf \wedge \geq 1)_c)$$
$$\subseteq \mathcal{L}(\mathrm{CD}, \mathrm{CF}[-\lambda], (sf \wedge \geq k)_c) \subseteq \mathcal{L}(\mathrm{P}, \mathrm{CF}[-\lambda], \mathrm{ac}),$$

for $k \geq 1$. We leave open whether these inclusions are strict or not.

Finally, if the sf-mode is combined with the t-mode, then a new characterization of the context-free languages is obtained in [2], that is,

$$\mathcal{L}(\mathrm{CD}, \mathrm{CF}[-\lambda], (sf \wedge t)) = \mathcal{L}(\mathrm{CF}).$$

A careful analysis reveals that exactly the same proof can be used to show the next theorem only replacing $(sf \wedge t)$-mode by $(sf \wedge t)_c$-mode in all the arguments.

Theorem 6. $\mathcal{L}(\mathrm{CD}, \mathrm{CF}[-\lambda], (sf \wedge t)_c) = \mathcal{L}(\mathrm{CF})$. $\qquad \square$

There is still a last combined sf-mode, namely the $(sf \wedge \bar{t})$-mode, which was not formally defined, but its meaning should be clear. Here x derives y with component P_i of the underlying CDGS in the $\bar{t}$-mode of derivation, i.e., $x \Rightarrow_i^{\bar{t}} y$, if and only if $x \Rightarrow_i^* y$ and there is z such that $y \Rightarrow_i z$. Trivially, the latter mode does not generate anything, since it can not terminate by the $\bar{t}$-mode property regardless whether the cut- or non-cut-mode version is considered. Therefore $\mathcal{L}(\mathrm{CD}, \mathrm{CF}[-\lambda], (sf \wedge \bar{t})) = \mathcal{L}(\mathrm{CD}, \mathrm{CF}[-\lambda], (sf \wedge \bar{t})_c) = \{\emptyset\}$.

4   Conclusions

Our findings on cut-mode derivations of combined sf-modes for context-free CDGSs (gray shaded) together with the results on cut-modes for ordinary and combined t-modes for context-free CDGSs are summarized in Table 2—if in the table only a lower bound for the language family under consideration is given and an upper bound is not stated, then one can safely assume that the upper bound is $\mathcal{L}(\mathrm{P}, \mathrm{CF}[-\lambda], \mathrm{ac})$. In most cases exact characterizations are obtained. However, for the stated inclusion results, we have to leave open, whether they are strict or not. Moreover, we also have to leave open, whether the inclusions

$$\mathcal{L}(\mathrm{CD}, \mathrm{CF}[-\lambda], (sf \wedge \geq k)_c) \subseteq \mathcal{L}(\mathrm{CD}, \mathrm{CF}[-\lambda], (sf \wedge \geq k \cdot \ell)_c),$$

for $k, \ell \geq 1$ are proper or not. Here it is worth mentioning that in the non-cut mode case, that is, the ordinary $(sf \wedge \geq k)$- and $(sf \wedge \geq k \cdot \ell)$-mode of derivation, we have the situation that the induced language families coincide. Thus, we have $\mathcal{L}(\mathrm{CD}, \mathrm{CF}[-\lambda], (sf \wedge \geq k)) = \mathcal{L}(\mathrm{CD}, \mathrm{CF}[-\lambda], (sf \wedge \geq k \cdot \ell))$, for $k, \ell \geq 1$, which in turn is equal to $\mathcal{L}(\mathrm{P}, \mathrm{CF}[-\lambda], \mathrm{ac})$.

Table 2. Generative capacity of CD grammar systems working in the cut-mode versions of the classical modes, the combined t-modes, and the combined sf-modes of derivation compared. The gray shaded parts mark the results obtained by our investigations.

Derivation mode f	$\mathcal{L}(CD, CF[-\lambda], f_c)$	$\mathcal{L}(CD, CF[-\lambda], (t \wedge f)_c)$	$\mathcal{L}(CD, CF[-\lambda], (sf \wedge f)_c)$
$*$			$\mathcal{L}(RP, CF[-\lambda], ac) \subseteq \cdot$
≤ 1			
$= 1$	$\mathcal{L}(CF)$		
≥ 1			$\mathcal{L}(RP, CF[-\lambda], ac) \subseteq \cdot$
$\leq k$ with $k \geq 2$			$\mathcal{L}(P, CF[-\lambda], ac)$
$= k$ with $k \geq 2$	$\mathcal{L}(CF) \subset \cdot \subseteq \mathcal{L}(P, CF[-\lambda])$	$\mathcal{L}(CF) \subset \cdot$ $\mathcal{L}_{fin}(P, CF[-\lambda]) \subseteq \cdot$	
$\geq k$ with $k \geq 2$			$\mathcal{L}(RP, CF[-\lambda], ac) \subseteq \cdot$
t		$\mathcal{L}(CF)$	
sf	$\mathcal{L}(RP, CF[-\lambda], ac) \subseteq \cdot$		$\mathcal{L}(RP, CF[-\lambda], ac) \subseteq \cdot$

References

1. Bordihn, H., Holzer, M.: Cooperating distributed grammar systems as models of distributed problem solving, revisited. Fund. Inform. 76(3), 255–270 (2007)
2. Bordihn, H., Holzer, M.: A note on cooperating distributed grammar systems working in combined modes. Inform. Process. Lett. 108(1), 10–14 (2008)
3. Csuhaj-Varjú, E., Dassow, J.: On cooperating/distributed grammar systems. J. Inform. Proc. Cybernet (formerly: EIK) 26(1-2), 49–63 (1990)
4. Csuhaj-Varjú, E., Dassow, J., Kelemen, J., Păun, G.: Grammar Systems: A Grammatical Approach to Distribution and Cooperation. Gordon and Breach (1994)
5. Dassow, J., Păun, G.: Regulated Rewriting in Formal Language Theory. EATCS Monographs in Theoretical Computer Science, vol. 18. Springer (1989)
6. Dassow, J., Păun, G., Rozenberg, G.: Grammar Systems. In: Handbook of Formal Languages. Linear Modeling: Background and Application, vol. 2, pp. 155–214. Springer (1997)
7. Fernau, H., Freund, R., Holzer, M.: Hybrid modes in cooperating distributed grammar systems: internal versus external hybridization. Theoret. Comput. Sci. 259(1-2), 405–426 (2001)
8. Fernau, H., Freund, R., Holzer, M.: Hybrid modes in cooperating distributed grammar systems: combining the t-mode with the modes $\leq k$ and $= k$. Theoret. Comput. Sci. 299(1-3), 633–662 (2003)
9. Kuroda, S.Y.: Classes of languages and linear bounded automata. Inform. Control 7(2), 207–223 (1964)
10. Meersman, R., Rozenberg, G.: Cooperating Grammar Systems. In: Winkowski, J. (ed.) MFCS 1978. LNCS, vol. 64, pp. 364–374. Springer, Heidelberg (1978)
11. von Solms, S.H.: Some notes on ET0L-languages. Int. J. Comput. Math. 5, 285–296 (1976)

Equations in the Partial Semigroup
of Words with Overlapping Products[*]

Mari Huova and Juhani Karhumäki

Department of Mathematics and TUCS,
University of Turku, 20014 Turku, Finland
{mahuov,karhumak}@utu.fi

Abstract. We consider an overlapping product of words as a partial operation where the product of two words is defined when the former ends with the same letter as the latter starts, and in this case the product is obtained by merging these two occurrences of letters, for example $aba \bullet ab = abab$. Some basic results on equations of words are established by reducing them to corresponding results of ordinary word equations.

Keywords: combinatorics on words, overlapping product, equations.

1   Introduction

Motivated by bio-operations, or more formally, DNA computing, see [18], we consider an operation of overlapping product of words defined as follows. For two words ua and bv, with a and b letters, we define their overlapping product

$$ua \bullet bv = \begin{cases} uav & \text{if } a = b \ , \\ \text{undefined} & \text{if } a \neq b \ . \end{cases}$$

Consequently, the operation is locally controlled, and clearly a (partial) associative operation on the set of nonempty words Σ^+.

Recently the descriptional complexity of this operation was analyzed in the case of regular languages, see [10]. The same operation and its extensions have been studied in a number of articles motivated by bio-operations in DNA strands, see, e.g., [4], [5], [6], [7], [11], [16] and [17]. We consider this operation in connection with word equations. It turns out that many questions on equations can be transformed, and finally solved, by translating these to related problems on ordinary word equations. The translation is made because, for example, the simple operation of cancellation does not work. Thus, for example, $x \bullet y = x \bullet z \bullet x$ is not equal to $y = z \bullet x$, as explained more closely in Section 3.

More concretely, we solve a few basic equations over overlapping product, introduce a general translation of such equations to a Boolean system of ordinary equations, and as a consequence establish, e.g., that the fundamental result of solvability of the satisfiability problem extends to these new types of equations.

---

[*] Supported by the Academy of Finland under the grant 121419.

H. Bordihn, M. Kutrib, and B. Truthe (Eds.): Dassow Festschrift 2012, LNCS 7300, pp. 99–110, 2012.

2   Preliminaries

Let Σ be a finite alphabet. We denote by Σ^+ the set of all nonempty words over Σ and view it as the free semigroup with respect to the product of words. Notation Σ^* is also used refering to the monoid $\Sigma^+ \cup \{1\}$, where 1 denotes the empty word. As general references of the combinatorics on words we refer to [13] and [9].

We define a new *partial* binary operation, so-called *overlapping product*, on Σ^+ as follows: For two words ua and bv, with $a, b \in \Sigma$, we set

$$ua \bullet bv = \begin{cases} uav & \text{if } a = b \text{ ,} \\ \text{undefined} & \text{if } a \neq b \text{ .} \end{cases}$$

Clearly, the operation $\bullet$ is associative (partial) operation so that we have

Fact 1. $(\Sigma^+, \bullet)$ is a partial semigroup.

Actually, in $(\Sigma^+, \bullet)$ letters (considered as words of length 1) constitute partial (nonunique) left and right units. Indeed, $a \bullet u$ with any $u \in \Sigma^+$, is equal to u if defined.

Due to the associativity it is justified to write the product without parenthesis:

$$\alpha = \alpha_1 \bullet \alpha_2 \bullet \cdots \bullet \alpha_n \text{ , for any } \alpha_i \in \Sigma^+. \tag{1}$$

The word α, if defined, as an element of Σ^+ is deduced from (1) as follows. We need one additional notation. For any word $u = a_1 a_2 \cdots a_k$, with $a_i \in \Sigma$, the notation $u(a_k)^{-1}$ refers to the word $a_1 a_2 \cdots a_{k-1}$ and correspondingly $(a_1)^{-1}u$ to the word $a_2 a_3 \cdots a_k$. In order for α to be defined, for each $i = 1, \ldots, n-1$, necessarily

$$\text{last } \alpha_i = \text{first } \alpha_{i+1} \text{ ,}$$

and then

$$\begin{aligned} \alpha &= \alpha_1 (\text{last } \alpha_1)^{-1} \alpha_2 (\text{last } \alpha_2)^{-1} \cdots \alpha_{n-1} (\text{last } \alpha_{n-1})^{-1} \alpha_n \\ &= \alpha_1 (\text{first } \alpha_2)^{-1} \alpha_2 (\text{first } \alpha_3)^{-1} \cdots \alpha_{n-1} (\text{first } \alpha_n)^{-1} \alpha_n \text{ .} \end{aligned}$$

On the other hand any word

$$\alpha = \alpha_1 \alpha_2 \cdots \alpha_n \text{ with } \alpha_i \in \Sigma^+$$

can be written as an element of the partial semigroup $(\Sigma^+, \bullet)$ as follows:

$$\alpha = \alpha_1 (\text{first } \alpha_2) \bullet \alpha_2 (\text{first } \alpha_3) \bullet \cdots \bullet \alpha_{n-1} (\text{first } \alpha_n) \bullet \alpha_n \text{ .}$$

It is worth noting that the latter translation is always defined.

These considerations make our goal to consider the theory of word equations over the overlapping product feasible - as described in details in Section 4.

3   Examples of Basic Equations

Common tools for solving word equations such as Levi's Lemma, splitting of equation and length argument are not so straightforward to use with equations containing overlapping products. Problems for using these tools arise from the facts that for overlapping products to be defined the last and the first letters of the adjacent factors have to coincide and when a product is conducted these two letters are unified to a single letter. For example, the first of these reasons causes the following problem.

Example 1. Consider an equation $x \bullet y = x \bullet z \bullet x$ with overlapping products and an equation $xy = xzx$. Equation $xy = xzx$ can be reduced into the form $y = zx$, accordingly we could suppose that $x \bullet y = x \bullet z \bullet x$ equals with equation $y = z \bullet x$. However, for example, $y = abb, z = ab, x = bb$ is a solution for $y = z \bullet x$ but not for the original equation because the overlapping product $x \bullet y = bb \bullet abb$ is not defined.

Example 1 shows that we cannot use Levi's Lemma straightforwardly to eliminate the leftmost or the rightmost unknowns. The same problem arises if we split an equation. Again we may loose the information of the requirements that originated from the overlapping product that was located at the point of splitting.

Unification of the last and the first letters of the adjacent factors complicates the use of length argument. The total length of an expression containing overlapping products depends on the lengths of the factors and on the number of factors, i.e. $|x_1 \bullet \cdots \bullet x_k| = |x_1| + \cdots + |x_k| - (k - 1)$. For some equations it may, nevertheless, be easy to detect, for example, the middle of both sides as for example in the equation $x \bullet y \bullet y \bullet x = z \bullet z$. From this we can conclude that $x \bullet y = z, y \bullet x = z$, but the consequences of splitting the equation have to be taken into account.

We proceed by solving some basic equations over the partial semigroup with overlapping product. First we consider the equation $x \bullet y = y \bullet x$, which corresponds to *commutation*.

Example 2. To solve the equation $x \bullet y = y \bullet x$ we first assume that $|x|, |y| > 1$. For the overlapping product to be defined we can assume that $x = ax'a$ and $y = ay'a$, where $a \in \Sigma$ and $x', y' \in \Sigma^*$. Now we can reduce the equation $x \bullet y = y \bullet x$ into an ordinary word equation $x \bullet y = ax'a \bullet ay'a = ax'ay'a = ay'ax'a = ay'a \bullet ax'a = y \bullet x$. From the equation $ax'ay'a = ay'ax'a$ we can notice that $ax'ay' = ay'ax'$, and hence ax' and ay' commute. Now we can write $ax' = t^i$ and $ay' = t^j$, where $t = a\alpha$ with $\alpha \in \Sigma^*$ and $i, j > 0$. From this we get $x = ax'a = t^ia = (a\alpha)^ia$ and $y = ay'a = t^ja = (a\alpha)^ja$, where $a \in \Sigma, \alpha \in \Sigma^*$ and $i, j > 0$. In the case that $|x| = 1$ (resp. $|y| = 1$) we have $x = a$ (resp. $y = a$), with $a \in \Sigma$ and $y = a\alpha a$ or $y = a$ (resp. $x = a\alpha a$ or $x = a$), with $\alpha \in \Sigma^*$. Thus the equation $x \bullet y = y \bullet x$ has solutions

$$\begin{cases} x = (a\alpha)^ia \\ y = (a\alpha)^ja \end{cases}, \text{ where } a \in \Sigma, \alpha \in \Sigma^* \text{ and } i, j \geq 0.$$

We remark that the answer of the equation of the previous example could also be written with the help of the overlapping product. For example, if $x = (a\alpha)^2 a, y = (a\alpha)^3 a$ we could also write $x = (a\alpha a) \bullet (a\alpha a), y = (a\alpha a) \bullet (a\alpha a) \bullet (a\alpha a)$. Thus, the words that are solutions of this equation refering to commutation are, in fact, overlapping products of words of the form $a\alpha a$ or letters as a special case.

The second equation we will examine is associated with *conjugation*, i.e. $xz = zy$.

Example 3. We first check two special cases for equation $x \bullet z = z \bullet y$. If $x = a$, with $a \in \Sigma$, then $y = b$, $b \in \Sigma$, and $z = a\alpha b$, $\alpha \in \Sigma^*$, or if $a = b$, then $z = a$ is possible, too. If $x = aa$, with $a \in \Sigma$, then $y = aa$ and $z = a^i$, where $i > 0$.

Now we can assume that $|x|, |y|, |z| > 2$ in equation $x \bullet z = z \bullet y$. As in Example 2 we may assume $x = ax'a$, $y = by'b$ and $z = az'b$, where $a, b \in \Sigma$ and $x', y', z' \in \Sigma^+$. These assumptions are due to the facts that overlapping products have to be defined and x and z have a common first letter and y has a common last letter with z. Reduction now gives now $x \bullet z = ax'az'b = az'by'b = z \bullet y$. From the word equation $x'az' = z'by'$ we can conclude that $x'a$ and by' conjugate. The conjugation property gives that there exist $p, q \in \Sigma^*$ so that $x'a = pq'$, $by' = q'p$ and $z' = p(q'p)^i$, where $i \geq 0$ and in addition if $q' \neq 1$ then $q' = bqa$ with $q \in \Sigma^*$. Now with these assumptions we have a solution

$$\begin{cases} x = ax'a = apq' = apbqa \\ y = by'b = q'pb = bqapb \\ z = az'b = ap(q'p)^i b = ap(bqap)^i b \end{cases},$$

where $a, b \in \Sigma, p, q \in \Sigma^*$ and $i \geq 0$.

If $q' = 1$ then $p = bp'a$, where $p' \in \Sigma^*$ and solutions are of the form

$$\begin{cases} x = ax'a = ap = abp'a \\ y = by'b = pb = bp'ab \\ z = az'b = a(p)^{i+1}b = a(bp'a)^{i+1}b \end{cases},$$

where $a, b \in \Sigma, p, p' \in \Sigma^*$ and $i \geq 0$.

In fact, these latter solutions are included in the upper formula. Thus equation $x \bullet z = z \bullet y$ has solutions

$$\begin{cases} x = apbqa \\ y = bqapb \\ z = ap(bqap)^i b \end{cases}, \quad \text{where } a, b \in \Sigma, p, q \in \Sigma^* \text{ and } i \geq 0$$

and special solutions

$$\begin{cases} x = a \\ y = b \\ z = a\alpha b \end{cases}, \quad \begin{cases} x = a \\ y = a \\ z = a^i \end{cases} \text{ and } \begin{cases} x = aa \\ y = aa \\ z = a^i \end{cases},$$

where $a, b \in \Sigma, \alpha \in \Sigma^*, i > 0$.

The third basic equation we consider asks when the product of two squares is a square, a problem first studied in [14]. In the case of word equation $x^2y^2 = z^2$ the answer is that the equation has only periodic solutions. If we consider the equation with overlapping product we get a corresponding result.

Example 4. We first assume that $|x|, |y|, |z| > 1$ in the equation $x \bullet x \bullet y \bullet y = z \bullet z$. Because overlapping products have to be defined we can again assume that $x = ax'a$, $y = ay'a$ and $z = az'a$, where $a \in \Sigma$ and $x', y', z' \in \Sigma^*$. Reduction of overlapping products into usual word products gives an equation $ax'ax'ay'ay'a = az'az'a$ from which we get a simpler equation $x'ax'ay'ay' = z'az'$. The length argument gives now that $|x'ay'| = |z'|$, and thus by comparing the beginnings and the ends of both sides on the equation $x'ax'ay'ay' = z'az'$ we conclude $z' = x'ay'$. Now the equation has the form $x'ax'ay'ay' = x'ay'ax'ay'$ which leads to an equation $ax'ay' = ay'ax'$ showing that ax' and ay' commute. From this observation we can conclude that $ax' = t^i$ and $ay' = t^j$ with $t = a\alpha$, $\alpha \in \Sigma^*$ and $i, j > 0$ and hence $x = ax'a = (a\alpha)^i a$, $y = ay'a = (a\alpha)^j a$ and $z = az'a = ax'ay'a = (a\alpha)^{i+j}a$. Again if some of the unknowns equal a letter, then the solution is gained from the following general formula by allowing $i, j \geq 0$. The equation $x \bullet x \bullet y \bullet y = z \bullet z$ has solutions

$$\begin{cases} x = (a\alpha)^i a \\ y = (a\alpha)^j a \\ z = (a\alpha)^{i+j} a \end{cases} , \quad \text{where } a \in \Sigma, \alpha \in \Sigma^* \text{ and } i, j \geq 0.$$

We yet give one example of a basic equation which leads us to analyze the defect property.

Example 5. To solve an equation $x \bullet y = u \bullet v$ we may assume $x = x'a$, $y = ay'$, $u = u'b$ and $v = bv'$ where $a, b \in \Sigma$ and $x', y', u', v' \in \Sigma^*$. With these assumptions we have an ordinary word equation $x'ay' = u'bv'$. We consider only the case $|x'| < |u'|$, the case $|u'| < |x'|$ is symmetric and $|x'| = |u'|$ is clear. The equation $x'ay' = u'bv'$ has now a solution $x' = \alpha$, $y' = \beta b\gamma$, $u' = \alpha a\beta$ and $v' = \gamma$ where $\alpha, \beta, \gamma \in \Sigma^*$. The solution for the original equation with the assumption $|x| < |u|$ can now be given:

$$\begin{cases} x = \alpha a \\ y = a\beta b\gamma \\ u = \alpha a\beta b \\ v = b\gamma \end{cases} , \quad \text{where } a, b \in \Sigma, \alpha, \beta, \gamma \in \Sigma^*.$$

We remark that these four words x, y, u and v of the previous example can be expressed in the form $x = \alpha a$, $y = a\beta b \bullet b\gamma$, $u = \alpha a \bullet a\beta b$ and $v = b\gamma$, thus they can be formed from three words by overlapping product. This implies, as stated in the next theorem, that a so-called *defect property*, see [3], is also valid in $(\Sigma^+, \bullet)$.

Theorem 1. *Let X be a set of n words with $X \cap \Sigma = \emptyset$. If X satisfies a nontrivial equation with overlapping products, then these words can be expressed with $n - 1$ words by using overlapping products.*

Proof. Let $x_1 \bullet x_2 \bullet \cdots \bullet x_k = y_1 \bullet y_2 \bullet \cdots \bullet y_l$ be a nontrivial equation such that $x_i, y_j \in X$ for all $i = 1, \ldots, k$ and $j = 1, \ldots, l$. We may assume that $|x_1| < |y_1|$ and hence y_1 can be written in the form $y_1 = x_1 \bullet (\text{last } x_1) \, y_1'$. Thus, the words of the set X can be expressed with words $X_1 = (X - \{y_1\}) \cup \{(\text{last } x_1) \, y_1'\}$. The number of words in X_1 is clearly at most n and $X_1 \cap \Sigma = \emptyset$. Now the equation corresponding to the original equation can be reduced at least from the beginning with a factor x_1 and hence, the new (nontrivial) equation will be shorter in terms of the total length of an expression which is given by $|x_1 \bullet \cdots \bullet x_k| = |x_1| + \cdots + |x_k| - (k-1)$. We divide the analyzis into two cases.

Case 1. Inductively with respect to the length of the nontrivial equation we will proceed into an equation $u = v_1 \bullet \cdots \bullet v_m$ with $u, v_1, \ldots v_m$ words from the processed set of at most n words. Now it is clear that the word u may be removed from the set and the original words can be expressed with $n-1$ words as claimed.

Case 2. If in some point of the procedure described above the equation will reduce into a trivial equation, the constructed set of words corresponding to that situation contains already at most $n-1$ words. This follows from the fact that the reduction from a nontrivial equation into a trivial equation is possible only if some factor replacing an old word already exists in the considered set of words.   □

As a conclusion, the above examples and the theorem show that results for word equations over overlapping product are often similar, but not exactly the same, as in the case of ordinary word equations. Moreover, the proofs reduce to that of ordinary words - as further explained in the next section.

4   Reduction into Word Equations

In this section the reduction of equations over overlapping products to that of ordinary word equations is analyzed in general. The reduction leads to a Boolean combination of word equations, as we shall see in the next result.

Theorem 2. *Let Σ be a finite alphabet, X be the set of unknowns and $e : u = v$ be an equation over X with overlapping products. Then the equation e can be reduced into a Boolean combination of ordinary word equations.*

Proof. Consider the equation $u = x_1 \bullet x_2 \bullet \cdots \bullet x_l = y_1 \bullet y_2 \bullet \cdots \bullet y_m = v$, where $x_i, y_j \in X$ for all $i = 1, \ldots, l$ and $j = 1, \ldots, m$.

Part 1. Assume that the solutions u_i for x_i and v_j for y_j have $|u_i|, |v_j| > 1$, for all $i = 1, \ldots, l$ and $j = 1, \ldots, m$, and hence we can mark the first and the last letters of the words and write

$$x_1 = a_1 x_1' a_2 \ , \quad x_2 = a_2 x_2' a_3 \ , \quad \ldots \ , \quad x_l = a_l x_l' a_{l+1} \ ,$$

$$y_1 = b_1 y_1' b_2 \ , \quad y_2 = b_2 y_2' b_3 \ , \quad \ldots \ , \quad y_m = b_m y_m' b_{m+1} \ ,$$

where $a_i, b_j \in \Sigma$ and x_i' and y_j' are new unknowns from the set X'.

Now we have some restrictions for choosing the letters a_i, b_j. If $x_i = x_j$ then $a_i = a_j$ and $a_{i+1} = a_{j+1}$, and similarly if $y_i = y_j$, then $b_i = b_j$ and $b_{i+1} = b_{j+1}$. Comparing unknowns of the equation e on both sides we have that if $x_i = y_j$, then $a_i = b_j$ and $a_{i+1} = b_{j+1}$, and in addition, $a_1 = b_1$ and $a_{l+1} = b_{m+1}$ always hold.

With these assumptions and markings we have a reduced word equation e' : $u' = v'$ without overlapping products where u' and v' are defined as follows:

$$u = x_1 \bullet x_2 \bullet \cdots \bullet x_l = a_1 x_1' a_2 x_2' a_3 \cdots a_l x_l' a_{l+1} = u'$$

$$v = y_1 \bullet y_2 \bullet \cdots \bullet y_m = b_1 y_1' b_2 y_2' b_3 \cdots b_m y_m' b_{m+1} = v' \ .$$

In fact, to solve the original equation e we have to solve the reduced equation e' with all possible combinations of values for letters a_i and b_j from the set Σ. In other words, the set of solutions of the original equation $u = v$ equals the set of solutions of a Boolean set of equations which is a disjunction of equations without overlapping products.

Part 2. In Part 1 we assumed that each unknown corresponds to a word of length at least two. Now we assume that at least one of the unknowns corresponds to a letter. We proceed as in Part 1 but with a bit different markings. Let $x_i = a_{i,1} x_i' a_{i,2}$ or $x_i = a_{i,12}$, with $a_{i,1}, a_{i,2}, a_{i,12} \in \Sigma$, depending on the length of the solution corresponding to x_i. Because overlapping products have to be defined we have $a_{i,2} = a_{i+1,1}$ or $a_{i,2} = a_{i+1,12}$ and $a_{i,12} = a_{i+1,1}$ or $a_{i,12} = a_{i+1,12}$. We process similarly with y_j's and b's. As in Part 1, we have some apparent additional restrictions for letters a's and b's depending on equation e. With these assumptions and markings we can again form a corresponding reduced word equation $e' : u' = v'$ without overlapping products.

To solve the original equation with assumptions of Part 2 we have again a Boolean combination of word equations to solve. This set is a disjunction of equations of the form e' with all possible combinations such that at least one unknown corresponds to a letter and values of corresponding a's and b's vary in the set Σ.

Part 3. In Part 1 and Part 2 we have only discussed the cases of constant free equations. If some factors in the equation $u = x_1 \bullet x_2 \bullet \cdots \bullet x_l = y_1 \bullet y_2 \bullet \cdots \bullet y_m = v$ are constants we proceed as previously in Parts 1 and 2 but with the additional knowledge of constants. If, for example, x_i is a constant in e and we have marked $x_i = a_i x_i' a_{i+1}$ we treat a_i, a_{i+1} and x_i' in equation e' as constants, too.

As a conclusion we remark that the considered Boolean sets are finite and the set of solutions of the original equation e is the set of solutions of a disjunction of Boolean sets of Part 1 and Part 2, the observations of the third part taken into account if necessary. Equations in this combined Boolean set do not contain overlapping products, and this proves the claim.   □

We remark that regardless of equation e having constants or not the equations in the constructed Boolean set have constants because the given reduction takes into consideration the fact that overlapping products have to be defined. The property that the overlapping product is only partially defined also makes the conversion of equations to the other direction difficult. As mentioned in Section 2 it is easy to write a word as the element of this partial semigroup $(\Sigma^+, \bullet)$. But if we try to convert, for example, an equation $xy = z$ we cannot just write $x \bullet y = z$. Instead, the equation $x \bullet y' = z$ with requirements $y' = ay, x = x'a$, with $a \in \Sigma$, would correspond the original equation.

5   Consequences of the Reduction

It is known that any Boolean combination of word equations can be transformed into a single equation, see [12], [3] or [2] as the original source. Another well known result concerning word equations is the satisfiability problem, that is decidability of whether a word equation has a solution or not. The satisfiability problem is shown to be decidable by Makanin [15], see also [19]. We will show that corresponding results are also valid for equations with overlapping products.

Theorem 3. *For any Boolean combination of equations with overlapping products we can construct a single equation without overlapping products such that the sets of solutions of the Boolean combination and the single equation are equal when restricted to unknowns of the original equations.*

Proof. The result of the previous section shows that an equation with overlapping products can be reduced into a Boolean combination of usual word equations. From this it follows that any Boolean combination of equations with overlapping products can be reduced into another Boolean combination of ordinary word equations. This, in turn, as stated above can be transformed into a single equation without overlapping products. ☐

We remind that combining a conjunction of two word equations into a single equation does not require any extra unknowns but in a case of disjunction two additional unknowns are required in the construction given in [12], see also [3]. Thus, the single equation constructed from the Boolean combination of equations is likely to contain many more unknowns than the original equations because of the disjunctions derived from the reduction method.

We next slightly modificate this old proof for the result of [12] concerning a disjunction of two equations. The new result shows that, in fact, two additional unknowns are enough to combine a disjunction of a finite set of equations into a single equation.

Theorem 4. *Let $e_1 : u_1 = v_1, \ldots, e_n : u_n = v_n$ be a finite set of equations. A disjunction of these equations, i.e. the property expressible by e_1 or e_2 or $\ldots$ or e_n, can be transformed into a single equation with only two additional unknowns.*

Proof. We may assume that the right hand sides of the equations are the same because the disjunctions of the equations of the following two sets S_1 and S_2 are equivalent:

$$S_1 : \begin{array}{c} u_1 = v_1 \\ u_2 = v_2 \\ \vdots \\ u_n = v_n \end{array} \quad \text{and} \quad S_2 : \begin{array}{c} u_1 v_2 v_3 \cdots v_n = v_1 v_2 \cdots v_n \\ v_1 u_2 v_3 \cdots v_n = v_1 v_2 \cdots v_n \\ \vdots \\ v_1 v_2 \cdots v_{n-1} u_n = v_1 v_2 \cdots v_n \end{array} .$$

Thus, we may assume that $v_1 = v_2 = \cdots = v_n = v$ holds for equations $e_1, \ldots, e_n$.

To complete the proof we will outline the necessary constructions, the justifications can be deduced as in [12]. First we define a function $\langle \ \rangle$ by

$$\langle \alpha \rangle = \alpha a \alpha b , \quad \text{where } a, b \in \Sigma, \ a \neq b.$$

We will use the properties that for each α the shortest period of $\langle \alpha \rangle$ is longer than half of its length and $\langle \alpha \rangle$ is primitive. We remark that now $\langle \alpha \rangle$ can occur in $\langle \alpha \rangle^2$ only as a prefix and a suffix. Let us denote $u_1 \cdots u_n = u$. With these observations we may deduce that

$$u_1 = v \text{ or } u_2 = v \text{ or } \cdots \text{ or } u_n = v \ \Leftrightarrow \ \exists Z, Z' : \ X = ZYZ' ,$$

where

$$Y = \langle u \rangle^2 v \langle u \rangle v \langle u \rangle^2$$

and

$$X = \langle u \rangle^2 u_1 \langle u \rangle u_1 \langle u \rangle^2 u_2 \langle u \rangle u_2 \langle u \rangle^2 \cdots \langle u \rangle^2 u_n \langle u \rangle u_n \langle u \rangle^2 .$$

The proof of the previous equivalence is based on the facts that the word $\langle u \rangle^2$ is a prefix and a suffix of Y and that it occurs in X in exactly $n+1$ places. We concentrate on the nontrivial part of the proof. Thus, if $X = ZYZ'$ holds there are essentially two possibilities for $v \langle u \rangle v$:

$$v \langle u \rangle v = u_i \langle u \rangle u_i , \quad \text{for some } i$$

or

$$v \langle u \rangle v = u_i \langle u \rangle u_i \langle u \rangle^2 u_{i+1} \langle u \rangle u_{i+1} \langle u \rangle^2 \cdots$$
$$u_{j-1} \langle u \rangle u_{j-1} \langle u \rangle^2 u_j \langle u \rangle u_j , \quad \text{for some } i \text{ and } j \text{ with } i < j.$$

In the first case $v = u_i$ as required. In the second case we can use the positions of factors $\langle u \rangle$ and $\langle u \rangle^2$ to conclude that this case is not possible, which completes the proof. We separate the analyzis into two cases depending on whether $v \langle u \rangle v$ equals to an expression containing an odd number of factors $\langle u \rangle^2$ or an even number of those. The following two examples illustrate the argumentation in each case. We leave it to the reader to apply corresponding arguments for the other values of i and j.

Let $w = u_1 \langle u \rangle u_1 \langle u \rangle^2 u_2 \langle u \rangle u_2$ and assume $v \langle u \rangle v = w$. Now the factor $\langle u \rangle$ in the middle of $v \langle u \rangle v$ has to overlap with the factor $\langle u \rangle^2$ of w, otherwise one of the v's would contain a factor $\langle u \rangle^2$. In a general case the overlapping concerns the centermost occurrence of factors $\langle u \rangle^2$. Now the factor preceding (or succeeding) the mentioned $\langle u \rangle$ has the length at least $2|u_1| + 2|\langle u \rangle|$ (or $2|u_2| + 2|\langle u \rangle|$). We may assume $|v| \geq 2|u_1| + 2|\langle u \rangle|$, the other case being similar. Now $|v \langle u \rangle v| \geq 4|u_1| + 5|\langle u \rangle| > |w|$ because $|\langle u \rangle| > 2|u_2|$. This gives a contradiction.

Let $w' = u_1 \langle u \rangle u_1 \langle u \rangle^2 u_2 \langle u \rangle u_2 \langle u \rangle^2 u_3 \langle u \rangle u_3$ and assume $v \langle u \rangle v = w'$. Now the factor $\langle u \rangle^2$ has to be located in the same place on both occurrences of v in the word $v \langle u \rangle v$. This gives $v = u_1 \langle u \rangle u_1 \langle u \rangle^2 u_3 \langle u \rangle u_3$ and thus $|v \langle u \rangle v| = 9|\langle u \rangle| + 4|u_1| + 4|u_3| > |w'|$ giving a contradiction. □

With a *positive* Boolean combination we refer to a Boolean combination that does not contain any negations, e.g. a Boolean combination of equations without inequalities. Now we can show that the conversion of a finite positive Boolean combination of equations over overlapping products into a single ordinary word equation requires only two extra unknowns.

Theorem 5. *For any finite positive Boolean combination of equations with overlapping products we can construct a single ordinary word equation with two additional unknowns such that the sets of solutions of the Boolean combination and the single equation are equal for some choice of these additional unknowns.*

Proof. For each equation over overlapping products we have a corresponding finite disjunction of ordinary equations based on reduction of Theorem 2. Thus, any finite positive Boolean combination of equations with overlapping products can be transformed into a finite positive Boolean combination of ordinary word equations. We may write the constructed Boolean combination in a disjunctive normal form and replace each conjuction of equations by a single equation. Thus, we have formed a finite disjunction of word equations without any additional unknowns. By Theorem 4 we can transform this disjunction into a single equation with two additional unknowns which proves the claim. □

The compactness theorem for words says that each system of equations over Σ^+ and with a finite number of unknowns is equivalent to some of its finite subsystems, see [1], [8] and also [9]. We remark that the analogical result concerning equations with overlapping products is not as obvious a consequence of the reduction as the satisfiability theorem analyzed in the end of this section. If we use the reduction on an infinite system of equations with overlapping products in order to be able to use the compactness theorem of ordinary word equations, we will end up with an infinite number of finite systems of disjunctions connected with conjunctions. Although, a finite positive Boolean combination of equations over overlapping products can be reduced into a single ordinary word equation with only two additional unknowns, a corresponding reduction of an infinite Boolean combination would require an infinite number of unknowns. Thus, we cannot use the original compactness theorem because of the infinite number of unknowns and the question about validity of the compactness theorem for equations over overlapping products remains open.

The decidability result for equations with overlapping products is instead obtained easily.

Theorem 6. *The satisfiability problem for a finite positive Boolean combination of equations with overlapping products is decidable.*

Proof. Theorem 3 shows that an equation with overlapping products can be reduced into a single equation without overlapping products. With Makanin's algorithm we can decide whether this equation without overlapping products has solutions or not and the existence of solutions is not affected by the additional unknowns in a sense that they would restrict the existence. Thus, we can straightforwardly decide the existence of solutions of the original equation with overlapping products, too.   □

Acknowledgements. We would like to thank the anonymous referee for constructive comments, especially on Theorem 4.

References

1. Albert, M.H., Lawrence, J.: A proof of Ehrenfeucht's conjecture. Theoret. Comput. Sci. 41, 121–123 (1985)
2. Büchi, J.R., Senger, S.: Coding in the existential theory of concatenation. Arch. Math. Logik Grundlag. 26, 101–106 (1986/1987)
3. Choffrut, C., Karhumäki, J.: Combinatorics of words. In: Rozenberg, G., Salomaa, A. (eds.) Handbook of Formal Languages, vol. 1, pp. 329–438. Springer (1997)
4. Csuhaj-Varjú, E., Petre, I., Vaszil, G.: Self-assembly of strings and languages. Theoret. Comput. Sci. 374, 74–81 (2007)
5. Cărăuşu, A., Păun, G.: String intersection and short concatenation. Revue Roumaine de Mathématiques Pures et Appliquées 26, 713–726 (1981)
6. Domaratzki, M.: Semantic Shuffle on and Deletion Along Trajectories. In: Calude, C.S., Calude, E., Dinneen, M.J. (eds.) DLT 2004. LNCS, vol. 3340, pp. 163–174. Springer, Heidelberg (2004)
7. Domaratzki, M.: Minimality in Template-Guided Recombination. Information and Computation 207, 1209–1220 (2009)
8. Guba, V.S.: Equivalence of infinite systems of equations in free groups and semigroups to finite subsystems. Mat. Zametki 40, 321–324 (1986) (in Russian)
9. Harju, T., Karhumäki, J., Plandowski, W.: Independent systems of equations. In: Lothaire, M. (ed.) Algebraic Combinatorics on Words, ch. 13, pp. 443–472. Cambridge University Press (2002)
10. Holzer, M., Jakobi, S.: Chop Operations and Expressions: Descriptional Complexity Considerations. In: Mauri, G., Leporati, A. (eds.) DLT 2011. LNCS, vol. 6795, pp. 264–275. Springer, Heidelberg (2011)
11. Ito, M., Lischke, G.: Generalized periodicity and primitivity. Mathematical Logic Quarterly 53, 91–106 (2007)
12. Karhumäki, J., Mignosi, F., Plandowski, W.: The expressibility of languages and relations by word equations. J. ACM 47, 483–505 (2000)
13. Lothaire, M.: Combinatorics on words. Addison-Wesley (1983)
14. Lyndon, R.C., Schützenberger, M.P.: The equation $a^M = b^N c^P$ in a free group. Michigan Math. J. 9, 289–298 (1962)

15. Makanin, G.S.: The problem of the solvability of equations in a free semigroup. Mat. Sb. (N.S.) 103, 147–236 (1997) (in Russian), English translation in: Math. USSR-Sb. 32, 129–198 (1977)
16. Mateescu, A., Păun, G., Rozenberg, G., Salomaa, A.: Simple splicing systems. Discrete Applied Mathematics 84, 145–162 (1998)
17. Mateescu, A., Salomaa, A.: Parallel composition of words with re-entrant symbols. Analele Universităţii Bucureşti Mathematică-Informatică 45, 71–80 (1996)
18. Păun, G., Rozenberg, G., Salomaa, A.: DNA Computing, New Computing Paradigms. Springer (1998)
19. Plandowski, W.: Satisfiability of word equations with constants is in PSPACE. J. ACM 51, 483–496 (2004)

On CD-Systems of Stateless Deterministic Two-Phase RR(1)-Automata

Martin Kutrib[1] and Friedrich Otto[2]

[1] Institut für Informatik, Universität Giessen,
Arndtstr. 2, 35392 Giessen, Germany
kutrib@informatik.uni-giessen.de
[2] Fachbereich Elektrotechnik/Informatik, Universität Kassel,
34109 Kassel, Germany
otto@theory.informatik.uni-kassel.de

Abstract. We study stateless deterministic two-phase RR-automata of window size one: stl-det-2-RR(1)-automata. While general deterministic RR-automata of window size one characterize the regular languages, it turns out that the class of languages accepted by the stateless two-phase variants is subregular. Therefore we combine stl-det-2-RR(1)-automata into computationally stronger cooperating distributed systems, obtaining the stl-det-local-CD-2-RR(1)-systems. By limiting their inherent non-determinism, two further variants are derived. The relations between the different classes and some well-known language families are investigated, and it is shown that the classes defined here form a finite hierarchy whose levels are incomparable to several well-known language families. Further, closure properties and decision problems are studied for these classes.

1 Introduction

One of the fundamental concepts of computing models and automata is that of internal states which evolve at discrete time steps. Accordingly, the number of these states can be seen as a parameter of such systems. By reducing this number as much as possible, we obtain types of automata that only have a single internal state. Thus, the behavior of these automata does not depend on their internal state at all and, therefore, these devices are called *stateless*. It is easily seen that the computational power of *stateless* finite automata is strictly weaker than that of general finite automata. On the other hand, it is well known that already stateless nondeterministic pushdown automata accept all context-free languages [5]. Thus, for nondeterministic pushdown automata, the resource 'pushdown store' can compensate for the absence of states. Generally speaking, it is a natural and interesting question of how resources given to finite automata relate to the absence or presence of internal states. Given some computational model, are states necessary at all?

Inspired by biologically motivated models of computing related studies were initiated in [6,18], as it is difficult and even unrealistic to maintain a global state for a massively parallel group of objects appearing in natural phenomena of cell

H. Bordihn, M. Kutrib, and B. Truthe (Eds.): Dassow Festschrift 2012, LNCS 7300, pp. 111–137, 2012.

evolutions and chemical reactions. The study of stateless multi-head finite automata and stateless multi-counter systems in [18] and the successor paper [6] shows that the resource 'heads' cannot compensate for the absence of states. Recently, also stateless two-pushdown automata have been investigated [7], and it has been shown that for shrinking as well as for length-reducing deterministic and nondeterministic two-pushdown automata states are not needed. Further, also stateless variants of restarting automata have been studied. In [7] so-called R-automata with combined rewrite/restart operations are considered, while in [8] restarting automata which, after executing a rewrite step may continue to read their tape before performing a restart, so-called RR-automata, are of main interest. Thus, even after executing a rewrite step an RR-automaton has still the option to accept or to reject instead of performing a restart. In particular, in [8] the *two-phase* RR-*automaton* has been introduced, which is a stateless RR-automaton that can distinguish between the two parts of each cycle: the first part, which ends with an application of a rewrite (that is, delete) operation, and the second part, which ends with an execution of a restart operation.

Here we study the influence of the size k of the read/write window on the expressive power of *stateless deterministic two-phase* RR-*automata*, abbreviated as stl-det-2-RR(k)-automata. We will see that based on the size k, we obtain an infinite strict hierarchy of language classes that, however, are incomparable to the class REG of regular languages with respect to inclusion. In particular, it turns out that the class of languages accepted by the stateless two-phase RR-automata of window size one is subregular, while general deterministic RR-automata of window size one characterize the regular languages [9].

Then, in analogy to the work presented in [14,15] we introduce *cooperating distributed systems* (CD-systems) of stl-det-2-RR(1)-automata, the so-called stl-det-local-CD-2-RR(1)-systems. These systems are an adaptation of the notion of *cooperating distributed grammar system* with external control (see, for example, [1,3]) to the setting of stl-det-2-RR(1)-automata. As it turns out these systems are strictly more expressive than the CD-systems of stateless deterministic R(1)-automata (the so-called stl-det-local-CD-R(1)-systems) studied in [14]. On the other hand, the class of languages $\mathscr{L}_{=1}$(stl-det-local-2-RR(1)) accepted by the stl-det-local-CD-2-RR(1)-systems is incomparable under inclusion to the classes of (deterministic) context-free languages, linear languages, Church-Rosser languages and growing context-sensitive languages.

Although all the component automata of a stl-det-local-CD-2-RR(1)-system are deterministic, the system itself is not. Therefore, also two types of *deterministic* CD-systems of stl-det-2-RR(1)-automata are defined: the *strictly deterministic* CD-systems and the *globally deterministic* CD-systems. We compare the resulting classes of languages to each other and to the class of regular languages, and we establish closure and non-closure properties for them.

The paper is organized as follows. First we describe in short the two-phase restarting automaton and derive a few fundamental results on them. In Section 3, CD-systems of stateless deterministic 2-RR(1)-automata are introduced and investigated. Then the two variants without nondeterminism are defined and studied in Section 4. It turns out that the strictly deterministic CD-systems define a language class that forms a non-reversal and non-intersection closed anti-AFL, which is quite surprising for a deterministic automaton model. Although anti-AFLs are sometimes referred to as "unfortunate families of languages," there is linguistical evidence that such language families might be of crucial importance, since in [2] it was shown that the family of natural languages is an anti-AFL. Decidability problems are the main aspect of Section 5. The results on the relations between the different language classes are summarized in Figure 1, and Table 1 summarizes the closure and non-closure properties. Finally, we conclude and present some open and untouched questions in Section 6.

2   Two-Phase Restarting Automata

A *stateless deterministic two-phase* RR-*automaton*, stl-det-2-RR-automaton for short, is described by a 6-tuple $M = (\Sigma, \math022, \$, k, \delta_1, \delta_2)$, where Σ is a finite input alphabet, $\math022$ and $\$$ are additional symbols that serve as markers for the left and right border of the input tape, $k \geq 1$ is the size of the read/write window, and δ_1 and δ_2 are the transition functions that associate a *transition step* to each possible content u of the window. There are four types of transition steps: A *move-right step* (MVR) causes M to shift the window one position to the right. However, the window cannot be shifted beyond the right border marker $\$$. A *rewrite step* causes M to delete at least one and at most all symbols of the content u of the window, thereby replacing u by v and shortening the tape. Subsequently, the window is placed immediately to the right of v. Some additional restrictions apply in that the border markers $\math022$ and $\$$ must not disappear from the tape. Hence, if u ends with the symbol $\$$, then so does v, and in this situation the window is placed on the $\$$. An *accept step* causes M to halt and accept, and a *restart step* causes M to place the window again over the left end of the tape, so that the first symbol it contains is the left border marker $\math022$. If the transition step is undefined for the current situation, then M necessarily halts and rejects.

A computation of M consists of cycles followed by a tail computation. A *cycle* begins with the window scanning the left border marker. It consists of a sequence of MVR steps which is followed by a rewrite step that completes the first phase of the cycle. The behavior of M during the first phase is determined by δ_1. After the rewrite step, the second phase controlled by δ_2 starts. It consists of further MVR steps followed by a restart step that completes the cycle. A computation of M ends by a *tail computation*, which is an incomplete cycle ending with an accept step or a reject. Accept instructions can occur in both δ_1 and δ_2.

With M we associate two languages – the *simple language*

$$S(M) = \{\, w \in \Sigma^* \mid M \text{ accepts } w \text{ in a tail computation} \,\}$$

and the *language*

$$L(M) = \{\, w \in \Sigma^* \mid \exists z \in S(M) : w \vdash_M^{c^*} z \,\}$$

of words accepted by M. Here $\vdash_M^c$ denotes the reduction relation on Σ^* that is induced by the cycles of M. In order to clarify our notion we give a first short example.

Example 1. The non-regular language $\{\, w \in \{a,b\}^* \mid |w|_a = |w|_b \,\}$ is accepted by the stateless deterministic two-phase RR-automaton $M = (\{a,b\}, \mathcal{c}, \$, 2, \delta_1, \delta_2)$, where $\delta_1(\mathcal{c}\$) = \mathsf{Accept}$, $\delta_1(u) = \varepsilon$ for all $u \in \{ab, ba\}$, $\delta_1(u) = \mathsf{MVR}$ for all $u \in \{\mathcal{c}a, \mathcal{c}b, aa, bb\}$, and $\delta_2(u) = \mathsf{Restart}$ for all $u \in \{ab, ba, aa, bb, a\$, b\$, \$\}$. □

In the example above, the stl-det-2-RR(2)-automaton restarts after a rewrite in any case. This particular behavior led to the definition of the so-called R-*automata* that cannot continue to read the input after a rewrite, that is, rewrite and restart steps are combined. Therefore, for these automata the transition function δ_2 can be omitted.

For each $k \geq 1$, stl-det-2-RR(k) denotes the class of stateless deterministic two-phase RR-automata with window of size k, and $\mathcal{L}$(stl-det-2-RR(k)) denotes the class of languages that are accepted by stl-det-2-RR(k)-automata. Similarly for R-automata. For devices with states it is evident that RR-automata are at least as powerful as R-automata. But this cannot be derived from the definition for stateless variants. Nevertheless, we have the following result.

Lemma 2. *For each $k \geq 1$ and each stl-det-R(k)-automaton M, there exists a stl-det-2-RR(k)-automaton M' such that the reduction relations $\vdash_M^c$ and $\vdash_{M'}^c$ coincide, $S(M) = S(M')$ and, thus, $L(M) = L(M')$.*

Proof. Let $M = (\Sigma, \mathcal{c}, \$, k, \delta)$ be a stl-det-R(k)-automaton. We obtain a stl-det-2-RR(k)-automaton $M' = (\Sigma, \mathcal{c}, \$, k, \delta_1, \delta_2)$ by taking $\delta_1 = \delta$ and $\delta_2(u) = \mathsf{Restart}$ for all u that can occur as the contents of the window of M. Then the cycles of M' and of M correspond to each other, and $S(M') = S(M)$ holds. □

Example 3. For $k \geq 1$ and $\Sigma = \{a,b\}$, we define the language $L_k = b^* \cdot (a^k \cdot b^+)^*$.

Claim. $L_k \in \mathcal{L}$(stl-det-2-RR(k)).

Proof (of claim). We define a stl-det-2-RR(k)-automaton $M_k = (\Sigma, \mathcal{c}, \$, k, \delta_1, \delta_2)$ as follows:

$$
\begin{aligned}
\delta_1(\mathcal{c}b^i\$) &= \mathsf{Accept}, \text{ for all } 0 \leq i \leq k-2, \\
\delta_1(\mathcal{c}b^i a^{k-1-i}) &= \mathsf{MVR}, \quad \text{for all } 0 \leq i \leq k-1, \\
\delta_1(b^i a^{k-i}) &= \mathsf{MVR}, \quad \text{for all } 1 \leq i \leq k, \\
\delta_1(a^k) &= \varepsilon, \\
\delta_1(b^{k-1}\$) &= \mathsf{Accept}; \\
\delta_2(b^i a^{k-i}) &= \mathsf{Restart}, \text{ for all } 1 \leq i \leq k, \\
\delta_2(b^i\$) &= \mathsf{Restart}, \text{ for all } 1 \leq i \leq k-1.
\end{aligned}
$$

Then $S(M) = b^*$, and $b^i a^k u \vdash_M^c b^i u$ for all $i \geq 0$ and all words u such that $u \in b^+$ or $u = b^r a^s u'$ for some $r, s \geq 1$ such that $r + s \geq k$. Thus, it is easily seen that $L(M) = L_k$ holds. □

Claim. $L_k \notin \mathcal{L}$(stl-det-R(k)).

Proof (of claim). Assume to the contrary that $M = (\Sigma, \mathfrak{c}, \$, k, \delta)$ is a stl-det-R(k)-automaton such that $L(M) = L_k$ holds. The word $w_1 = a^k b$ belongs to L_k, that is, M accepts on input w_1. Now we consider the function δ. Obviously, $\delta(\mathfrak{c}a^{k-1})$ must be defined. It cannot be an accept instruction, and it cannot be a rewrite instruction, as the prefix a^{k-1} of w_1 cannot be replaced by any shorter word without obtaining a word that is not a member of L_k. Thus, it follows that $\delta(\mathfrak{c}a^{k-1}) = \mathsf{MVR}$. If $\delta(a^k) = \mathsf{MVR}$, then we must consider $\delta(a^{k-1}b)$. The suffix $a^{k-1}b$ of w_1 cannot be replaced by a shorter word without obtaining a word that is not a member of L_k, and so it follows that $\delta(a^{k-1}b) = \mathsf{MVR}$. Finally, $\delta(a^{k-2}b\$)$ must be an accept instruction. However, then together with w_1, M also accepts the word $a^{k+1}b \notin L_k$. This contradiction shows that $\delta(a^k) = \varepsilon$ must hold. But then $a^k a^k b \vdash_M^c a^k b \vdash_M^* \mathsf{Accept}$, and M accepts $a^k a^k b \notin L_k$. $\square$

Together with Lemma 2, Example 3 yields the following proper inclusions.

Corollary 4. *For all $k \geq 1$, $\mathscr{L}(\mathsf{stl\text{-}det\text{-}R}(k)) \subsetneq \mathscr{L}(\mathsf{stl\text{-}det\text{-}2\text{-}RR}(k))$.*

In [14] it is shown that the regular language $L'_k = \{ (ab^k)^i \mid i \geq 0 \}$ separates the language class $\mathscr{L}(\mathsf{stl\text{-}det\text{-}R}(k))$ from the class $\mathscr{L}(\mathsf{stl\text{-}det\text{-}R}(k+1))$. From Lemma 2 we see that L'_k is also accepted by a stl-det-2-RR($k+1$)-automaton.

Lemma 5. *The language L'_k is not accepted by any stl-det-2-RR(k)-automaton.*

Proof. Assume to the contrary that $M = (\Sigma, \mathfrak{c}, \$, k, \delta_1, \delta_2)$ is a stl-det-2-RR(k)-automaton that accepts the language L'_k. Then on input $ab^k ab^k$, M will have to accept. However, as M has a window of size k only, it cannot accept the word $ab^k ab^k$ in a tail computation without accepting some word not belonging to L'_k. Hence, the accepting computation of M on input $ab^k ab^k$ begins with a cycle $ab^k ab^k \vdash_M^c z$. Then $|z| < 2k + 2$, and as M can delete at most k symbols in a single cycle, we have $|z| \geq k + 2$. This implies, however, that $z \notin L'_k$. So, M cannot accept $ab^k ab^k$ without accepting z as well. It follows that L'_k is not accepted by any stl-det-2-RR(k)-automaton. $\square$

Recall from [9] that $\mathscr{L}(\mathsf{det\text{-}RR}(1))$ coincides with the class of regular languages, and from Example 1 that $\mathscr{L}(\mathsf{stl\text{-}det\text{-}2\text{-}RR}(2))$ includes a non-regular language. Thus, together with Lemma 5 this yields the following results.

Corollary 6. (a) *For all $k \geq 1$, $\mathscr{L}(\mathsf{stl\text{-}det\text{-}2\text{-}RR}(k)) \subsetneq \mathscr{L}(\mathsf{stl\text{-}det\text{-}2\text{-}RR}(k+1))$.*
(b) *The class $\mathscr{L}(\mathsf{stl\text{-}det\text{-}2\text{-}RR}(1))$ is properly contained in the class* REG *of regular languages.* (c) *For all $k \geq 2$, the class $\mathscr{L}(\mathsf{stl\text{-}det\text{-}2\text{-}RR}(k))$ is incomparable under inclusion to the class* REG.

3   CD-Systems of stl-det-2-RR(1)-Automata

Cooperating distributed systems (CD-systems) of restarting automata were introduced and studied in [12]. Here we study CD-systems of stateless deterministic 2-RR(1)-automata, comparing them in particular to the CD-systems of stateless deterministic R(1)-automata of [14].

A CD-system of stateless deterministic 2-RR(1)-automata consists of a finite collection $\mathcal{M} = ((M_i, \sigma_i)_{i \in I}, I_0)$ of stateless deterministic 2-RR(1)-automata $M_i = (\Sigma, \mathfrak{c}, \$, 1, \delta_1^{(i)}, \delta_2^{(i)})$ $(i \in I)$, *successor relations* $\sigma_i \subseteq I$ $(i \in I)$, and a subset $I_0 \subseteq I$ of *initial indices*. Here it is required that $I_0 \neq \emptyset$, and that $\sigma_i \neq \emptyset$ for all $i \in I$. For the CD-systems of stl-det-R(1)-automata introduced in [14] it was required in addition that $i \notin \sigma_i$ for all $i \in I$, but this requirement is easily met by using two isomorphic copies of each component automaton. Therefore, we abandon it here in order to simplify the presentation.

Various modes of operation have been introduced and studied for CD-systems of restarting automata, but here we are only interested in mode = 1 computations. A computation of $\mathcal{M}$ in mode = 1 on an input word w proceeds as follows. First an index $i_0 \in I_0$ is chosen nondeterministically. Then the 2-RR-automaton M_{i_0} starts the computation with the initial configuration $\mathfrak{c}w\$$, and executes a single cycle. Thereafter an index $i_1 \in \sigma_{i_0}$ is chosen nondeterministically, and M_{i_1} continues the computation by executing a single cycle. This continues until, for some $l \geq 0$, the automaton M_{i_l} accepts. Such a computation will be denoted as $(i_0, w) \vdash_{\mathcal{M}}^c (i_1, w_1) \vdash_{\mathcal{M}}^c \cdots \vdash_{\mathcal{M}}^c (i_l, w_l) \vdash_{M_{i_l}}^* \mathsf{Accept}$. Should at some stage the chosen automaton M_{i_l} be unable to execute a cycle or to accept, then the computation fails. By $L_{=1}(\mathcal{M})$ we denote the language that the system $\mathcal{M}$ accepts in mode = 1, and by $\mathscr{L}_{=1}(\mathsf{stl\text{-}det\text{-}local\text{-}CD\text{-}2\text{-}RR(1)})$ we denote the class of languages that are accepted by mode = 1 computations of stl-det-local-CD-2-RR(1)-systems, that is, by CD-systems of stateless deterministic 2-RR(1)-automata.

From Lemma 2 we immediately obtain that $\mathscr{L}_{=1}(\mathsf{stl\text{-}det\text{-}local\text{-}CD\text{-}R(1)})$ is contained in $\mathscr{L}_{=1}(\mathsf{stl\text{-}det\text{-}local\text{-}CD\text{-}2\text{-}RR(1)})$. Below we will see that this inclusion is actually a proper one.

Recall from [4] or from [14] that a language $L \subseteq \Sigma^*$ is called a *rational trace language* if there exists a reflexive and transitive binary relation D on Σ (a *dependency relation*) such that $L = \bigcup_{w \in R} [w]_D$ for some regular language R on Σ. Here $[w]_D$ denotes the congruence class of w with respect to the congruence $\equiv_D = \{ (uabv, ubav) \mid u, v \in \Sigma^*, a, b \in \Sigma, (a, b) \notin D \}$. In [14] it is shown that the stl-det-local-CD-R(1)-systems accept all rational trace languages. Thus, we see that also the stl-det-local-CD-2-RR(1)-systems accept all rational trace languages. Further, it is shown in [14] that one can extract a finite-state acceptor A from a stl-det-local-CD-R(1)-system $\mathcal{M}$ such that A accepts a sublanguage of $L_{=1}(\mathcal{M})$ that is letter-equivalent to $L_{=1}(\mathcal{M})$. Below we prove that this result does not carry over to stl-det-local-CD-2-RR(1)-systems.

Example 7. Let $\mathcal{M} = ((M_i, \sigma_i)_{i \in \{1,2\}}, \{1\})$ be the CD-system of stl-det-2-RR(1)-automata on $\Sigma = \{a, b\}$ that is specified by $\sigma_1 = \{2\}$, $\sigma_2 = \{1\}$, and

$$M_1 : \delta_1^{(1)} : \mathfrak{c} \mapsto \mathsf{MVR}, a \mapsto \varepsilon, \$ \mapsto \mathsf{Accept}; \delta_2^{(1)} : a \mapsto \mathsf{Restart}, b \mapsto \mathsf{Restart};$$

$$M_2 : \delta_1^{(2)} : \mathfrak{c} \mapsto \mathsf{MVR}, b \mapsto \varepsilon, a \mapsto \mathsf{MVR}; \quad \delta_2^{(2)} : b \mapsto \mathsf{MVR}, \quad \$ \mapsto \mathsf{Restart}.$$

Then $\mathcal{M}$ accepts the empty word. If $w \in \Sigma^+$ is accepted, then we see from the definition of $\mathcal{M}$ that $w = a^n b^m$ for some $n, m \geq 1$. In fact, as M_1

and M_2 alternate in every computation of $\mathcal{M}$, we have $n = m$ and, therefore, $L_{=1}(\mathcal{M}) = \{\, a^n b^n \mid n \geq 0 \,\}$.   □

Actually, the following stronger result can be derived.

Proposition 8. *For all* $m \geq 1$,

$$L_m = \{\, a_1^n a_2^n \dots a_m^n \mid n \geq 0 \,\} \in \mathcal{L}_{=1}(\mathsf{stl\text{-}det\text{-}local\text{-}CD\text{-}2\text{-}RR(1)}).$$

The language L_m $(m \geq 2)$ does not contain a regular sublanguage that is letter-equivalent to L_m. It follows that this language is not accepted by any stl-det-local-CD-R(1)-systems. So we obtain the following proper inclusion.

Corollary 9. $\mathcal{L}_{=1}(\mathsf{stl\text{-}det\text{-}local\text{-}CD\text{-}R(1)}) \subsetneqq \mathcal{L}_{=1}(\mathsf{stl\text{-}det\text{-}local\text{-}CD\text{-}2\text{-}RR(1)}).$

In order to determine the computational capacity of stl-det-local-CD-2-RR(1)-systems we continue with an example that shows that these systems accept a language that is not even growing context-sensitive.

Example 10. Let $\Sigma = \{a, b, \tilde{a}, \tilde{b}\}$. For any word $w = x_1 x_2 \cdots x_n \in \{a, b\}^*$, we set $\tilde{w} = \tilde{x}_1 \tilde{x}_2 \cdots \tilde{x}_n \in \{\tilde{a}, \tilde{b}\}^*$, and consider $L_{tc} = \{\, a w \tilde{a} \tilde{w} \mid w \in \{a, b\}^* \,\}$ over Σ.

The language L_{tc} is not growing context-sensitive, as the growing context-sensitive languages are closed under union and ε-free homomorphisms, and the copy language is not growing context-sensitive [10]. However, it is accepted by the stl-det-local-CD-2-RR(1)-system $\mathcal{M} = ((M_i, \sigma_i)_{i \in \{0,1,2,3,4\}}, \{0\})$ that is specified by $\sigma_0 = \{1\}$, $\sigma_1 = \{0, 2, 4\}$, $\sigma_2 = \{3\}$, $\sigma_3 = \{0, 2, 4\}$, $\sigma_4 = \{4\}$, and

$$\delta_1^{(0)} : \math鐷 \mapsto \mathsf{MVR}, \quad a \mapsto \varepsilon;$$
$$\delta_2^{(0)} : a \mapsto \mathsf{Restart}, \quad b \mapsto \mathsf{Restart}, \quad \tilde{a} \mapsto \mathsf{Restart};$$
$$\delta_1^{(1)} : \math鐷 \mapsto \mathsf{MVR}, \quad a \mapsto \mathsf{MVR}, \quad b \mapsto \mathsf{MVR}, \quad \tilde{a} \mapsto \varepsilon;$$
$$\delta_2^{(1)} : \tilde{a} \mapsto \mathsf{Restart}, \quad \tilde{b} \mapsto \mathsf{Restart}, \quad \$ \mapsto \mathsf{Restart};$$
$$\delta_1^{(2)} : \math鐷 \mapsto \mathsf{MVR}, \quad b \mapsto \varepsilon;$$
$$\delta_2^{(2)} : a \mapsto \mathsf{Restart}, \quad b \mapsto \mathsf{Restart}, \quad \tilde{b} \mapsto \mathsf{Restart};$$
$$\delta_1^{(3)} : \math鐷 \mapsto \mathsf{MVR}, \quad a \mapsto \mathsf{MVR}, \quad b \mapsto \mathsf{MVR}, \quad \tilde{b} \mapsto \varepsilon;$$
$$\delta_2^{(3)} : \tilde{a} \mapsto \mathsf{Restart}, \quad \tilde{b} \mapsto \mathsf{Restart}, \quad \$ \mapsto \mathsf{Restart};$$
$$\delta_1^{(4)} : \math鐷 \mapsto \mathsf{MVR}, \quad \$ \mapsto \mathsf{Accept}.$$

Initially, component 0 deletes the first input symbol if it is an a, otherwise the input is rejected. Then component 1 searches for the first occurrence of an input letter from $\{\tilde{a}, \tilde{b}\}$. It is deleted if it is $\tilde{a}$, otherwise the input is rejected. In subsequent cycles corresponding symbols a and $\tilde{a}$ or b and $\tilde{b}$ are deleted by the components 0 and 1 or 2 and 3. After deleting an a, component 0 rejects if the next input symbol is $\tilde{b}$ or $\$$. In all other cases it restarts. The following component 1 deletes the first occurrence of an input letter from $\{\tilde{a}, \tilde{b}\}$ if it is $\tilde{a}$, otherwise the input is rejected. Moreover, component 1 restarts only if the deleted symbol is followed by another symbol from $\{\tilde{a}, \tilde{b}\}$ or by $\$$. Similarly, for the components 2 and 3, and b and $\tilde{b}$. Whenever a pair of corresponding symbols has been deleted, system $\mathcal{M}$ guesses of which type the next pair is, or whether all pairs have been deleted. In the first case either component 0 or 2 is chosen to continue the computation. In the latter case, component 4 is used to verify that in fact all symbols have been deleted. Only in this situation it accepts. It follows that $L_{=1}(\mathcal{M}) = L_{tc}$. □

The power of stl-det-local-CD-2-RR(1)-systems is only deployed for languages over an alphabet with at least two symbols. For unary languages we have the following characterization.

Theorem 11. *A language $L \subseteq \{a\}^*$ is accepted by a stl-det-local-CD-2-RR(1)-system if and only if it is regular.*

Proof. As already stl-det-local-CD-R(1)-systems accept all regular languages [14], we see from Corollary 9 that the implication from right to left holds. To prove the reverse implication let $\mathcal{M} = ((M_i, \sigma_i)_{i \in I}, I_0)$ be a CD-system of stateless deterministic 2-RR(1)-automata on $\Sigma = \{a\}$. For all $i \in I$, if $\delta_1^{(i)}(\text{¢})$ is undefined, then each computation of $\mathcal{M}$ that activates M_i fails, and if $\delta_1^{(i)}(\text{¢}) = \mathsf{Accept}$, then each computation of $\mathcal{M}$ that activates M_i accepts. Thus, in the former case M_i can be seen as a trap "state," while in the latter case it can be seen as an accepting "state" that keeps on digesting a's. Now assume that $\delta_1^{(i)}(\text{¢}) = \mathsf{MVR}$. If also $\delta_1^{(i)}(a) = \mathsf{MVR}$, then M_i can only execute tail computations. In fact, either M_i accepts all words from Σ^* in tail computations, and this is the case if $\delta_1^{(i)}(\$) = \mathsf{Accept}$, or it rejects all words from Σ^* in tail computations, and this is the case if $\delta_1^{(i)}(\$)$ is undefined. Also if $\delta_1^{(i)}(a) = \varepsilon$ and $\delta_2^{(i)}(a) \neq \mathsf{Restart}$ and $\delta_2^{(i)}(\$) \neq \mathsf{Restart}$, then M_i can only execute tail computations. Hence, again M_i can be seen as a trap "state" or as an accepting "state."

Now we can construct a finite-state acceptor $A = (Q, \Sigma, S, F, \delta_A)$ from $\mathcal{M}$ that accepts the language $L = L_{=1}(\mathcal{M})$. Essentially the states of A correspond to the component automata M_i of $\mathcal{M}$, with certain component automata becoming trap states and others becoming accepting states. For each $i \in I$, if M_i can execute a cycle (that is, $\delta_1^{(i)}(\text{¢}) = \mathsf{MVR}$, $\delta_1^{(i)}(a) = \varepsilon$, and $\delta_2^{(i)}(a) = \mathsf{Restart}$ or $\delta_2^{(i)}(a) = \mathsf{MVR}$ and $\delta_2^{(i)}(\$) = \mathsf{Restart}$), then A has an a-transition from the state corresponding to M_i to all states that correspond to component automata M_j with $j \in \sigma_i$. It is now easy to set up the transition relation δ_A in such a way that $L(A) = L_{=1}(\mathcal{M})$ holds. □

Since, for example, the unary language $\{a^{2^n} \mid n \geq 0\}$ belongs to the class of Church-Rosser languages [11], which in turn is a proper subset of the growing context-sensitive languages, we obtain the following incomparability results.

Corollary 12. *The language class $\mathscr{L}_{=1}(\mathsf{stl\text{-}det\text{-}local\text{-}CD\text{-}2\text{-}RR(1)})$ is incomparable under inclusion to the classes CRL of Church-Rosser languages and GCSL of growing context-sensitive languages.*

Now we turn to consider the closure properties of the language class $\mathscr{L}_{=1}(\text{stl-det-local-CD-2-RR}(1))$. Closure under certain operations indicates a certain robustness of the language families considered, while non-closure properties may serve, for example, as a valuable basis for extensions. We start to explore the closure properties under the Boolean operations union, intersection, and complementation. The first result is immediate.

Lemma 13. *The class $\mathscr{L}_{=1}(\text{stl-det-local-CD-2-RR}(1))$ is closed under union.*

Proof. Given two stl-det-local-CD-2-RR(1)-systems $\mathcal{M} = ((M_i, \sigma_i)_{i \in I}, I_0)$ and $\mathcal{M}' = ((M_i', \sigma_i')_{i \in I'}, I_0')$, we can assume without loss of generality that I and I' are disjoint. So, it suffices to construct a new stl-det-local-CD-2-RR(1)-system $\mathcal{M}'' = ((M_i, \sigma_i)_{i \in I \cup I'}, I_0 \cup I_0')$ that consists of the components of $\mathcal{M}$ and $\mathcal{M}'$. Initially, $\mathcal{M}''$ guesses a starting component from the union $I_0 \cup I_0'$, that is, whether to simulate $\mathcal{M}$ or $\mathcal{M}'$.   □

In order to show non-closure under intersection with regular sets we give the following example.

Example 14. Let D_1 denote the Dyck language on $\Sigma = \{a, b\}$, $\varphi : \Sigma^* \to \Sigma^*$ be the homomorphism that is induced by $a \mapsto a$ and $b \mapsto ba$, and $D_\varphi = \varphi(D_1)$. Then $w \in \Sigma^+$ belongs to D_φ if and only if $w \in \{a, ba\}^+$ and there exists an $n \geq 1$ such that $(w = a^n baz) \wedge (a^{n-1}z \in D_\varphi)$.

The stl-det-local-CD-2-RR(1)-system $\mathcal{M} = ((M_i, \sigma_i)_{i \in \{1,2,3\}}, \{1\})$ is specified by $\sigma_1 = \{2\}$, $\sigma_2 = \{3\}$, $\sigma_3 = \{1\}$, and

$$
\begin{array}{lll}
M_1 : \delta_1^{(1)}(\textcent) = \text{MVR}, & M_2 : \delta_1^{(2)}(\textcent) = \text{MVR}, & M_3 : \delta_1^{(3)}(\textcent) = \text{MVR}, \\
\quad \delta_1^{(1)}(a) = \varepsilon, & \quad \delta_1^{(2)}(a) = \text{MVR}, & \quad \delta_1^{(3)}(a) = \varepsilon; \\
\quad \delta_1^{(1)}(\$) = \text{Accept}; & \quad \delta_1^{(2)}(b) = \varepsilon; & \quad \delta_2^{(3)}(a) = \text{Restart}, \\
\quad \delta_2^{(1)}(a) = \text{Restart}, & \quad \delta_2^{(2)}(a) = \text{Restart}; & \quad \delta_2^{(3)}(b) = \text{Restart}, \\
\quad \delta_2^{(1)}(b) = \text{Restart}; & & \quad \delta_2^{(3)}(\$) = \text{Restart}.
\end{array}
$$

Obviously, $\mathcal{M}$ accepts on input ε. Now let $w = a^n baz$ such that $a^{n-1}z \in D_\varphi$. Then on input w, $\mathcal{M}$ proceeds as follows:

$$(1, w) = (1, a^n baz) \vdash_{M_1}^c (2, a^{n-1}baz) \vdash_{M_2}^c (3, a^{n-1}az) \vdash_{M_3}^c (1, a^{n-1}z).$$

By induction it follows that $\mathcal{M}$ accepts input $a^{n-1}z$, which shows $w \in L_{=1}(\mathcal{M})$. Thus, $D_\varphi \subseteq L_{=1}(\mathcal{M})$.

Conversely, assume that $w \in L_{=1}(\mathcal{M})$. If $w = \varepsilon$, then $w \in D_\varphi$. Otherwise, the accepting computation of $\mathcal{M}$ on input w looks as follows:

$$(1, w) \vdash_{M_1}^c (2, w_1) \vdash_{M_2}^c (3, w_2) \vdash_{M_3}^c (1, w_3) \vdash_{\mathcal{M}}^* \text{Accept},$$

where $w = aw_1$ for $w_1 \neq \varepsilon$, $w_1 = a^m baz$ for some $z \in \Sigma^*$, $w_2 = a^m az$, and $w_3 = a^m z \in L_{=1}(\mathcal{M})$. By induction it follows that $a^m z \in D_\varphi$, which implies that $w = aw_1 = a^{m+1}baz$ belongs to D_φ. Thus, $L_{=1}(\mathcal{M}) = D_\varphi$ holds.   □

Theorem 15. *The class $\mathscr{L}_{=1}$(stl-det-local-CD-2-RR(1)) is not closed under intersection (with regular sets), complementation, and ε-free homomorphisms.*

Proof. By Example 14, the language D_φ is accepted by a stl-det-local-CD-2-RR(1)-system. We take $R = a^* \cdot (ba)^*$ and show that the intersection $D_\varphi \cap R$ does not belong to the class $\mathscr{L}_{=1}$(stl-det-local-CD-2-RR(1)).

Claim. $D_\varphi \cap R \notin \mathscr{L}_{=1}$(stl-det-local-CD-2-RR(1)).

Proof (of claim). Assume that $\mathcal{M} = ((M_i, \sigma_i)_{i \in I}, I_0)$ is a stl-det-local-CD-2-RR(1)-system such that $L_{=1}(\mathcal{M}) = D_\varphi \cap R$. Then, for each $n \geq 1$, $\mathcal{M}$ has an accepting computation on input $w_{n+1} = a^{n+1}(ba)^{n+1}$. Let $n > |I|$, and let

$$(i_0, w_{n+1}) \vdash^c_{M_{i_0}} (i_1, z_1) \vdash^c_{M_{i_1}} \cdots \vdash^c_{M_{i_{m-1}}} (i_m, z_m) \vdash^*_{M_{i_m}} \text{Accept}$$

be an accepting computation of $\mathcal{M}$ on input w_{n+1}. We now analyze this computation. Assume that there exists an index $k < n$ such that

$$(i_0, w_{n+1}) \vdash^{c^k}_{\mathcal{M}} (i_k, a^{n+1-k}(ba)^{n+1}) \vdash^c_{M_{i_k}} (i_{k+1}, a^{n+1-k}a(ba)^n) \vdash^*_{\mathcal{M}} \text{Accept}$$

holds, that is, in each of the first $k < n$ cycles, an occurrence of the letter a is deleted, while in the $(k+1)$-st cycle the first occurrence of the letter b is deleted. Then $\mathcal{M}$ would also perform the following computation:

$$(i_0, a^n baa(ba)^n) \vdash^{c^k}_{\mathcal{M}} (i_k, a^{n-k}baa(ba)^n) \vdash^c_{M_{i_k}} (i_{k+1}, a^{n-k}aa(ba)^n) \vdash^*_{\mathcal{M}} \text{Accept},$$

which shows that $\mathcal{M}$ accepts on input $a^n baa(ba)^n$ as well. However, since $a^n baa(ba)^n \notin D_\varphi \cap R$, this contradicts our assumption on $\mathcal{M}$.

Thus, during the first k cycles, in the accepting computation above the prefix a^k is deleted, for a $k \geq n$. As $n > |I|$, this means that there exist integers j and $\ell > 0$ such that $j + \ell \leq n$ and $i_j = i_{j+\ell}$. Hence, the accepting computation above has the following form:

$$(i_0, w_{n+1}) \vdash^{c^j}_{\mathcal{M}} (i_j, a^{n+1-j}(ba)^{n+1}) \vdash^{c^\ell}_{\mathcal{M}} (i_j, a^{n+1-j-\ell}(ba)^{n+1}) \vdash^*_{\mathcal{M}} \text{Accept}.$$

But then $\mathcal{M}$ will also execute the following accepting computation:

$$(i_0, a^{n+1-\ell}(ba)^{n+1}) \vdash^{c^j}_{\mathcal{M}} (i_j, a^{n+1-j-\ell}(ba)^{n+1}) \vdash^*_{\mathcal{M}} \text{Accept},$$

which shows that it accepts on input $a^{n+1-\ell}(ba)^{n+1} \notin D_\varphi \cap R$. Again this contradicts our assumption on $\mathcal{M}$. It follows that $D_\varphi \cap R$ is not accepted by any stl-det-local-CD-2-RR(1)-system working in mode $= 1$. $\qquad\square$

So the class $\mathscr{L}_{=1}$(stl-det-local-CD-2-RR(1)) is not closed under intersection even with regular sets. By Lemma 13 it is closed under union. Since closure under complementation and union implies closure under intersection, it cannot be closed under complementation, either.

By Example 7, the language $L_2 = \{ a^n b^n \mid n \geq 0 \}$ is accepted by a stl-det-local-CD-2-RR(1)-system. Let $h : \{a,b\}^* \to \{a,b\}^*$ be the ε-free homomorphism defined by $h(a) = a$ and $h(b) = ba$. Then $h(L_2) = D_\varphi \cap R$ does not belong to $\mathscr{L}_{=1}$(stl-det-local-CD-2-RR(1)), which shows the non-closure under ε-free homomorphisms. $\qquad\square$

The proof of Theorem 15 together with Example 10 reveal further incomparabilities. Since the language $D_\varphi \cap R$ belongs to the intersection of deterministic and linear context-free languages, we have the following corollary.

Corollary 16. *The language class* $\mathscr{L}_{=1}(\mathsf{stl\text{-}det\text{-}local\text{-}CD\text{-}2\text{-}RR}(1))$ *is incomparable under inclusion to the classes* CFL *of context-free languages,* LIN *of linear, and* DCFL *of deterministic context-free languages.*

We continue with further (non-)closure properties.

Example 17. Let $\Sigma = \{a, b, c, d\}$ and define the language

$$L_{dc} = \{\, wc^m dc^n \mid w \in \{a, b\}^*, m = |w|_a, n = |w|_b \,\}.$$

The language L_{dc} is accepted by the stl-det-local-CD-2-RR(1)-system $\mathcal{M} = ((M_i, \sigma_i)_{i \in \{0,1,2,3,4,5\}}, \{0, 2\})$ that is specified by $\sigma_0 = \{1\}$, $\sigma_1 = \{0, 2\}$, $\sigma_2 = \{3, 5\}$, $\sigma_3 = \{4\}$, $\sigma_4 = \{3, 5\}$, $\sigma_5 = \{5\}$, and

$$\delta_1^{(0)} : \mathbb{c} \mapsto \mathsf{MVR}, \quad b \mapsto \mathsf{MVR}, \quad a \mapsto \varepsilon;$$
$$\delta_2^{(0)} : a \mapsto \mathsf{Restart}, \quad b \mapsto \mathsf{Restart}, \quad c \mapsto \mathsf{Restart};$$
$$\delta_1^{(1)} : \mathbb{c} \mapsto \mathsf{MVR}, \quad a \mapsto \mathsf{MVR}, \quad b \mapsto \mathsf{MVR}, \quad c \mapsto \varepsilon;$$
$$\delta_2^{(1)} : c \mapsto \mathsf{Restart}, \quad d \mapsto \mathsf{Restart};$$
$$\delta_1^{(2)} : \mathbb{c} \mapsto \mathsf{MVR}, \quad b \mapsto \mathsf{MVR}, \quad d \mapsto \varepsilon;$$
$$\delta_2^{(2)} : c \mapsto \mathsf{Restart}, \quad \$ \mapsto \mathsf{Restart};$$
$$\delta_1^{(3)} : \mathbb{c} \mapsto \mathsf{MVR}, \quad b \mapsto \varepsilon;$$
$$\delta_2^{(3)} : b \mapsto \mathsf{Restart}, \quad c \mapsto \mathsf{Restart};$$
$$\delta_1^{(4)} : \mathbb{c} \mapsto \mathsf{MVR}, \quad b \mapsto \mathsf{MVR}, \quad c \mapsto \varepsilon;$$
$$\delta_2^{(4)} : c \mapsto \mathsf{Restart}, \quad \$ \mapsto \mathsf{Restart};$$
$$\delta_1^{(5)} : \mathbb{c} \mapsto \mathsf{MVR}, \quad \$ \mapsto \mathsf{Accept}.$$

Basically, the idea of the construction is that components 0 and 1 are used to delete one a from the prefix w and, subsequently, one c from the first block of c's. When all a's and c's have been deleted, component 2 is used to delete the sole symbol d. The input is rejected if component 2 sees an a or a c before reaching the d. Next, components 3 and 4 are used to delete successively the remaining b's from the prefix and the c's from the second block. Finally, component 5 checks that all symbols have been deleted. Only in this situation it accepts. □

Theorem 18. *The class* $\mathscr{L}_{=1}(\mathsf{stl\text{-}det\text{-}local\text{-}CD\text{-}2\text{-}RR}(1))$ *is not closed under inverse homomorphisms.*

Proof. By Example 17, the language L_{dc} is accepted by a stl-det-local-CD-2-RR(1)-system. Let $h : \{a, c, d\}^* \to \{a, b, c, d\}^*$ be the homomorphism that is defined by $h(a) = ab$, $h(c) = c$, and $h(d) = d$. Then $h^{-1}(L_{dc}) = \{\, a^n c^n dc^n \mid n \geq 0 \,\}$.

Assume that $\mathcal{M} = ((M_i, \sigma_i)_{i \in I}, I_0)$ is a stl-det-local-CD-2-RR(1)-system such that $L_{=1}(\mathcal{M}) = h^{-1}(L_{dc})$.

First we note that in any accepting computation none of the c's following the sole d can be deleted as long as there is at least one c left before the d.

Let $n > |I|$. Clearly, $\mathcal{M}$ cannot accept the input $w_n = a^n c^n d c^n$ in a tail computation. So, there exist integers j and $\ell > 0$ with $j + \ell \le n$ and $i_j = i_{j+\ell}$, and integers k_1, k_2, ℓ_1, ℓ_2 with $k_1 + \ell_1 + k_2 + \ell_2 = j + \ell$ such that, assuming that the sole d is not deleted during the first $j + \ell$ cycles, the accepting computation on w_n has the following form:

$$(i_0, w_n) \vdash_{\mathcal{M}}^{c^j} (i_j, a^{n-k_1} c^{n-k_2} d c^n) \vdash_{\mathcal{M}}^{c^\ell} (i_j, a^{n-k_1-\ell_1} c^{n-k_2-\ell_2} d c^n) \vdash_{\mathcal{M}}^* \text{Accept.}$$

But then $\mathcal{M}$ will also execute the following accepting computation:

$$(i_0, a^{n-\ell_1} c^{n-\ell_2} d c^n) \vdash_{\mathcal{M}}^{c^j} (i_j, a^{n-k_1-\ell_1} c^{n-k_2-\ell_2} d c^n) \vdash_{\mathcal{M}}^* \text{Accept,}$$

which shows that it accepts the input $a^{n-\ell_1} c^{n-\ell_2} d c^n$ not belonging to $h^{-1}(L_{dc})$.

Now assume that the sole d is deleted during the first $j + \ell$ cycles. Then we obtain immediately a contradiction since the input $a^n c^{n+1} d c^{n-1} \notin h^{-1}(L_{dc})$ is accepted as well. It follows that $h^{-1}(L_{dc})$ is not accepted by any stl-det-local-CD-2-RR(1))-system, which proves the non-closure under inverse homomorphisms.   □

Theorem 19. $\mathcal{L}_{=1}(\text{stl-det-local-CD-2-RR(1)})$ *is not closed under reversal.*

Proof. By Example 14, the language D_φ is accepted by a stl-det-local-CD-2-RR(1)-system. We show the theorem by proving that the reversal D_φ^R does not belong to $\mathcal{L}_{=1}(\text{stl-det-local-CD-2-RR(1)})$.

In contrast to the assertion assume that $\mathcal{M} = ((M_i, \sigma_i)_{i \in I}, I_0)$ is a stl-det-local-CD-2-RR(1)-system such that $L_{=1}(\mathcal{M}) = D_\varphi^R$. We consider accepting computations on inputs of the form $w_n = (ab)^n a^n$, for n large enough.

First we note that each component that deletes a symbol has to delete the leftmost occurrence of that symbol. Therefore, none of the components can delete an a from the suffix a^n as long as there is at least one a left in the prefix $(ab)^n$. Moreover, it is not hard to see that $\mathcal{M}$ cannot accept without deleting some a's from the suffix. Consider the tape inscription before the cycle in which the first symbol a from the suffix is deleted. It must be of the form $b^k a^n$, and k is determined by the prefix $(ab)^n$. Furthermore, for a fixed k, there are less than $|I|$ different values n such that $b^k a^n$ is the tape inscription in that situation. This implies that k is not bounded, that is, for any $k \ge 0$ we can find an n such that $(ab)^n a^n$ is transformed into $b^{k'} a^n$, where $k' \ge k$. We choose an n large enough such that k is large enough as well. Therefore, during the computation on the prefix there must occur two cycles in which the same component deletes an a such that the number of b's preceding the a is larger in the second cycle. More precisely, there exist integers j and $\ell > 0$ with $j + \ell \le n$ and $i_j = i_{j+\ell}$, and integers $k_1 \ge 0$, k_2, m_1, m_2 with $m_2 - m_1 \ge 1$, $k_2 - m_2 - 1 \ge 0$ such that the accepting computation on $w_n = (ab)^n a^n$ has the following form:

$$(i_0, w_n) \vdash_{\mathcal{M}}^{c^j} (i_j, b^{k_1}(ab)^{k_2} a^n)$$
$$\vdash_{\mathcal{M}}^{c} (i_{j+1}, b^{k_1} b(ab)^{k_2-1} a^n) \vdash_{\mathcal{M}}^{c^{\ell-1}} (i_j, b^{k_1-m_1+m_2}(ab)^{k_2-m_2} a^n)$$
$$\vdash_{\mathcal{M}}^{c} (i'_{j+1}, b^{k_1-m_1+m_2} b(ab)^{k_2-m_2-1} a^n) \vdash_{\mathcal{M}}^{*} \text{Accept}.$$

But then $\mathcal{M}$ will also execute the following accepting computation:

$$(i_0, (ab)^{n-k_2} abb^{m_2-m_1}(ab)^{k_2-m_2-1} a^n) \vdash_{\mathcal{M}}^{c^j} (i_j, b^{k_1} abb^{m_2-m_1}(ab)^{k_2-m_2-1} a^n)$$
$$\vdash_{\mathcal{M}}^{c} (i'_{j+1}, b^{k_1} bb^{m_2-m_1}(ab)^{k_2-m_2-1} a^n) \vdash_{\mathcal{M}}^{*} \text{Accept},$$

which shows that it accepts the input $(ab)^{n-k_2+1} b^{m_2-m_1}(ab)^{k_2-m_2-1} a^n \notin D_\varphi^R$. It follows that D_φ^R is not accepted by any stl-det-local-CD-2-RR(1)-system. □

4   Deterministic CD-Systems of stl-det-2-RR(1)-Automata

Although all the component automata of a stl-det-local-CD-2-RR(1)-system are deterministic, the system itself is not. Indeed, the initial component with which to begin a particular computation is chosen nondeterministically from the set I_0 of all initial components, and after each cycle the component for executing the next cycle is chosen nondeterministically from among all the successors of the previously active component. Here we define two types of *deterministic* CD-systems of stl-det-2-RR(1)-automata: the *strictly deterministic* CD-systems and the *globally deterministic* CD-systems.

4.1   Strictly Deterministic CD-Systems of stl-det-2-RR(1)-Automata

Here we introduce and study a first type of CD-system of stateless deterministic 2-RR(1)-automata that is completely deterministic. The idea and the notation is taken from [13], where a corresponding notion was introduced for CD-systems of general restarting automata.

A CD-system $\mathcal{M} = ((M_i, \sigma_i)_{i \in I}, I_0)$ of stateless deterministic 2-RR(1)-automata is called *strictly deterministic* if $|I_0| = 1$ and $|\sigma_i| = 1$ for all $i \in I$. Then, for each word $w \in \Sigma^*$, $\mathcal{M}$ has a unique computation that begins with the initial configuration corresponding to input w. By $\mathcal{L}_{=1}$(stl-det-strict-CD-2-RR(1)) we denote the class of languages that are accepted by strictly deterministic stateless CD-2-RR(1)-systems. Note that the CD-systems in Examples 7, 14, and Proposition 8 are strictly deterministic. On the other hand, we have the following simple but useful observation on the weakness of stl-det-strict-CD-2-RR(1)-systems.

Lemma 20. *Let* $\mathcal{M} = ((M_i, \sigma_i)_{i \in I}, \{i_0\})$ *be a* stl-det-strict-CD-2-RR(1)-*system that accepts a language over the alphabet* Σ, *where* $\delta_1^{(i_0)}(\mathdollar) = $ MVR. *For all* $w \in \Sigma^*$ *and all* $x, y \in \Sigma$, *if* $\delta_1^{(i_0)}(x) = \delta_1^{(i_0)}(y) = \varepsilon$, *then* $xw \in L_{=1}(\mathcal{M})$ *if and only if* $yw \in L_{=1}(\mathcal{M})$.

Lemma 21. *The finite language $L_0 = \{aaa, bb\}$ is not accepted by any strictly deterministic stateless* CD-2-RR(1)-*system.*

Proof. Assume that $\mathcal{M} = ((M_i, \sigma_i)_{i \in I}, I_0)$ is a strictly deterministic stateless CD-2-RR(1)-system such that $L_{=1}(\mathcal{M}) = L_0$, and let $I_0 = \{i_0\}$. Since L_0 is neither $\{a, b\}^*$ nor empty, we have $\delta_1^{(i_0)}(\mathfrak{c}) = $ MVR. Similarly, $L_0 \cap a^+$ is neither a^+ nor empty and, thus, we see that $\delta_1^{(i_0)}(a) = \varepsilon$. Analogously it follows that $\delta_1^{(i_0)}(b) = \varepsilon$. So we see from Lemma 20 that $aaa \in L_{=1}(\mathcal{M})$ if and only if $baa \in L_{=1}(\mathcal{M})$, a contradiction. $\square$

We obtain the following consequences.

Corollary 22. $\mathscr{L}_{=1}(\text{stl-det-strict-CD-2-RR}(1))$ *is incomparable under inclusion to the language classes* FIN *of finite languages,* REG *of regular languages, and* CFL *of context-free languages. In particular, it follows that the inclusion* $\mathscr{L}_{=1}(\text{stl-det-strict-CD-2-RR}(1)) \subseteq \mathscr{L}_{=1}(\text{stl-det-local-CD-2-RR}(1))$ *is proper.*

Further, we see that $\mathscr{L}_{=1}(\text{stl-det-strict-CD-2-RR}(1))$ is incomparable under inclusion to the language class $\mathscr{L}_{=1}(\text{stl-det-local-CD-R}(1))$. For future reference we consider another finite example language.

Lemma 23. *The finite language $L_0' = \{aaaa, abb\}$ is not accepted by any strictly deterministic stateless* CD-2-RR(1)-*system.*

Proof. Assume that $\mathcal{M} = ((M_i, \sigma_i)_{i \in I}, I_0)$ is a strictly deterministic stateless CD-2-RR(1)-system such that $L_{=1}(\mathcal{M}) = L_0'$, let $I_0 = \{i_0\}$, and let $\sigma_{i_0} = \{i_1\}$ and $\sigma_{i_1} = \{i_2\}$. Obviously, we have $\delta_1^{(i_0)}(\mathfrak{c}) = $ MVR, and $\delta_1^{(i_0)}(a) = \varepsilon$. Further, it holds that $\delta_1^{(i_1)}(\mathfrak{c}) = $ MVR, and $\delta_1^{(i_1)}(a) = \delta_1^{(i_1)}(b) = \varepsilon$. Now $(i_0, aaaa) \vdash_{\mathcal{M}}^c (i_1, aaa) \vdash_{\mathcal{M}}^c (i_2, aa)$, which leads to acceptance, while $(i_0, abaa) \vdash_{\mathcal{M}}^c (i_1, baa) \vdash_{\mathcal{M}}^c (i_2, aa)$ should lead to rejection, which is a contradiction. Thus, L_0' is not accepted by any strictly deterministic stateless CD-2-RR(1)-system working in mode $= 1$. $\square$

From Lemma 21 we immediately obtain several non-closure properties for the class $\mathscr{L}_{=1}(\text{stl-det-strict-CD-2-RR}(1))$. In fact, we can derive the following result.

Theorem 24. *The language class $\mathscr{L}_{=1}(\text{stl-det-strict-CD-2-RR}(1))$ is not closed under union, intersection with regular sets, ε-free homomorphisms, and inverse homomorphisms.*

Proof. The languages $\{aaa\}$, $\{bb\}$, and $\{a, b\}^*$ are all accepted by stl-det-strict-CD-2-RR(1)-systems. As $\{aaa\} \cup \{bb\} = \{aaa, bb\} = \{aaa, bb\} \cap \{a, b\}^*$, Lemma 21 shows that this language class is neither closed under union nor under intersection with regular sets.

The languages $\{c, d\}$ and $\{c^6\}$ are accepted by stl-det-strict-CD-2-RR(1)-systems. Let $h_1 : \{c, d\}^* \to \{a, b\}^*$ be the homomorphism defined by $c \mapsto aaa$ and $d \mapsto bb$, and let $h_2 : \{a, b\}^* \to \{c\}^*$ be the homomorphism defined by $a \mapsto c^2$ and $b \mapsto c^3$. Then $h_1(\{c, d\}) = \{aaa, bb\} = h_2^{-1}(\{c^6\})$, and hence, Lemma 21 shows that this language class is neither closed under ε-free homomorphisms nor under inverse homomorphisms. $\square$

Proposition 25. *The class* $\mathscr{L}_{=1}$(stl-det-strict-CD-2-RR(1)) *is* (a) *closed under complementation and* (b) *not closed under intersection.*

Proof. (a) Let $\mathcal{M} = ((M_i, \sigma_i)_{i \in I}, \{i_0\})$ be a stl-det-strict-CD-2-RR(1)-system on Σ such that $L_{=1}(\mathcal{M}) = L$. By interchanging accept transitions and undefined transitions within each function $\delta_1^{(i)}$ and $\delta_2^{(i)}$, we obtain a stl-det-strict-CD-2-RR(1)-system $\mathcal{M}' = ((M_i', \sigma_i)_{i \in I}, \{i_0\})$ that executes exactly the same cycles as $\mathcal{M}$, but for each index $i \in I$, the accepting tail computations of M_i' correspond to rejecting tail computations of M_i, and vice versa. Hence, it follows that $L_{=1}(\mathcal{M}') = \Sigma^* \setminus L = \overline{L}$.
(b) Closure under complementation and non-closure under union yield immediately that $\mathscr{L}_{=1}$(stl-det-strict-CD-2-RR(1)) is not closed under intersection.   □

Proposition 26. *The class* $\mathscr{L}_{=1}$(stl-det-strict-CD-2-RR(1)) *is* (a) *not closed under commutative closure and* (b) *not closed under reversal.*

Proof. (a) From Example 7 we know that the language $L_2 = \{a^n b^n \mid n \geq 0\}$ is accepted by a stl-det-strict-CD-2-RR(1)-system. Its commutative closure is the language $L_= = \{w \in \{a, b\}^* \mid |w|_a = |w|_b \geq 0\}$.
Assume that $\mathcal{M} = ((M_i, \sigma_i)_{i \in I}, I_0)$ is a stl-det-strict-CD-2-RR(1)-system accepting the language $L_=$, and assume that $I_0 = \{i_0\}$. Then $\delta_1^{(i_0)}(\mathfrak{c}) = \mathsf{MVR}$, and as $\varepsilon \in L_=$, we also have $\delta_1^{(i_0)}(\$) = \mathsf{Accept}$. As $a \notin L_=$, we see that $\delta_1^{(i_0)}(a) = \varepsilon$, and as $b \notin L_=$, we also have $\delta_1^{(i_0)}(b) = \varepsilon$. Further, it holds that $\delta_2^{(i_0)}(b) = \mathsf{Restart}$, or $\delta_2^{(i_0)}(b) = \mathsf{MVR}$ and $\delta_2^{(i_0)}(\$) = \mathsf{Restart}$, as $ab \in L_=$, while $abb \notin L_=$. Hence, $\mathcal{M}$ performs the following computations, where $\sigma_{i_0} = \{i_1\}$:

$$(i_0, ab) \vdash_{\mathcal{M}}^c (i_1, b) \vdash_{\mathcal{M}}^* \mathsf{Accept} \quad \text{and} \quad (i_0, bb) \vdash_{\mathcal{M}}^c (i_1, b) \vdash_{\mathcal{M}}^* \mathsf{Accept}.$$

As bb does not belong to $L_=$, this contradicts our assumption on $\mathcal{M}$. Hence, $L_=$ is not accepted by any stl-det-strict-CD-2-RR(1)-system, which means that $\mathscr{L}_{=1}$(stl-det-strict-CD-2-RR(1)) is not closed under the operation of commutative closure.
(b) Let $L = \{aaw \mid w \in \{a, b\}^*\}$. Then L is accepted by the following stl-det-strict-CD-2-RR(1)-system $\mathcal{M} = ((M_0, \{1\}), (M_1, \{1\}), \{0\})$, where M_0 and M_1 are defined as follows:

$$M_0 : \delta_1^{(0)} : \mathfrak{c} \mapsto \mathsf{MVR}, a \mapsto \varepsilon; \delta_2^{(0)} : a \mapsto \mathsf{Restart};$$
$$M_1 : \delta_1^{(1)} : \mathfrak{c} \mapsto \mathsf{MVR}, a \mapsto \varepsilon; \delta_2^{(1)} : a \mapsto \mathsf{MVR}, \quad b \mapsto \mathsf{MVR}, \$ \mapsto \mathsf{Accept}.$$

Assume that $\mathcal{M} = ((M_i, \sigma_i)_{i \in I}, I_0)$ is a stl-det-strict-CD-2-RR(1)-system such that $L_{=1}(\mathcal{M}) = L^R$. Without loss of generality we can assume that $I = \{0, 1, \ldots, m\}$, that $I_0 = \{0\}$ and that $\sigma_0 = \{1\}$. Obviously, $\delta_1^{(0)}(\mathfrak{c}) = \mathsf{MVR}$, and $\delta_1^{(0)}(a) = \varepsilon$, as $a \notin L^R$, while $a^2 \in L^R$. If $\delta_1^{(0)}(b) = \varepsilon$ as well, then with $aa \in L^R$, $\mathcal{M}$ would also accept $ba \notin L^R$. Hence, $\delta_1^{(0)}(b) = \mathsf{MVR}$. It remains to consider the function $\delta_2^{(0)}$.

If $\delta_2^{(0)}(a) = $ Accept, then $\mathcal{M}$ would accept the word $aab \notin L^R$. Also if $\delta_2^{(0)}(\$) = $ Accept, then $\mathcal{M}$ would accept the word $a \notin L^R$. Hence, $\delta_2^{(0)}(a) = $ Restart, or $\delta_2^{(0)}(a) = $ MVR and $\delta_2^{(0)}(\$) = $ Restart. Further, as $abaa \in L^R$, while $abab \notin L^R$, it follows that $\delta_2^{(0)}(b) \in \{$MVR, Restart$\}$ holds, too. But then $\mathcal{M}$ executes the following computations:

$$(0, baa) \vdash_{\mathcal{M}}^c (1, ba) \vdash_{\mathcal{M}}^* \text{ Accept and } (0, aba) \vdash_{\mathcal{M}}^c (1, ba) \vdash_{\mathcal{M}}^* \text{ Accept.}$$

As $aba \notin L^R$, this again contradicts our assumption on $\mathcal{M}$. Thus, L^R is not accepted by any stl-det-strict-CD-2-RR(1)-system, and it follows that $\mathscr{L}_{=1}$(stl-det-strict-CD-2-RR(1)) is not closed under reversal. □

For showing that the class $\mathscr{L}_{=1}$(stl-det-strict-CD-2-RR(1)) is an anti-AFL, it remains to be proven that this class is not closed under concatenation and Kleene star, either. Let L_p be the language $L_p = a^+ \cdot b \cdot a^+$ on $\Sigma_2 = \{a, b\}$.

Lemma 27. $L_p \in \mathscr{L}_{=1}$(stl-det-strict-CD-2-RR(1)).

Proof. The language L_p is accepted by the stl-det-strict-CD-2-RR(1)-system $\mathcal{M}_p = ((M_i, \sigma_i)_{i \in \{0,1,2\}}, \{0\})$, where $\sigma_0 = \{1\}$, $\sigma_1 = \{2\}$, $\sigma_2 = \{0\}$, and the stl-det-2-RR(1)-automata M_0, M_1 and M_2 are defined as follows:

$M_0 : \delta_1^{(0)} : \mathcal{c} \mapsto $ MVR, $a \mapsto \varepsilon$; $\delta_2^{(0)} : a \mapsto $ Restart, $b \mapsto $ Restart;

$M_1 : \delta_1^{(1)} : \mathcal{c} \mapsto $ MVR, $a \mapsto $ MVR, $b \mapsto \varepsilon$; $\delta_2^{(1)} : a \mapsto $ Restart;

$M_2 : \delta_1^{(2)} : \mathcal{c} \mapsto $ MVR, $a \mapsto $ MVR, $\$ \mapsto $ Accept. □

Non-closure under concatenation for $\mathscr{L}_{=1}$(stl-det-strict-CD-2-RR(1)) will follow from the following negative result.

Lemma 28. $L_p \cdot L_p \notin \mathscr{L}_{=1}$(stl-det-strict-CD-2-RR(1)).

Proof. Obviously,

$$L_p^2 = L_p \cdot L_p = a^+ \cdot b \cdot a \cdot a^+ \cdot b \cdot a^+ = \{a^m b a^n b a^p \mid m, p \geq 1, n \geq 2\}.$$

We claim that the language L_p^2 is not accepted by any stl-det-strict-CD-2-RR(1)-system.

Assume to the contrary that $\mathcal{M} = ((M_i, \sigma_i)_{i \in I}, I_0)$ is a stl-det-strict-CD-2-RR(1)-system such that $L_{=1}(\mathcal{M}) = L_p^2$. Without loss of generality we may assume that $I = \{0, 1, \ldots, r\}$, and that $I_0 = \{0\}$.

We first analyze the transition functions of M_0. Obviously, $\delta_1^{(0)}(\mathcal{c}) = $ MVR. If $\delta_1^{(0)}(a) = $ MVR, then $\delta_1^{(0)}(b) = \varepsilon$, as $L_{=1}(\mathcal{M})$ is neither empty nor the set Σ_2^*. But then $\mathcal{M}$ cannot distinguish between the input $abaaba \in L_p^2$ and the input $aababa \notin L_p^2$, contradicting our assumption above. Hence, we conclude that $\delta_1^{(0)}(a) = \varepsilon$.

If $\delta_1^{(0)}(b) = $ Accept, then $\mathcal{M}$ would accept all words beginning with the letter b. If $\delta_1^{(0)}(b) = $ MVR, then $\mathcal{M}$ cannot distinguish between the input $abaaba \in L_p^2$

and the input $baaaba \notin L_p^2$. It follows that $\delta_1^{(0)}(b) = \emptyset$. Further, as in the second phase of the first cycle, M_0 cannot possibly ensure that the remaining tape contents is of the form $a^* \cdot b \cdot a \cdot a^+ \cdot b \cdot a^+$, we see that M_0 executes a restart operation after deleting the first a.

Let $\sigma_0 = \{1\}$. We continue by analyzing the transition functions of M_1. Obviously, $\delta_1^{(1)}(\mathbb{c}) = \mathsf{MVR}$.

Assume first that $\delta_1^{(1)}(a) = \varepsilon$. If also $\delta_1^{(1)}(b) = \varepsilon$, then $\mathcal{M}$ cannot distinguish between the input $aabaaba \in L_p^2$ and the input $abbaaba \notin L_p^2$, which contradicts our assumption above. If $\delta_1^{(1)}(b) = \mathsf{MVR}$, then we have the following partial computations of $\mathcal{M}$, where $\sigma_1 = \{2\}$ is taken:

$$(0, abaaba) \vdash_{\mathcal{M}}^c (1, baaba) \vdash_{\mathcal{M}}^c (2, baba),$$

and

$$(0, aababa) \vdash_{\mathcal{M}}^c (1, ababa) \vdash_{\mathcal{M}}^c (2, baba).$$

Hence, $\mathcal{M}$ cannot distinguish between the input $abaaba \in L_p^2$ and the input $aababa \notin L_p^2$. It follows that $\delta_1^{(1)}(a) = \mathsf{MVR}$. But then $\delta_1^{(1)}(b) = \varepsilon$ follows, which in turn means that $\mathcal{M}$ executes the following partial computations:

$$(0, abaaba) \vdash_{\mathcal{M}}^c (1, baaba) \vdash_{\mathcal{M}}^c (2, aaba)$$

and

$$(0, aababa) \vdash_{\mathcal{M}}^c (1, ababa) \vdash_{\mathcal{M}}^c (2, aaba).$$

This again shows that $\mathcal{M}$ cannot distinguish between the input $abaaba \in L_p^2$ and the input $aababa \notin L_p^2$. In conclusion we see that L_p^2 is not accepted by any stl-det-strict-CD-2-RR(1)-system.   □

In fact, it can be shown that each stl-det-strict-CD-2-RR(1)-system that accepts all words from L_p^2 also accepts some words from $(a^* \cdot b \cdot a^*)^*$ that do not belong to the language L_p^*. Hence, it follows that L_p^* (and also L_p^+) is not accepted by any stl-det-strict-CD-2-RR(1)-system. Thus, we have the following additional non-closure results.

Corollary 29. *The language class $\mathscr{L}_{=1}$(stl-det-strict-CD-2-RR(1)) is not closed under concatenation, Kleene plus and Kleene star.*

Thus, we see that $\mathscr{L}_{=1}$(stl-det-strict-CD-2-RR(1)) is an anti-AFL.

4.2 Globally Deterministic CD-Systems of stl-det-2-RR(1)-Automata

In a globally deterministic CD-system of stateless deterministic R(1)-automata, each rewrite operation of each component automaton is associated with a particular successor index. Thus, if M_{i_1} is the active component, and if it executes a cycle involving the deletion of the letter $a \in \Sigma$, then the component $i_2 \in \sigma_{i_1}$

that is associated with the delete operation $\delta_{i_1}(a) = \varepsilon$ is activated. Hence, the choice of the successor component is based on the symbol deleted.

In a computation of a CD-system of stateless deterministic 2-RR(1)-automata, a successor component is chosen whenever the active component executes a restart operation. Accordingly, for these CD-systems we associate a particular successor index with each restart operation.

Let $((M_i, \sigma_i)_{i \in I}, I_0)$ be a CD-system of stateless deterministic 2-RR(1)-automata over Σ such that $|I_0| = 1$. For each $i \in I$, let $\Sigma_{\text{rs}}^{(i)}$ be the set of symbols that cause the component automaton M_i to perform a restart operation, that is,

$$\Sigma_{\text{rs}}^{(i)} = \{ a \in \Sigma \mid \delta_2^{(i)}(a) = \text{Restart} \} \cup \{ \$ \mid \delta_2^{(i)}(\$) = \text{Restart} \}.$$

Further, let $\delta : \bigcup_{i \in I}(\{i\} \times \Sigma_{\text{rs}}^{(i)}) \to I$ be a mapping that assigns to each pair $(i, a) \in \{i\} \times \Sigma_{\text{rs}}^{(i)}$ an element $j \in \sigma_i$. Then δ is called a *global successor function*. It assigns a successor component $j \in \sigma_i$ to the active component i based on the symbol $a \in \Sigma_{\text{rs}}^{(i)}$ that causes M_i to perform a restart operation in the current cycle. It follows that, for each input word $w \in \Sigma^*$, the system $\mathcal{M} = ((M_i, \sigma_i)_{i \in I}, I_0, \delta)$ has a unique computation that starts from the initial configuration corresponding to input w. Accordingly we call $\mathcal{M}$ a *globally deterministic stateless* CD-2-RR(1)-*system*, and by $\mathscr{L}_{=1}(\text{stl-det-global-CD-2-RR(1)})$ we denote the class of languages that are accepted by these systems.

Obviously, each strictly deterministic stateless CD-2-RR(1)-system is globally deterministic. However, the globally deterministic stateless CD-2-RR(1)-systems are more expressive than the strictly deterministic ones.

Example 30. Let $\mathcal{M} = ((M_i, \sigma_i)_{i \in I}, I_0, \delta)$ be the globally deterministic CD-system of stateless deterministic 2-RR(1)-automata over $\Sigma = \{a, b\}$ that is defined as follows:
$I = \{0, 1, 2, 3, 4, 5\}$, $I_0 = \{0\}$, $\sigma_0 = \{1, 4\}$, $\sigma_1 = \{2\}$, $\sigma_2 = \{3\}$, $\sigma_3 = \{5\} = \sigma_4$, $\sigma_5 = \{1\}$, and M_0 to M_5 are the stateless deterministic 2-RR(1)-automata that are given by the following transition functions:

$$M_0 : \delta_1^{(0)} : \math{¢} \mapsto \text{MVR},\ a \mapsto \varepsilon;\ \delta_2^{(0)} : a \mapsto \text{Restart},\ b \mapsto \text{Restart};$$
$$M_1 : \delta_1^{(1)} : \math{¢} \mapsto \text{MVR},\ a \mapsto \varepsilon;\ \delta_2^{(1)} : a \mapsto \text{Restart};$$
$$M_2 : \delta_1^{(2)} : \math{¢} \mapsto \text{MVR},\ a \mapsto \varepsilon;\ \delta_2^{(2)} : a \mapsto \text{Restart};$$
$$M_3 : \delta_1^{(3)} : \math{¢} \mapsto \text{MVR},\ a \mapsto \varepsilon;\ \delta_2^{(3)} : \$ \mapsto \text{Accept};$$
$$M_4 : \delta_1^{(4)} : \math{¢} \mapsto \text{MVR},\ b \mapsto \varepsilon;\ \delta_2^{(4)} : b \mapsto \text{Restart};$$
$$M_5 : \delta_1^{(5)} : \math{¢} \mapsto \text{MVR},\ b \mapsto \varepsilon;\ \delta_2^{(5)} : \$ \mapsto \text{Accept}.$$

and δ is defined by $\delta(0, a) = 1$, $\delta(0, b) = 4$, $\delta(1, a) = 2$, $\delta(2, a) = 3$, $\delta(4, b) = 5$. Then it is easily seen that $L_{=1}(\mathcal{M}) = \{aaaa, abb\}$, which is not accepted by any strictly deterministic stateless CD-2-RR(1)-system by Lemma 23. □

Thus, we have the following proper inclusion.

Corollary 31

$$\mathscr{L}_{=1}(\text{stl-det-strict-CD-2-RR(1)}) \subsetneq \mathscr{L}_{=1}(\text{stl-det-global-CD-2-RR(1)}).$$

Further, we relate the stl-det-global-CD-2-RR(1)-systems to the stl-det-local-CD-2-RR(1)-systems.

Proposition 32

$$\mathscr{L}_{=1}(\text{stl-det-global-CD-2-RR(1)}) \subseteq \mathscr{L}_{=1}(\text{stl-det-local-CD-2-RR(1)}).$$

Proof Let $\mathcal{M} = ((M_i, \sigma_i)_{i \in I}, \{i_0\}, \delta)$ be a stl-det-global-CD-2-RR(1)-system on alphabet Σ and, for each $i \in I$, $\Sigma_{\text{rs}}^{(i)}$ as defined above. From $\mathcal{M}$ we now construct a stl-det-local-CD-2-RR(1)-system $\mathcal{M}' = ((M_j', \sigma_j')_{j \in J}, J_0)$ satisfying $L_{=1}(\mathcal{M}') = L_{=1}(\mathcal{M})$. For all $i \in I$, let $S^{(i)} = \Sigma_{\text{rs}}^{(i)}$, if $\Sigma_{\text{rs}}^{(i)} \neq \emptyset$, and $S^{(i)} = \{+\}$, otherwise. Now let $J = \{ (i,a) \mid i \in I, a \in S^{(i)} \}$, let $J_0 = \{ (i_0, a) \mid a \in S^{(i_0)} \}$, and for all $i \in I$, take

$$\sigma_{(i,a)}' = \{ (j,b) \mid j = \delta(i,a),\ b \in S^{(j)} \} \text{ for all } a \in \Sigma_{\text{rs}}^{(i)},$$
$$\sigma_{(i,+)}' = J_0, \qquad\qquad\qquad \text{if } \Sigma_{\text{rs}}^{(i)} = \emptyset.$$

Finally, we define the stateless deterministic 2-RR(1)-automata $M_{(i,a)}'$ as follows, where $i \in I$, $a \in S^{(i)}$, and $b \in \Sigma$:

$$M_{(i,a)}' : \delta_1^{(i,a)}(x) = \delta_1^{(i)}(x) \quad \text{for all } x \in \Sigma \cup \{\text{¢}, \$\};$$
$$\delta_2^{(i,a)}(x) = \delta_2^{(i)}(x) \quad \text{for all } x \in (\Sigma \cup \{\$\}) \smallsetminus \Sigma_{\text{rs}},$$
$$\delta_2^{(i,a)}(a) = \text{Restart, if } a \in \Sigma_{\text{rs}},$$
$$\delta_2^{(i,a)}(b) = \emptyset, \qquad \text{for all } b \in \Sigma_{\text{rs}} \smallsetminus \{a\}.$$

Let $w = a_1 a_2 \cdots a_n \in \Sigma^*$, where $n \geq 0$ and $a_1, \ldots, a_n \in \Sigma$. Assume that the computation of $\mathcal{M}$ on input w has the following form:

$$(i_0, w) = (i_0, u_0 b_0 v_0) \vdash_{\mathcal{M}}^c \quad (i_1, u_0 v_0) \quad = (i_1, u_1 b_1 v_1) \vdash_{\mathcal{M}}^c \cdots$$
$$\vdash_{\mathcal{M}}^c (i_r, u_{r-1} v_{r-1}) = \quad (i_r, w_r),$$

and that starting with the configuration (i_r, w_r), the component automaton M_{i_r} performs a tail computation. Then $\mathcal{M}'$ can simulate this sequence of cycles by guessing, in each step, on which letter the next restart operation of $\mathcal{M}$ will be executed. Thus, we conclude that $L_{=1}(\mathcal{M}) \subseteq L_{=1}(\mathcal{M}')$ holds.

Conversely, if $\mathcal{M}'$ has an accepting computation on input $w \in \Sigma^*$, then it follows easily from the above construction of $\mathcal{M}'$ that $\mathcal{M}$ will also accept on input w. Thus, we see that $L_{=1}(\mathcal{M}') = L_{=1}(\mathcal{M})$, which completes the proof. $\square$

Since all rational trace languages are accepted by stl-det-local-CD-2-RR(1)-systems, the inclusion result above raises the question of whether all rational trace languages are accepted by stl-det-global-CD-2-RR(1)-systems as well. The following result answers this question in the negative.

Proposition 33. *The rational trace language*

$$L_\vee = \{\, w \in \{a, b\}^* \mid \exists n \geq 0 : |w|_a = n \text{ and } |w|_b \in \{n, 2n\} \,\}$$

is not accepted by any globally deterministic stateless CD-2-RR(1)-*system.*

Proof. As $L_\vee$ is the commutative closure of the regular language $(ab)^* \cup (abb)^*$, it is obviously a rational trace language.

It remains to be proven that $L_\vee \notin \mathcal{L}_{=1}(\text{stl-det-global-CD-2-RR(1)})$. Assume to the contrary that $\mathcal{M} = ((M_i, \sigma_i)_{i \in I}, I_0, \delta)$ is a stl-det-global-CD-2-RR(1)-system such that $L_{=1}(\mathcal{M}) = L_\vee$. Without loss of generality we can assume that $I = \{0, 1, \ldots, m-1\}$ and that $I_0 = \{0\}$.

Let $n > 2m$, and let $w = a^n b^n \in L_\vee$. Then the computation of $\mathcal{M}$ on input w is accepting, that is, it is of the form

$$(0, a^n b^n) \vdash^c_{\mathcal{M}} (i_1, w_1) \vdash^c_{\mathcal{M}} \cdots \vdash^c_{\mathcal{M}} (i_r, w_r) \vdash^*_{M_{i_r}} \text{Accept},$$

where M_{i_r} accepts the tape contents $\text{\textcent} w_r \$$ in a tail computation. Let $i = |w_r|_a$ and $j = |w_r|_b$.

If $j > 1$, then M_{i_r} would also accept the tape contents $w_r b^k = a^i b^{j+k}$ for any $k > 0$, and therewith $\mathcal{M}$ would accept the input $wb^{2n} = a^n b^{3n}$. As this word is not contained in $L_\vee$, this contradicts our assumption that $L_{=1}(\mathcal{M}) = L_\vee$. Hence, we conclude that $j = |w_r|_b \leq 1$.

Analogously, if $i > 1$, then M_{i_r} would also accept the tape contents $a^{i+k} b^j$ for any $k > 0$, and therewith $\mathcal{M}$ would accept the input $a^n w = a^{2n} b^n \notin L_\vee$. Hence, we conclude that $i = |w_r|_a \leq 1$. Thus, $|w_r| = i + j \leq 2$, which shows that in the above computation at least the first $n - 1$ occurrences of the letter a and the first $n - 1$ occurrences of the letter b are deleted letter by letter, and then M_{i_r} accepts the word w_r of length at most two.

As $n > m$, there exists an index $i \in I$ such that the component automaton M_i is used twice within the above sequence of cycles. Thus, there are integers $s, t, k, \ell \geq 0$, $m \geq s + t \geq 0$ and $m \geq k + \ell > 0$, such that the above computation can be written as follows:

$$(0, a^n b^n) \vdash^{c^*}_{\mathcal{M}} (i, a^{n-s} b^{n-t}) \vdash^{c^+}_{\mathcal{M}} (i, a^{n-s-k} b^{n-t-\ell}) \vdash^{c^*}_{\mathcal{M}} (i_r, w_r) \vdash^*_{M_{i_r}} \text{Accept}.$$

Obviously, $\mathcal{M}$ will also execute the following shortened computation:

$$(0, a^{n-k} b^{n-\ell}) \vdash^{c^*}_{\mathcal{M}} (i, a^{n-s-k} b^{n-t-\ell}) \vdash^{c^*}_{\mathcal{M}} (i_r, w_r) \vdash^*_{M_{i_r}} \text{Accept},$$

that is, $\mathcal{M}$ accepts on input $a^{n-k} b^{n-\ell}$. From our assumption that $L_{=1}(\mathcal{M}) = L_\vee$ we can therefore conclude that $k = \ell$, as $n > 2m$.

Now consider the computation of $\mathcal{M}$ on input $a^n b^{2n}$. As $a^n b^{2n} \in L_\vee$, this computation is accepting, that is, it has the following form:

$$(0, a^n b^{2n}) \vdash^{c^*}_{\mathcal{M}} (i, a^{n-s} b^{2n-t}) \vdash^{c^+}_{\mathcal{M}} (i, a^{n-s-k} b^{2n-t-k}) \vdash^{c^*}_{\mathcal{M}} (i', z') \vdash^*_{M_{i'}} \text{Accept}$$

for some $i' \in I$ and some word $z' \in \Sigma^*$. But then $\mathcal{M}$ will also execute the following computation:

$$(0, a^{n-k} b^{2n-k}) \vdash^{c^*}_{\mathcal{M}} (i, a^{n-s-k} b^{2n-t-k}) \vdash^{c^*}_{\mathcal{M}} (i', z') \vdash^*_{M_{i'}} \text{Accept},$$

that is, it accepts on input $a^{n-k}b^{2n-k} \notin L_\vee$. It follows that $L_{=1}(\mathcal{M}) \neq L_\vee$, that is, $L_\vee$ is not accepted by any globally deterministic stateless CD-2-RR(1)-system working in mode $= 1$. □

This yields the following consequence.

Corollary 34.

$$\mathscr{L}_{=1}(\text{stl-det-global-CD-2-RR}(1)) \subsetneq \mathscr{L}_{=1}(\text{stl-det-local-CD-2-RR}(1)).$$

The Dyck language D_1 is not a rational trace language, but it is accepted by a strictly deterministic stateless CD-2-RR(1)-system as can be shown easily (see Example 14). Thus, we have the following consequence.

Corollary 35. *The two language classes* $\mathscr{L}_{=1}(\text{stl-det-strict-CD-2-RR}(1))$ *and* $\mathscr{L}_{=1}(\text{stl-det-global-CD-2-RR}(1))$ *are incomparable under inclusion to the class of rational trace languages.*

In a stl-det-global-CD-R(1)-system, the choice of the successor component is based on the letter removed in the current cycle, while in a stl-det-global-CD-2-RR(1)-system, this choice is based on the letter on which the currently active component automaton executes the restart that completes the current cycle. This raises the question of whether each stl-det-global-CD-R(1)-system can be simulated by a stl-det-global-CD-2-RR(1)-system. In order to answer this question we first note that Lemma 20 applies also to stl-det-global-CD-2-RR(1)-systems. Hence, from Lemma 21 we adapt the following negative result.

Corollary 36. *The finite language* $L_0 = \{aaa, bb\}$ *is not accepted by any globally deterministic stateless* CD-2-RR(1)-*system.*

Since all regular languages are accepted by stl-det-global-CD-R(1)-systems, the corollary implies that the class $\mathscr{L}_{=1}(\text{stl-det-global-CD-R}(1))$ is not contained in $\mathscr{L}_{=1}(\text{stl-det-global-CD-2-RR}(1))$. On the other hand, Example 7 shows that already stl-det-strict-CD-2-RR(1)-systems accept some languages that are not accepted by stl-det-local-CD-R(1)-systems. Hence, we have the following incomparability results.

Corollary 37. *The language classes*

$$\mathscr{L}_{=1}(\text{stl-det-strict-CD-2-RR}(1)) \ \text{and} \ \mathscr{L}_{=1}(\text{stl-det-global-CD-2-RR}(1))$$

are incomparable under inclusion to the classes

$$\mathscr{L}_{=1}(\text{stl-det-global-CD-R}(1)) \ \text{and} \ \mathscr{L}_{=1}(\text{stl-det-local-CD-R}(1)).$$

Even though stl-det-global-CD-2-RR(1)-systems cannot accept all finite languages, they seem to be powerful devices. In particular, the language of Example 10, which is not even growing context-sensitive, can be shown to belong to the class $\mathscr{L}_{=1}(\text{stl-det-global-CD-2-RR}(1))$ by adding a corresponding global successor function to the stl-det-local-CD-2-RR(1))-system of Example 10.

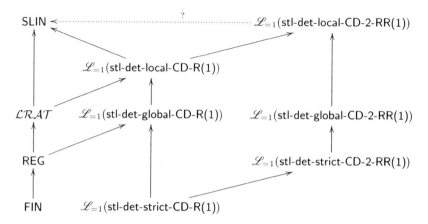

Fig. 1. Hierarchy of language classes accepted by various types of CD-R(1)- and CD-2-RR(1)-systems. Here SLIN denotes the class of *semi-linear languages*, $\mathcal{LRAT}$ is the class of *rational trace languages*, and FIN is the class of all finite languages. Each arrow represents a proper inclusion, the dotted arrow represents an inclusion that is still open, and all other classes that are not connected by a sequence of arrows are incomparable under inclusion.

Theorem 38. *The language class $\mathcal{L}_{=1}$(stl-det-global-CD-2-RR(1)) is incomparable under inclusion to the classes* GCSL *of growing context-sensitive languages,* CRL *of Church-Rosser languages,* CFL *of context-free languages,* LIN *of linear languages,* DCFL *of deterministic context-free languages,* REG *of regular languages as well as to the class* FIN *of finite languages.*

The diagram in Figure 1 summarizes our inclusion results. We have the following results on closure and non-closure properties for the language class $\mathcal{L}_{=1}$(stl-det-global-CD-2-RR(1)).

Proposition 39. *The class $\mathcal{L}_{=1}$(stl-det-global-CD-2-RR(1)) is (a) closed under complementation, (b) not closed under union or intersection, (c) not closed under intersection with regular languages, concatenation, ε-free homomorphisms, and inverse homomorphisms, and (d) not closed under commutative closure and reversal.*

Proof. (a) Let $\mathcal{M} = ((M_i, \sigma_i)_{i \in I}, \{i_0\}, \delta)$ be a stl-det-global-CD-2-RR(1)-system on Σ such that $L_{=1}(\mathcal{M}) = L$. By interchanging accept transitions and undefined transitions within each function $\delta_1^{(i)}$ and $\delta_2^{(i)}$, we obtain a stl-det-global-CD-2-RR(1)-system $\mathcal{M}' = ((M_i', \sigma_i)_{i \in I}, \{i_0\}, \delta)$ that executes exactly the same cycles as $\mathcal{M}$, but for each index $i \in I$, the accepting tail computations of M_i' correspond to rejecting tail computations of M_i, and vice versa. Hence, it follows that $L_{=1}(\mathcal{M}') = \Sigma^* \setminus L = \overline{L}$.

(b) By Corollary 36 the finite language $L_0 = \{aaa, bb\}$ is not accepted by any stl-det-global-CD-2-RR(1)-system. It is easily seen that the languages

$L_{0,1} = \{aaa\}$ and $L_{0,2} = \{bb\}$ are accepted by such systems and, thus, the class $\mathscr{L}_{=1}(\mathsf{stl\text{-}det\text{-}global\text{-}CD\text{-}2\text{-}RR(1)})$ is not closed under union. Together with closure under complementation this also yields non-closure under intersection.

(c) As $\{a, b\}^*$ is accepted by a $\mathsf{stl\text{-}det\text{-}global\text{-}CD\text{-}2\text{-}RR(1)}$-system, and as L_0 is regular, we see from the claim above and the fact that $L_0 = \{a, b\}^* \cap L_0$ that the class $\mathscr{L}_{=1}(\mathsf{stl\text{-}det\text{-}global\text{-}CD\text{-}2\text{-}RR(1)})$ is not closed under intersection with regular languages.

It is not hard to see that the languages $\{\varepsilon, aa\}$ and $\{\varepsilon, bb\}$ are accepted by $\mathsf{stl\text{-}det\text{-}global\text{-}CD\text{-}2\text{-}RR(1)}$-systems. By an application of Lemma 20 their concatenation $\{\varepsilon, aa\} \cdot \{\varepsilon, bb\} = \{\varepsilon, aa, bb, aabb\}$ is not.

The languages $\{c, d\}$ and $\{c^6\}$ are accepted by $\mathsf{stl\text{-}det\text{-}global\text{-}CD\text{-}2\text{-}RR(1)}$-systems. Let $h_1 : \{c, d\}^* \to \{a, b\}^*$ be the homomorphism defined by $c \mapsto aaa$ and $d \mapsto bb$, and let $h_2 : \{a, b\}^* \to \{c\}^*$ be the homomorphism defined by $a \mapsto c^2$ and $b \mapsto c^3$. Then $h_1(\{c, d\}) = \{aaa, bb\} = h_2^{-1}(\{c^6\})$, and hence, Corollary 36 shows that the language class $\mathscr{L}_{=1}(\mathsf{stl\text{-}det\text{-}global\text{-}CD\text{-}2\text{-}RR(1)})$ is neither closed under ε-free homomorphisms nor under inverse homomorphisms.

(d) Since the regular language $(ab)^* \cup (abb)^*$ can be accepted by some $\mathsf{stl\text{-}det\text{-}global\text{-}CD\text{-}2\text{-}RR(1)}$-system, Proposition 33 implies that the language class $\mathscr{L}_{=1}(\mathsf{stl\text{-}det\text{-}global\text{-}CD\text{-}2\text{-}RR(1)})$ is not closed under commutative closure.

From Example 30 we know that the language $L_0' = \{aaaa, abb\}$ is accepted by a $\mathsf{stl\text{-}det\text{-}global\text{-}CD\text{-}2\text{-}RR(1)}$-system. In analogy it can be shown that also the language $L_1' = \{caaa, cbb\}$ is accepted by a $\mathsf{stl\text{-}det\text{-}global\text{-}CD\text{-}2\text{-}RR(1)}$-system. Here we claim that the language $L_1'^{R} = \{aaac, bbc\}$ is not accepted by any $\mathsf{stl\text{-}det\text{-}global\text{-}CD\text{-}2\text{-}RR(1)}$-system.

Assume that $\mathcal{M} = ((M_i, \sigma_i)_{i \in I}, I_0, \delta)$ is a $\mathsf{stl\text{-}det\text{-}global\text{-}CD\text{-}2\text{-}RR(1)}$-system accepting the language $L_1'^{R}$. Without loss of generality we can assume that $I = \{0, 1, \ldots, n\}$, and that $I_0 = \{0\}$. Obviously, $\delta_1^{(0)}(\mathfrak{c}) = \mathsf{MVR}$. We now consider various cases.

(i) If $\delta_1^{(0)}(a) = \delta_1^{(0)}(b) = \mathsf{MVR}$, then necessarily $\delta_1^{(0)}(c) = \varepsilon$ and $\delta_2^{(0)}(\$) = \mathsf{Restart}$ follow. Let $\delta(0, \$) = 1$. Then the system $\mathcal{M}' = ((M_i, \sigma_i)_{i \in I}, \{1\}, \delta)$ accepts the language $L_0 = \{aaa, bb\}$. This, however, contradicts Corollary 36.

(ii) If $\delta_1^{(0)}(a) = \delta_1^{(0)}(b) = \varepsilon$, Lemma 20 shows that $L_1'^{R}$ is not accepted by $\mathcal{M}$.

(iii) If $\delta_1^{(0)}(a) = \mathsf{MVR}$ and $\delta_1^{(0)}(b) = \varepsilon$, then $\delta_1^{(0)}(c) = \varepsilon$ and $\delta_2^{(0)}(\$) = \mathsf{Restart}$. Then $\mathcal{M}$ executes the following accepting computation:

$$(0, aaac) \vdash_{\mathcal{M}}^{c} (1, aaa) \vdash_{\mathcal{M}}^{*} \mathsf{Accept},$$

where $\delta(0, \$) = 1$. But then $\mathcal{M}$ also executes the following computation:

$$(0, aaab) \vdash_{\mathcal{M}}^{c} (1, aaa) \vdash_{\mathcal{M}}^{*} \mathsf{Accept},$$

which again contradicts our assumption on $\mathcal{M}$, as $aaab \notin L_1'^{R}$.

(iv) If $\delta_1^{(0)}(a) = \varepsilon$ and $\delta_1^{(0)}(b) = \mathsf{MVR}$, then $\delta_1^{(0)}(c) = \varepsilon$ and $\delta_2^{(0)}(\$) = \mathsf{Restart}$. Then $\mathcal{M}$ executes the following accepting computation:

$$(0, bbc) \vdash_{\mathcal{M}}^{c} (1, bb) \vdash_{\mathcal{M}}^{*} \mathsf{Accept},$$

where $\delta(0, \$) = 1$. But then $\mathcal{M}$ also executes the following computation:

$$(0, bba) \vdash_{\mathcal{M}}^{c} (1, bb) \vdash_{\mathcal{M}}^{*} \text{Accept},$$

which again contradicts our assumption on $\mathcal{M}$, as $bba \notin L_1'^{R}$.

As this covers all cases, we see that indeed $L_1'^{R}$ is not accepted by any stl-det-global-CD-2-RR(1)-system.   □

5   Decidability Problems

In this section we turn to investigate decidability problems for the classes $\mathscr{L}_{=1}(\text{stl-det-local-CD-R}(1))$ and $\mathscr{L}_{=1}(\text{stl-det-local-CD-2-RR}(1))$. Basically it turns out that all undecidable problems are not even semidecidable. For these problems, it suffices to show the results for $\mathscr{L}_{=1}(\text{stl-det-local-CD-R}(1))$. We will use a reduction of Post's Correspondence Problem (PCP) (see, for example, [17]).

Let Σ be an alphabet. An instance of the PCP is given by two lists $\alpha = \alpha_1, \alpha_2, \ldots, \alpha_n$ and $\beta = \beta_1, \beta_2, \ldots, \beta_n$ of words from Σ^+. It is well known that it is undecidable whether a PCP has a solution [16], that is, whether there is a nonempty finite sequence of indices $i_1, i_2, \ldots, i_k$ such that $\alpha_{i_1} \alpha_{i_2} \cdots \alpha_{i_k} = \beta_{i_1} \beta_{i_2} \cdots \beta_{i_k}$. In particular, it is semidecidable whether a PCP has a solution, but is *not* semidecidable whether it has *no* solution. In the sequel we call $i_1, i_2, \ldots, i_k$ as well as $\alpha_{i_1} \alpha_{i_2} \cdots \alpha_{i_k}$ a solution of the PCP.

Theorem 40. *Regularity, context-freeness, equivalence, and inclusion are not semidecidable for $\mathscr{L}_{=1}(\text{stl-det-local-CD-R}(1))$ and $\mathscr{L}_{=1}(\text{stl-det-local-CD-2-RR}(1))$.*

Proof. Let an instance of the PCP be given by the lists $\alpha = \alpha_1, \alpha_2, \ldots, \alpha_n$ and $\beta = \beta_1, \beta_2, \ldots, \beta_n$ of nonempty words over some alphabet $\Sigma = \{a_1, a_2, \ldots, a_m\}$. Further, let $H = \{1, 2, \ldots, n\}$, $\alpha_j = \alpha_{j,1} \alpha_{j,2} \cdots \alpha_{j,|\alpha_j|}$, $\beta_j = \beta_{j,1} \beta_{j,2} \cdots \beta_{j,|\beta_j|}$, define $\tilde{\Sigma} = \{\tilde{a} \mid a \in \Sigma\}$ to be a disjoint copy of Σ, and set $\tilde{\beta}_j = \tilde{\beta}_{j,1} \tilde{\beta}_{j,2} \cdots \tilde{\beta}_{j,|\beta_j|}$, for $1 \leq j \leq n$.

In order to construct a language that meets our purposes we start with the set $E = \{x_1 x_1' x_2 x_2' \cdots x_\ell x_\ell' \mid \ell \geq 0, x_i \in \Sigma, x_i' \in \tilde{\Sigma}, 1 \leq i \leq \ell, \text{ and } x_i' = \tilde{x}_i\}$, that is, the set of words in which symbols from Σ and $\tilde{\Sigma}$ occur alternatingly (!) so that each symbol from $\tilde{\Sigma}$ is the copy of its left neighbor from Σ. Now define

$$L_1 = ((\Sigma \cup \tilde{\Sigma})^* \setminus E) \sqcup H^*,$$

that is, the complement of E with respect to $\Sigma \cup \tilde{\Sigma}$ shuffled with indices from H, and

$$L_2 = (E \sqcup H^*) \cap \{(w \sqcup \tilde{w}) \sqcup v \mid v = v_1 v_2 \cdots v_k \in H^+$$
$$\text{such that } w = \alpha_{v_1} \alpha_{v_2} \cdots \alpha_{v_k}, \tilde{w} = \tilde{\beta}_{v_1} \tilde{\beta}_{v_2} \cdots \tilde{\beta}_{v_k}\}.$$

To conclude the construction of the language set $L_P = L_1 \cup L_2$. It can be shown that L_P is accepted by a stl-det-local-CD-R(1)-system $\mathcal{M}$. Essentially $\mathcal{M}$ is the union of two subsystems, the one of which accepts the language L_1, while the other accepts a superset of the language L_2.

Now assume that the PCP has no solution. Then L_2 is empty and $L_P = L_1$ is regular and, thus, context-free. Conversely, if L_P is context-free, then $L_P \cap (E \amalg H^*) = L_2$ is context free. A straightforward application of the pumping lemma on words of the form $(\Sigma \cup \tilde{\Sigma})^* \cdot H^*$ that belong to L_2 shows a contradiction. Therefore, the non-semidecidability of regularity and context-freeness follows by the non-semidecidability of the unsolvability of the PCP.

As mentioned above, $\mathcal{M}$ includes a subsystem for accepting L_1. Therefore, the semidecidability of equivalence implies the semidecidability of whether $\mathcal{M}$ accepts L_1, that is, whether L_2 is empty and, thus, the semidecidability of the unsolvability of the PCP. Finally, if inclusion is semidecidable then so is equivalence. $\square$

Theorem 41.   *Universality and cofiniteness are not semidecidable for the classes* $\mathscr{L}_{=1}$(stl-det-local-CD-R(1)) *and* $\mathscr{L}_{=1}$(stl-det-local-CD-2-RR(1)).

Proof. Given an instance of the PCP, a stl-det-local-CD-2-RR(1)-system can be constructed for the language $\overline{L_2}$ (which includes L_1). Since L_2 is empty if and only if the PCP has no solution, $\overline{L_2} = (\Sigma \cup \tilde{\Sigma} \cup H)^*$ if and only if the PCP has no solution. Therefore, universality is non-semidecidable.

A PCP has either no solution or infinitely many solutions. So $\overline{L_2}$ is cofinite if and only if the PCP has no solution, which completes the proof. $\square$

6   Conclusions

We have investigated cooperating distributed systems of stateless deterministic two-phase RR-automata of window size one. The main interest was on the computational power and the closure properties of the language classes induced by the systems considered. The proven inclusion relations are depicted in Figure 1, while Table 1 summarizes the closure and non-closure properties obtained. Moreover, we considered decidability problems for the classes $\mathscr{L}_{=1}$(stl-det-local-CD-R(1)) and $\mathscr{L}_{=1}$(stl-det-local-CD-2-RR(1)). However, several questions remain unanswered: (1) Do the systems in question only accept semilinear languages? (2) Is the class $\mathscr{L}_{=1}$(stl-det-local-CD-2-RR(1)) not closed under concatenation or Kleene plus? Similarly, is $\mathscr{L}_{=1}$(stl-det-global-CD-2-RR(1)) not closed under Kleene plus? (3) What are the remaining algorithmic properties of the language class $\mathscr{L}_{=1}$(stl-det-local-CD-2-RR(1))? Is emptiness or finiteness decidable? How about the decidability problems for $\mathscr{L}_{=1}$(stl-det-strict-CD-2-RR(1)) and $\mathscr{L}_{=1}$(stl-det-global-CD-2-RR(1))?

Table 1. "+" denotes the fact that the corresponding class is closed under the given operation, "−" denotes the fact that it is not closed, and "?" indicates that the status of this property is still open. $\cup$ denotes union, $^{-}$ complementation, $\cap$ intersection, $\cap_{REG}$ intersection with regular languages, $\cdot$ concatenation, $^{+}$ Kleene plus, h_{ε} ε-free homomorphism, h^{-1} inverse homomorphism, *com* commutative closure, and ^{R} denotes reversal.

Types of CD-Systems	Operations									
	$\cup$	$^{-}$	$\cap$	$\cap_{REG}$	$\cdot$	$^{+}$	h_{ε}	h^{-1}	*com*	R
stl-det-local-CD-2-RR(1)	+	−	−	−	?	?	−	−	?	−
stl-det-global-CD-2-RR(1)	−	+	−	−	−	?	−	−	−	−
stl-det-strict-CD-2-RR(1)	−	+	−	−	−	−	−	−	−	−

References

1. Csuhaj-Varjú, E., Dassow, J., Kelemen, J., Păun, G.: Grammar Systems. A Grammatical Approach to Distribution and Cooperation. Gordon and Breach, London (1994)
2. Culy, C.: Formal properties of natural language and linguistic theories. Linguistics and Philosophy 19, 599–617 (1996)
3. Dassow, J., Păun, G., Rozenberg, G.: Grammar systems. In: Rozenberg, G., Salomaa, A. (eds.) Handbook of Formal Languages, vol. 2, pp. 155–213. Springer, Berlin (1997)
4. Diekert, V., Rozenberg, G.: The Book of Traces. World Scientific, Singapore (1995)
5. Harrison, M.A.: Introduction to Formal Language Theory. Addison-Wesley, Reading (1978)
6. Ibarra, O., Karhumäki, J., Okhotin, A.: On stateless multihead automata: Hierarchies and the emptiness problem. Theoret. Comput. Sci. 411, 581–593 (2009)
7. Kutrib, M., Messerschmidt, H., Otto, F.: On stateless two-pushdown automata and restarting automata. Int. J. Found. Comput. Sci. 21, 781–798 (2010)
8. Kutrib, M., Messerschmidt, H., Otto, F.: On stateless deterministic restarting automata. Acta Inform. 47, 391–412 (2010)
9. Kutrib, M., Reimann, J.: Succinct description of regular languages by weak restarting automata. Inform. Comput. 206, 1152–1160 (2008)
10. Lautemann, C.: One pushdown and a small tape. In: Dirk Siefkes zum 50. Geburtstag, pp. 42–47. TU Berlin and Universität Augsburg (1988)
11. McNaughton, R., Narendran, P., Otto, F.: Church-Rosser Thue systems and formal languages. J. ACM 35, 324–344 (1988)
12. Messerschmidt, H., Otto, F.: Cooperating distributed systems of restarting automata. Int. J. Found. Comput. Sci. 18, 1333–1342 (2007)
13. Messerschmidt, H., Otto, F.: Strictly Deterministic CD-Systems of Restarting Automata. In: Csuhaj-Varjú, E., Ésik, Z. (eds.) FCT 2007. LNCS, vol. 4639, pp. 424–434. Springer, Heidelberg (2007)
14. Nagy, B., Otto, F.: CD-Systems of Stateless Deterministic R(1)-Automata Accept All Rational Trace Languages. In: Dediu, A.-H., Fernau, H., Martín-Vide, C. (eds.) LATA 2010. LNCS, vol. 6031, pp. 463–474. Springer, Heidelberg (2010)

15. Nagy, B., Otto, F.: Globally Deterministic CD-Systems of Stateless R(1)-Automata. In: Dediu, A.-H., Inenaga, S., Martín-Vide, C. (eds.) LATA 2011. LNCS, vol. 6638, pp. 390–401. Springer, Heidelberg (2011)
16. Post, E.L.: A variant of a recursively unsolvable problem. Bull. AMS 52, 264–268 (1946)
17. Salomaa, A.: Formal Languages. Academic Press, New York (1973)
18. Yang, L., Dang, Z., Ibarra, O.: On stateless automata and P systems. In: Workshop on Automata for Cellular and Molecular Computing, pp. 144–157. MTA SZTAKI (2007)

The Boolean Formula Value Problem
as Formal Language

Klaus-Jörn Lange

WSI, Universität Tübingen,
Sand 13, D72076 Tübingen, Germany
`lange@informatik.uni-tuebingen.de`

Abstract. The Boolean formula value problem asks for the Boolean output value of a given input formula. We code it as a formal language $\mathcal{D}_+ \subset \{a,b\}^*$. $\mathcal{D}_+$ is a nonregular, visibly pushdown language. We give automata for $\mathcal{D}_+$ which enable us to derive some of its syntactic equations. It is unknown whether the given list of equations is complete. Using these equations some algebraic properties of the syntactic monoid of $\mathcal{D}_+$ are sketched.

Keywords: Boolean formula value problem, syntactic monoid, word equations, algebraic approach, TC^0 vs NC^1.

1 Introduction

The Boolean formula value problem (*BFVP*) consists in evaluating Boolean formulas. Compared to the *P*-complete circuit value problem it is of rather low complexity; depending in the coding of the input formula the evaluation can be done in NC^1 if the formula is given in a parenthesized way (or in polish normal form) or deterministically in logarithmic space if the input formula is given as a tree coded by nodes and edges. In this paper we will be interested in the first of these two possibilities. It is quite easy to show the NC^1-hardness of the Boolean evaluation problem for formulas given by parenthesized words or expressions in polish normal form. On the other hand,the construction to prove membership in NC^1 is quite involved [5].

The class NC^1 is quite attractive from the formal language viewpoint since Barrington showed, that each regular set, whose syntactic monoid contains an nonsolvable group is NC^1-complete ([3]). Thus there exist regular sets whose word problem is NC^1-complete.

The aim of this investigation is to consider the Boolean formula evaluation problem from the algebraic formal language view-point despite the fact that it is nonregular and hence its syntactic monoid is infinite. This short note comes without proofs. Just the constructions and some examples are given. Proofs for the correctness of the automata constructions can be found in the Studienarbeit of Bernd Brumm ([4]).

This paper is structured as follows: we first express (a special version of) the Boolean formula value problem via the Dyck language $\mathcal{D}$ over one pair of parenthesis. Then we present automata constructions. Finally, some first algebraic

H. Bordihn, M. Kutrib, and B. Truthe (Eds.): Dassow Festschrift 2012, LNCS 7300, pp. 138–144, 2012.
© Springer-Verlag Berlin Heidelberg 2012

properties of *BFVP* are given including a list of defining equations which are not known to be complete.

1.1 Preliminaries

Let $L \subseteq \{a, b\}^*$. We say that two words $x, y \in \{a, b\}^*$ are *congruent modulo L* if and only if we have

$$zxz' \in L \iff zyz' \in L$$

for all $z, z' \in \{a, b\}^*$. By $[x]_L$ we denote the congruence class of x modulo L. For the resulting notions and results concerning the *syntactic monoid* of L we refer to [6].

In our investigations of the Boolean formula value problem we will use the notion of visibly pushdown automata and languages as introduced by Alur ([1]). These were known before as input-driven languages and are characterized by the restriction that the modification of the stack in terms of push or pop moves are not dependent in the state but only in the input symbol. While the one-sided Dyck languages are visibly pushdown languages, the two-sided ones are not.

2 Coding the Boolean Formula Value Problem

The coding of a problem, i.e.: its representation as formal language containing words which code problem instances, can usually be done in different ways without affecting the complexity of the problem. But the resulting formal languages will differ signifcantly in their algebraic properties expressed in their syntactic monoids. For instance will a parenthesis-language contain a zero in its syntactic monoid, while in the corresponding polish normal form language arbitrary words are subwords of valid expressions, which means that there is no zero in the syntactic monoid.

Our aim is to choose a representation of BFVP which leads to a syntactic monoid as simple as possible. That is why we will represent boolean formulas by the *NAND*-operation, we will use a polish normal form instead of using parentheses, and we will represent binary trees by dyck words and not by the more usual Lukasiewicz words.

It is well known that every Boolean function can be expressed by the (binary) *NAND*-function together with the Boolean constant *TRUE*. The conversion is possible by replacing $AND(x, y)$ by $NAND(TRUE, NAND(x, y))$, $OR(x, y)$ by $NAND(NAND(TRUE, x), NAND(TRUE, y))$, and $NOT(x)$ by $NAND(TRUE, x)$. The size of the resulting $NAND$-formula is linear in the size of the original AND, OR-formula.

Hence we will consider as input formulas, which are to be evaluated, complete binary trees labelled with NAND-function, i.e.; all inner nodes have two predecessors and are labelled by the NAND-function, while the remaining nodes are leaves of indegree zero labelled by the Boolean constant TRUE.

As mentioned before, coding these formulas as graphs, with vertices and edges, leads to evaluation problems which are hard for deterministic logarithmic space.

The well-known alternative is to use parentheses to express the tree structure. Throughout of this paper we will represent the opening paranthesis by the letter a and the closing one by the letter b.

We code complete binary trees as follows: the tree consisting of a single (root) vertex is coded by the empty word λ. If a vertex has two outgoing edges leading to its predecessors the left edge is labelled by a and the right one by b. The tree is then read in-order from left to right. Thus the word $aabb$ represents the binary tree with 3 leaves, the left subree containing 2 leaves, and the right one containing one leave. Switching the left and right subtree yields the tree represented by $abab$. This gives a one-to-one corrspondence between complete binary trees and the Dyck language $\mathcal{D} \subset \{a, b\}^*$.

We decided in favour of labelling the edges and against the more usual labelling of the vertices, which would lead to the well known representation of complete binary trees by the Lukasiewicz language.

While (contextfree) grammars are in general easier to construct for the Lukasiewicz language, in the Dyck case the resulting syntactic (bicyclic) monoid is more simple. For instance $\mathcal{D}$ is generated by the single equation $ab = \lambda$ whereas the Lukasiewicz language results in the equations $aba = a, abb = b$, and $aab = a$.

A tree labelled by Boolean functions and constants evaluates either to TRUE or to FALSE. In this way the Dyck set $\mathcal{D}$ is divided into the two disjoint subsets $\mathcal{D} = \mathcal{D}_+ \cup \mathcal{D}_-$ where $\mathcal{D}_+$ consists in those elements of $\mathcal{D}$ which represent a tree (labelled with the NAND-function and the constant TRUE) which evaluates to TRUE and $\mathcal{D}_-$ contain those which evaluate to FALSE. Thus $\mathcal{D}_+$ is a special formulation of the Boolean formual value problem which makes $\mathcal{D}_+$ NC^1-complete.

In the following, we are going to investigate the properties of the formal language $\mathcal{D}_+$.

3   Properties of $\mathcal{D}_+$

It is easy to see that $\mathcal{D}_+$ is a context-free language. For instance, the set $\mathcal{D}_+ b^1$ is generated by the grammar with the rules $S \to aaSSS|aSaSS|aaSSaSS|b$.

Obviously, $\mathcal{D}_+$ is a *visibly push-down language* as defined by Alur([1]), i.e. for each element of the terminal alphabet it is determined whether the stack of an push-down automaton accepting $\mathcal{D}_+$ is pushed (here by the symbol a) or popped (here by the symbol b).

Alur et al. showed that a language is visibly push-down if and only if a certain congruence relation is of finite index ([2]).

A close inspection shows that the resulting congruence relation divides the set $\mathcal{D}$ into four classes

$$\mathcal{D} = F \cup N \cup P \cup T.$$

This was explicated in [4].

---

[1] This set might be regarded as the Lukasiewicz-version of $\mathcal{D}_+$

These four classes might be explained in the following way: T consists in those Dyck-words w which represent formulas, which evaluate to TRUE and if we add a suffix v such that wv is still a Dyck word, wv evaluates to TRUE, as well.

F consists of the dual class of words representing fomulae evaluating to FALSE regardless how they are completed by a Dyck suffix.

P and N are represent those formulas which evaluate to TRUE (respectively, FALSE) whose value can be changed by a suffix.

The shortest member of these four classes are $\lambda \in P, ab \in N, aabb \in T$, and, $abaabb \in F$. We have

$$\mathcal{D}_+ = T \cup P \text{ and } \mathcal{D}_- = F \cup N.$$

These four classes can be characterized in the following way: every $w \in \mathcal{D}$ admits a unique decomposition $w = aw_1 baw_2 \cdots aw_n b$ for some $n \geq 0$ and some $w_i \in \mathcal{D}$. We then have

- $w \in P$ iff n is even and for all $1 \leq j \leq n$ we have $w_j \in \mathcal{D}_+$,
- $w \in N$ iff n is odd and for all $1 \leq j \leq n$ we have $w_j \in \mathcal{D}_+$,
- $w \in T$ iff there exists some $i < n/2$ such that $w_{2i+1} \in \mathcal{D}_-$ and for all $1 \leq j \leq 2i$ we have $w_j \in \mathcal{D}_+$, and
- $w \in F$ iff there exists some $i < n/2$ such that $w_{2i} \in \mathcal{D}_-$ and for all $1 \leq j \leq 2i - 1$ we have $w_j \in \mathcal{D}_+$.

Thus $w \in \mathcal{D}_+$ iff w consists in a concatenation of an even number of words $avb, v \in \mathcal{D}_+$, followed by a word $aub, u \in \mathcal{D}_-$, or followed by nothing. If that number is odd, we have $w \in \mathcal{D}_-$.

3.1   Automata for $\mathcal{D}_+$

Alur showed how to construct a push-down automaton out of the congruence whose classes serve both as stack alphabet and as set of states. In the case of $\mathcal{D}_+$ the resulting automaton can be simplified by keeping as set of states $\{F, N, P, T\}$ but shrinking the stack alphabet to $\Gamma := \{P, N\}$ with the pushing transitions

$$T\ a \to F, P$$
$$P\ a \to P, P$$
$$N\ a \to P, N$$
$$F\ a \to F, N$$

and the popping transitions

$$T, P\ b \to N$$
$$P, P\ b \to N$$
$$N, P\ b \to T$$
$$F, P\ b \to T$$
$$T, N\ b \to P$$
$$P, N\ b \to P$$
$$N, N\ b \to F$$
$$F, N\ b \to F$$

The resulting automaton has a very regular structure like an infinite binary tree. It is thus possible to represent uniquely each combination of state and stack content of this automaton as a binary string in a way, that configurations connected by transitions are of a very similar shape. A pushing transition, i.e. reading an a, acts on a binary string ending in the bits xy, $x, y \in \{0, 1\}$, by appending the second to last bit x which makes the string now ending in xyx. A popping transition, i.e. reading a b, acts on a binary string ending in the bits xyz, $x, y, z \in \{0, 1\}$, by deleting the last bit z and then exchanging the remaining last two bits which makes the string now ending in yx. Interpreting these strings as binary numbers, the four pushing and eight popping rules can be given by the following rules: The automaton has infinitely many states labelled by natural numbers greater or equal to 4. The starting state is state 6. If the (one-way) input reads an a and the automaton is in state $4n + i$ for some n and some $i < 4$ it goes to state $8n + j$ where i and j are given by:

$$4n + 0 \rightarrow 8n + 0$$
$$4n + 1 \rightarrow 8n + 2$$
$$4n + 2 \rightarrow 8n + 5$$
$$4n + 3 \rightarrow 8n + 7$$

If a b is read we have the following rules:

$$8n + 0 \rightarrow 4n + 0$$
$$8n + 1 \rightarrow 4n + 0$$
$$8n + 2 \rightarrow 4n + 2$$
$$8n + 3 \rightarrow 4n + 2$$
$$8n + 4 \rightarrow 4n + 1$$
$$8n + 5 \rightarrow 4n + 1$$
$$8n + 6 \rightarrow 4n + 3$$
$$8n + 7 \rightarrow 4n + 3$$

Since $\mathcal{D}_+$ is NC^1-completeit is thus an NC^1-complete task to read a Dyck word and make according to the input letters the corresponding modulo computations and then to determine whether the result is 6 or 7 (corresponding to $\mathcal{D}_+$) or 4 or 5 (corresponding to $\mathcal{D}_-$).

4   The Syntactic Monoid of $\mathcal{D}_+$

We now investigate the infinite syntactic monoid of $\mathcal{D}_+$. To do so, we first consider equations fulfilled by the syntactic congruence of $\mathcal{D}_+$. After that we give a few algebraic properties of $\mathcal{D}_+$.

4.1   Equations

It is easy to check the validity of the following equations fulfilled in $\{a, b\}^*$ by $\mathcal{D}_+$ either directly or using the automata given in the previous section:

1. $abab = \lambda$,
2. $aabbb = b$,
3. $aabaabb = aabba$,
4. $abbabb = babb$,
5. $aabba^i ab = aabba^i$ for all $i \geq 0$, and
6. $abb^i abaabb = b^i abaabb$ for all $i \geq 1$.

The last two (sets of) equations express that to the right of the word in $aabba^*$ respectively to the left of the word in $b^* babaabb$ the $\mathcal{D}_+$-evaluation is simply just a $\mathcal{D}$-evaluation, i.e.: the reduction of the subword ab to λ.

It is unclear, whether these equations are complete, i.e.: whether there are new equations not implied by the given ones, or independent, i.e.; whether one of them is implied by the others.

Another interesting question is, whether these equations can be directed in either way (either from left to right or from right to left) such that each sequence of applications of the bidirectional equations could be simulated by a sequence of applications of the unidirectional versions of these equations.

The last question is closely connected to the search for rewriting systems converting each $w \in \{a, b\}^*$ into a normal form w' (for instance a shortest word w.r.t. some ordering) such that w' and w are congruent modulo $\mathcal{D}_+$. An application of the Knuth-Bendix-procedure to (unidirectional versions of) the given equations didn't give new ones.

4.2   Algebraic Properties of the $\mathcal{D}_+$

We finally give a few algebraic properties of the syntactic monoid $\mathcal{M}_{\mathcal{D}_+}$ of $\mathcal{D}_+$.

A standard tool to investigate the structure of a monoid are *Green's relations* (see for instance [6]). For finite monoids the D- and the J-relation allways coincide. This can hold in the infinite case, as well; an example is the syntactic monoid of the language $\mathcal{D}$ which is the *bicyclic monoid*. In contrast to that the syntactic monoid of $\mathcal{D}_+$ has one J-class (i.e. for all $x, y \in \{a, b\}^*$ there exist $z, z' \in \{a, b\}^*$ such that x is congruent zyz' modulo $\mathcal{D}_+$), but more then one D-class, since $aabb$ and the empty word λ are not D-equivalent, i.e.: there is no $x \in \{a, b\}^*$ such that both $[aabb]_L \mathcal{M}_{\mathcal{D}_+} = [x]_L \mathcal{M}_{\mathcal{D}_+}$ and $\mathcal{M}_{\mathcal{D}_+}[x]_L = \mathcal{M}_{\mathcal{D}_+}[\lambda]_L$.

We finally mention, that the syntactic monoid of $\mathcal{D}_+$ is regular. That is, for all $x \in \{a, b\}^*$ there exists some y (called an *inverse* of x) such that xyx is congruent with x and yxy with y modulo $\mathcal{D}_+$.

If that inverse element y is uniquely determined by x, such a monoid is called inverse. While the syntactic monoid of $\mathcal{D}$ is inverse, that of $\mathcal{D}_+$ is not; for instance the word bab has the two different inverses a and $aaabb$.

These algebraic differences between $\mathcal{D}$ and $\mathcal{D}_+$ express their differences in complexity. While $\mathcal{D}$ is in TC^0 (and complete w.r.t. Turing reducibilities), $\mathcal{D}_+$ is NC^1-complete(w.r.t. many-one reducibilities).

Compared to $\mathcal{D}_+$ a totally different NC^1-complete problem is the word problem of A_5 (or of any other regular set whose syntactic monoid contains a nonsolvable group ([3])). In the proof of this fact the action of a Boolean gate with inputs

x and y is in some sense simulated by evaluating the commutator $x^{-1}y^{-1}xy$ of the algebraic simulations of x and y. The proof makes uses of the fact, that a nonsolvable group contains arbitrarily long, nonvanishing commutatorchains. This leads to the question whether we can find in the syntactic monoid of $\mathcal{D}_+$, which is regular and has only one J-class, a similar algebraic simmulation of the action of Boolean gates, which would give a new proof of the NC^1-hardness of the Boolean formula value problem.

Acknowledgement. I would like to thank the referees for their careful reading of this note.

References

1. Alur, R.: Visibly Pushdown Languages. In: Proc. 36th ACM Symp. on Theory of Computing, pp. 202–211 (2004)
2. Alur, R., Kumar, V., Madhusudan, P., Viswanathan, M.: Congruences for Visibly Pushdown Languages. In: Caires, L., Italiano, G.F., Monteiro, L., Palamidessi, C., Yung, M. (eds.) ICALP 2005. LNCS, vol. 3580, pp. 1102–1114. Springer, Heidelberg (2005)
3. Barrington, D.A.: Bounded-width polynomial-size branching programs can recognize exactly those languages in NC^1. J. Comp. System Sci. 38, 150–164 (1989)
4. Brumm, B.: Das Auswertungsproblem als formale Sprache. Private Communication (2011)
5. Buss, S.R.: The Boolean formula value problem is in ALOGTIME. In: Proc. 19th Ann. ACM Symp. on Theory of Computing, pp. 123–131 (1987)
6. Pin, J.E.: Varieties of Formal Languages. Plenum, London (1986)

Hairpin Lengthening and Shortening
of Regular Languages*

Florin Manea[1], Robert Mercas[2], and Victor Mitrana[3]

[1] Institut für Informatik, Christian-Albrechts-Universität zu Kiel,
Christian-Albrechts-Platz 4, 24098 Kiel, Germany
`flm@informatik.uni-kiel.de`
[2] Facultät für Informatik, Otto-von-Guericke-Universität Magdeburg,
Postfach 4120, 39016 Magdeburg, Germany
`robertmercas@gmail.com`
[3] Department of Organization and Structure of Information,
University School of Informatics, Polytechnic University of Madrid,
Crta. de Valencia km. 7 - 28031 Madrid, Spain
`victor.mitrana@upm.es`

Abstract. We consider here two formal operations on words inspired by the DNA biochemistry: hairpin lengthening introduced in [15] and its inverse called hairpin shortening. We study the closure of the class of regular languages under the non-iterated and iterated variants of the two operations. The main results are: although any finite number of applications of the hairpin lengthening to a regular language may lead to non-regular languages, the iterated hairpin lengthening of a regular language is always regular. As far as the hairpin shortening operation is concerned, the class of regular languages is closed under bounded and unbounded iterated hairpin shortening.

1 Introduction

This paper is a continuation of a series of works started with [3] (based on some ideas from [1]), where a new, bio-inspired, formal operation on words, called hairpin completion, was introduced. The initial work was followed by a series of related papers ([5,7,12,14,16,17,13]), where both the hairpin completion, as well as its inverse operation, the hairpin reduction, were further investigated both from the algorithmic and the language theoretic points of view.

We briefly recall the biological motivation of this operation. Polymerase chain reaction (PCR) is an automated process which enables researchers to produce a huge number of copies of a specific DNA sequence. Although PCR starts with a test tube containing double-stranded DNA molecules (called template and primer), we consider here a pretty similar phenomenon involving single-stranded DNA (ssDNA) following the second and third step of PCR, namely annealing and extension, respectively. It is well known that ssDNA are composed by nucleotides which differ from each other by their bases: A (adenine), G

---

* Work supported by the Alexander von Humboldt Foundation.

H. Bordihn, M. Kutrib, and B. Truthe (Eds.): Dassow Festschrift 2012, LNCS 7300, pp. 145–159, 2012.

(guanine), C (cytosine), and T (thymine). Two single strands can bind to each other, forming the secondary structure of DNA, if they are pairwise *Watson-Crick complementary*: A is complementary to T, and C to G. The binding of two strands is usually called annealing. This process which appears at about 54^0C is due to the Hydrogen-bonds that are constantly formed and broken between the two ssDNA. More bonds last longer and allow the polymerase to attach new nucleotides [22]. Once a few bases are built in, the ionic bond between the two ssDNA becomes so strong, that it does not break anymore. The polymerase attaches the bases (complementary to the template) to the primer on the 3' side.

We now imagine the following situation in which the role of the two ssDNA (template and primer) in the PCR is played by only one ssDNA. An intramolecular base pairing, known as *hairpin*, is a pattern that can occur in single-stranded DNA or RNA molecules. In this case, the single-stranded molecule bends, and one part of the strand bonds to another part of the same strand. In this way, the role of template and primer is played by the prefix and suffix, or vice-versa, of the ssDNA. In our case the phenomenon produces a new molecule as follows (see Figure 1): one starts with a ssDNA molecule, such that one of its ends (a prefix or, respectively, a suffix) is annealed to another part of itself by Watson-Crick complementarity forming a hairpin, and a *polymerization buffer* with many copies of the four basic nucleotides. Then, the initial hairpin is lengthened by polymerases (thus adding a suffix or, respectively, a prefix), until a complete hairpin structure is obtained (the beginning of the strand is annealed to the end of the strand). Of course, all these phenomena are considered here in an idealized way. For instance, we allow polymerase to extend the strand at either end (usually denoted in biology with 3' and 5') despite that, due to the greater stability of 3' when attaching new nucleotides, DNA polymerase can act continuously only in the 5'$\longrightarrow$ 3' direction. However, polymerase can also act in the opposite direction, but in short "spurts" (Okazaki fragments). This is the source of inspiration for the hairpin completion operation introduced in [3]. The situation is schematically illustrated in Picture 1.

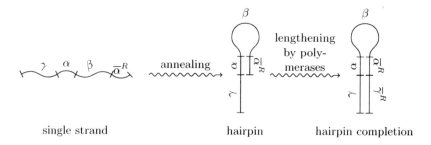

Fig. 1. Hairpin Completion

Hairpin or hairpin-free structures have numerous applications to molecular genetics and DNA-computing. In some DNA-based algorithms, these DNA molecules cannot be used in the subsequent computations. Therefore, it is important to design methods for constructing sets of DNA sequences which are unlikely to lead to such "bad" hybridizations. This problem was considered in a series of papers, see e.g. [4,8,10] and the references therein. On the other hand, molecules which may form a hairpin structure have been used as the basic feature of a computational model reported in [21], where an instance of the 3-SAT problem has been solved by a DNA algorithm whose second phase is mainly based on the elimination of hairpin structured molecules. Different types of hairpin and hairpin-free languages are defined in [2,19] and more recently in [11,9], where they are studied from a language theoretical point of view.

In [15] and later on in [18], a new variant of the hairpin completion, called hairpin lengthening, which seems more appropriate for possible bio-lab implementation, is considered. Informally, it seems more natural to consider that the prefix/suffix added by the hairpin completion cannot be arbitrarily long, since every step of a computation in a laboratory has to use a finite amount of resources and finite time. This variant concerns the prolongation of a strand which forms a hairpin, similarly to the process described for hairpin completion, but not necessarily until a complete hairpin structure is obtained. The main motivation in introducing this operation is that, in practice, it may be a difficult task to control the completion of a hairpin structure, and it seems easier to model only the case when such a structure is extended.

Both the hairpin completion and hairpin lengthening can be seen as formal operations by which one can generate a set of words, starting from a single word: for each possible pairing between a prefix and a complementary factor, or a suffix and a complementary factor, we can obtain a word by hairpin completion and several words by hairpin lengthening. As most of the unary operations on words, the hairpin completion and lengthening defined above can be extended canonically to operations on languages, and, then, their iterated version can be defined. We consider here two formal operations on words inspired by the DNA biochemistry: hairpin lengthening discussed above and a new operation which is its inverse, namely hairpin shortening. We study the closure of the class of regular languages under the non-iterated and iterated variants of the two operations.

In this paper we show that although any finite number of applications of the hairpin lengthening to a regular language may lead to non-regular languages, the iterated hairpin lengthening preserves the regularity of a language (in the final preparation of this paper we learned[1] that this result has been independently obtained by [6]). As far as the hairpin shortening operation is concerned, we prove that the class of regular languages is closed both under finitely-iterated and freely iterated hairpin shortening.

2 Preliminaries

We assume the reader to be familiar with the fundamental concepts of formal languages and automata theory, particularly with the notions of regular languages and finite automata; for all the related notions see [20].

---

[1] Volker Diekert, personal communication.

We denote by V^* and V^+ the set of all words over V including the empty word ε, and the set of all non-empty words over V, respectively. Given a word w over an alphabet V, we denote by $|w|$ its length, while $w[i..j]$ denotes the factor of w starting at position i and ending at position j, $1 \le i \le j \le |w|$. If $i = j$, then $w[i..j]$ is the i-th letter of w, which is simply denoted by $w[i]$.

Let Ω be a "superalphabet", that is an infinite set such that any alphabet considered in this paper is a subset of Ω. In other words, Ω is the *universe* of the languages in this paper, i.e., all words and languages are over alphabets that are subsets of Ω. An *involution* over a set S is a bijective mapping $\sigma : S \longrightarrow S$ such that $\sigma = \sigma^{-1}$. Any involution σ on Ω such that $\sigma(a) \ne a$ for all $a \in \Omega$ is said to be, in this paper's context, a *Watson-Crick involution*. Despite that this is nothing more than a fixed point-free involution, we prefer this terminology since the hairpin lengthening defined later is inspired by the DNA lengthening by polymerases, where the Watson-Crick complementarity plays an important role. Let $\bar{\cdot}$ be a Watson-Crick involution fixed for the rest of the paper. The Watson-Crick involution is extended to a morphism from Ω^* to Ω^* in the usual way. We say that the letters a and $\bar{a}$ are complementary to each other. For an alphabet V, we set $\overline{V} = \{\bar{a} \mid a \in V\}$. Note that V and $\overline{V}$ could be disjoint or intersect or be equal. We denote by $(\cdot)^R$ the mapping defined by $^R : V^* \longrightarrow V^*$, $(a_1 a_2 \ldots a_n)^R = a_n \ldots a_2 a_1$. Note that ^{R} is an involution and an *anti-morphism* $((xy)^R = y^R x^R$ for all $x, y \in V^*)$. Note also that the two mappings $\bar{\cdot}$ and $\cdot^R$ commute, namely, for any word x the equality $(\overline{x})^R = \overline{x^R}$ holds.

Let V be an alphabet, for any $w \in V^+$ we define the *k-hairpin lengthening* of w, denoted by $HL_k(w)$, for some $k \ge 1$, as follows:

- $HLP_k(w) = \{\overline{\delta^R} w \mid w = \alpha \beta \overline{\alpha}^R \gamma, |\alpha| = k, \alpha, \beta, \gamma \in V^+ \text{ and } \delta \text{ is a prefix of } \gamma\}$,
- $HLS_k(w) = \{w \overline{\delta^R} \mid w = \gamma \alpha \beta \overline{\alpha}^R, |\alpha| = k, \alpha, \beta, \gamma \in V^+ \text{ and } \delta \text{ is a suffix of } \gamma\}$,
- $HL_k(w) = HLP_k(w) \cup HLS_k(w)$.

The *hairpin lengthening* of w is defined by $HL(w) = \bigcup_{k \ge 1} HL_k(w)$. Clearly, $HL_{k+1}(w) \subseteq HL_k(w)$ for any $w \in V^+$ and $k \ge 1$. Therefore, one can easily note that $HL(w) = HL_1(w)$.

The k-hairpin lengthening is naturally extended to languages by $HL_k(L) = \bigcup_{w \in L} HL_k(w)$ for $k \ge 1$.

This operation is schematically illustrated in Figure 2.

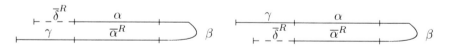

Fig. 2. Hairpin lengthening

The iterated version of the k-hairpin lengthening is defined as usual by:

$$HL_k^0(w) = \{w\}, \; HL_k^{n+1}(w) = HL_k(HL_k^n(w)), \; HL_k^*(w) = \bigcup_{n \geq 0} HL_k^n(w),$$

and $HL_k^*(L) = \bigcup_{w \in L} HL_k^*(w)$.

The iterated version of the hairpin lengthening is $HL^*(L) = HL_1^*(L)$.

Let us define the inversion operation of the hairpin lengthening, namely the hairpin shortening. Let V be an alphabet, and define for any $w \in V^+$ the k-hairpin shortening of w, denoted by $HS_k(w)$, for some $k \geq 1$, as follows:

- $HSP_k(w) = \{x|$ exists δ such that $\overline{\delta^R}x = \overline{\delta^R}\alpha\beta\overline{\alpha^R}\gamma = w, |\alpha| = k, \alpha, \beta, \gamma \in V^+$ and δ is a prefix of $\gamma\}$,
- $HSS_k(w) = \{x|$ exists δ such that $x\overline{\delta^R} = \gamma\alpha\beta\overline{\alpha^R}\overline{\delta^R} = w, |\alpha| = k, \alpha, \beta, \gamma \in V^+$ and δ is a suffix of $\gamma\}$,
- $HS_k(w) = HSP_k(w) \cup HSS_k(w)$.

The *hairpin shortening* of w is defined by $HS(w) = \bigcup_{k \geq 1} HS_k(w)$. The k-hairpin shortening is naturally extended to languages by $HS_k(L) = \bigcup_{w \in L} HS_k(w)$ for $k \geq 1$.

The iterated version of the k-hairpin shortening is defined as usual by:

$$HS_k^0(w) = \{w\}, \; HS_k^{n+1}(w) = HS_k(HS_k^n(w)), \; HS_k^*(w) = \bigcup_{n \geq 0} HS_k^n(w),$$

and $HS_k^*(L) = \bigcup_{w \in L} HS_k^*(w)$.

The iterated version of the hairpin lengthening is $HS^*(L) = HS_1^*(L)$.

3 Hairpin Lengthening of Regular Languages

The following results were reported in [18]:

Theorem 1. *A language is linear context-free if and only if it is the gsm image of the k-hairpin lengthening of a regular language for $k \geq 1$.*

Consequently,

Proposition 1. *The class of regular languages is not closed under k-hairpin lengthening for $k \geq 1$.*

A first natural question concerns the closure of the class of regular languages under the application for a finite number of times of the k-hairpin lengthening operation. In [18], we show that if we apply for a finite number of times the k-hairpin lengthening operation to a regular language we can obtain non-regular languages. The regular language $L = \{\diamond a^n c^k d\overline{c^k} \mid n \geq k\}$ proves this. Indeed, it is not hard to see that, for $m \geq 2$, we have

$$HL_k^m(L) = \{\diamond a^n c^k d\overline{c^k}a^s \mid n \geq k, 0 \leq s \leq n + (m-1)(n-k)\} \cup$$
$$\{\diamond a^n c^k d\overline{c^k}a^s\diamond \mid n \geq k, s \geq k, s \leq n + (m-1)(n-k)\}.$$

Thus, $HL_k^m(L) \cap \diamond V^* \bar{\diamond} = \{\diamond a^n c^k d\bar{c^k} a^s \diamond \mid n \geq k, s \geq k, s \leq n + (m-1)(n-k)\}$ and one can easily show that this language is not regular.

We now came to the second natural question: Is the class of regular languages closed under iterated k-hairpin lengthening? Somehow surprising (in the view of the above discussion), the answer is affirmative.

First, let us note that every iterated k-hairpin lengthening can be simulated by an iterated k-hairpin lengthening, where only one symbols is added in the right-hand or left-hand end of the current word. Let V be an alphabet and $k \geq 1$. For each letter $a \in V$ we define the two languages:

$$R(k,a) = \{x\bar{a}yz\bar{y}^R \mid x \in V^*, y, z \in V^+, |y| = k\},$$
$$L(k,a) = \{yz\bar{y}^R\bar{a}x \mid x \in V^*, y, z \in V^+, |y| = k\}.$$

Obviously, the two languages are regular.

We now define the mapping $\phi_k : V^* \longrightarrow V^*$ by:

$$\phi_k(w) = \{w\} \cup \{wa \mid w \in R(k,a), a \in V\} \cup \{aw \mid w \in L(k,a), a \in V\}.$$

This mapping is naturally extended to languages by $\phi_k(L) = \bigcup_{w \in L} \phi_k(w)$.

The iterated version of this mapping is defined as usual by:
$$\phi_k^0(w) = \{w\}, \qquad \phi_k^{n+1}(w) = \phi_k(\phi_k^n(w)), \qquad \phi_k^*(w) = \bigcup_{n \geq 0} \phi_k^n(w),$$

and $\phi_k^*(L) = \bigcup_{w \in L} \phi_k^*(w)$.

It is plain that for any language L and any $k \geq 1$, $HL_k^*(L) = \phi_k^*(L)$. The main result of this section is based on this simple remark.

Theorem 2. *The class of regular languages is closed under iterated k-hairpin lengthening for $k \geq 1$.*

Proof. We begin our proof with a simple remark. Let w and w' be two words over an alphabet V such that $w \in HL_k^*(w')$. According to the remarks made before this proof, there exist the words $w_0, w_1, \ldots, w_t$ such that $w_0 = w'$, $w_t = w$, $w_i \in HL_k(w_{i-1})$ and $|w_i| = |w_{i-1}| + 1$ for $1 \leq i \leq t$. Then, for $1 \leq i \leq t$:

- If $w_i \in HLS_k(w_{i-1})$ has the suffix αa with $a \in V$ and $\alpha \in V^k$, and w ends with α, then $wa \in HL_k^*(w')$.
- If $w_i \in HLP_k(w_{i-1})$ has the prefix $a\alpha$ with $a \in V$ and $\alpha \in V^k$, and w starts with α, then $aw \in HL_k^*(w')$.

Assume now that we want to determine whether a word w of length n can be obtained by iterated k-hairpin lengthening from one of its factors $w[i..j]$. More precisely, we determine whether w can be obtained from $w[i..j]$ by iterated k-hairpin lengthening such that in each step the current word is lengthened with exactly one symbol (which is equivalent to the general iterated k-hairpin lengthening, as we have already seen before in this proof).

Let $S_1 = (x_1, x_2, \ldots, x_t)$ be a list of the distinct words of length $k + 1$ that occur as factors of the word $w[1..i + k - 1]$, ordered increasingly according to the rightmost position where they appear in $w[1..i + k - 1]$ (i.e., the rightmost occurrence of x_ℓ in $w[1..i + k - 1]$ is to the left of the rightmost position where $x_{\ell+1}$ occurs in the same factor). Let p_ℓ be the maximum number such that $p_\ell \leq i - 1$ and $w[p_\ell..p_\ell + k] = x_\ell$, i.e., p_ℓ is the starting position of the rightmost occurrence of x_ℓ in $w[1..i + k - 1]$; clearly, $p_t = i - 1$. Finally, we define the list of numbers $S_1' = (n_1, \ldots, n_t)$ by $n_1 = \min(k, p_1 - 1)$ and $n_\ell = \min(k, p_\ell - p_{\ell-1} - 1)$ for $2 \leq \ell \leq t$.

It is not hard to see that all the factors $w[i'..i' + k]$ with $p_{\ell-1} + 1 \leq i' \leq i - 1$ are from the set $\{x_\ell, x_{\ell+1}, \ldots, x_t\}$. By the remark stated at the beginning of this proof, it follows that if $j' \geq j$ and $w[p_\ell..j']$ can be obtained by iterated k-hairpin lengthening from $w[i..j]$, then $w[p_{\ell-1} + 1..j']$ can be also obtained by iterated k-hairpin lengthening from $w[i..j]$. Moreover, $w[p_\ell..j']$ can be obtained from $w[p_\ell + 1..j]$ by k-hairpin lengthening if and only if $\overline{x}_\ell{}^R$ appears as a factor in $w[p_\ell + k + 1..j']$. Finally, $w[1..j']$ can be obtained from $w[p_1 + 1..j]$ by k-hairpin lengthening if and only if $\overline{x}_1{}^R$ appears as a factor in $w[p_1 + k + 1..j']$.

Analogously, let $S_2 = (y_1, y_2, \ldots, y_s)$ be a list of all the different words of length $k + 1$ that occur as factors of the word $w[j - k + 1..n]$, ordered increasingly according to the leftmost position where they appear in $w[j - k + 1..n]$ (i.e., the leftmost occurrence of y_ℓ in $w[j - k + 1..n]$ is to the left of the leftmost position where $y_{\ell+1}$ occurs in the same factor). Let q_ℓ be the minimum number such that $q_\ell \geq j + 1$ and $w[q_\ell - k..q_\ell] = y_\ell$, i.e., q_ℓ is the ending position of the leftmost occurrence of y_ℓ in $w[j - k + 1..n]$; clearly, $q_1 = j + 1$. We also define the list of numbers $S_2' = (m_1, \ldots, m_s)$ by $m_s = \min(k, n - q_s)$ and $m_\ell = \min(k, q_{\ell+1} - q_\ell + 1)$ for $1 \leq \ell \leq s - 1$.

It is immediate that all the factors $w[i' - k..i']$ with $j + 1 \leq i' \leq q_{\ell+1} - 1$ are from the set $\{y_1, y_2, \ldots, y_\ell\}$. Thus, if $i' \leq i$ and $w[i'..q_\ell]$ can be obtained by iterated k-hairpin lengthening from $w[i..j]$, then $w[i'..q_{\ell+1} - 1]$ can be also obtained by iterated k-hairpin lengthening from $w[i..j]$. Moreover, $w[i'..q_\ell]$ can be obtained from $w[i'..q_\ell - 1]$ by k-hairpin lengthening if and only if $\overline{y}_\ell{}^R$ appears as a factor in $w[i'..q_\ell - k - 1]$. Also, $w[i'..n]$ can be obtained from $w[i'..q_s - 1]$ by k-hairpin lengthening if and only if $\overline{y}_s{}^R$ appears as a factor in $w[i'..q_s - k - 1]$.

Further, assume that S is a list of the distinct factors of length $k + 1$ of $w[i..j - k]$ and S' a list of the distinct factors of length $k + 1$ of $w[i + k..j]$; also, let T and T' be two lists containing initially k times the word $\perp^{k+1}$, where $\perp \notin V$. It follows that $w[p_{t-1} + 1..j]$ can be obtained from $w[i..j]$ if and only if $\overline{x}_t^R$ appears in S'. Similarly, $w[i..q_2 - 1]$ can be obtained from $w[i..j]$ if and only if $\overline{y}_1^R$ appears in S. Now we are in the position of extending these words even more, in order to see if the whole word w can be generated. We update the lists $S_1, S_1', S_2, S_2', S, S'$ as well as T and T' as follows:

- If $w[p_{t-1} + 1..j]$ can be obtained from $w[i..j]$, we delete from S_1 the element x_t and from S_1' the element n_t. If $n_t = k$ we add x_t to S' and set T' to be the list with k elements $\perp^{k+1}$; otherwise, if $n_t = \ell < k$, we update the list T' as follows: we delete the last $\ell + 1$ elements of T and put in S' those that do not contain $\bot$, and add at the beginning of T' the word x_t and, then, ℓ times the word $\bot^{k+1}$.

- If $w[i..q_1 - 1]$ can be obtained from $w[i..j]$, we delete from S_2 the element y_1 and from S_2' the element m_1. If $m_1 = k$ we add y_1 to S' and set T to be the list with k elements $\bot^{k+1}$; otherwise, if $m_1 = \ell < k$, we update the list T as follows: we delete the last $\ell + 1$ elements of T and put in S those that do not contain $\bot$, and add at the beginning of T the word y_1 and, then, ℓ times the word $\bot^{k+1}$.

At this point we can repeat the process just described above and try to extend the newly obtained word ($w[p_{t-1} + 1..j]$ or $w[i..q_1 - 1]$) even more, to the left or to the right. We stop iterating this process either when both S_1 and S_2 are empty or when the current word cannot be extended (to the left or to the right) anymore. In the first case, when the two lists become empty, we decide that w can be obtained from $w[i..j]$ by iterated k-hairpin lengthening, while in the second case we decide that w can not be obtained from $w[i..j]$ by iterated k-hairpin lengthening. It is worth noting that our decision is taken by looking only at the lists S_1, S_1', S_2, S_2', S and S', and the total number of possible assignments for these lists depends only on k, not on w or on the values of i and j. Therefore, we call a 6-tuples of lists $(S_1, S_1', S_2, S_2', S, S')$ acceptable if the process described above ends with S_1 and S_2 empty; otherwise, they are called unacceptable. One clearly needs a constant amount of time (depending on k) to decide which 6-tuples of lists are acceptable, and which are unacceptable; this can be done by a preprocessing phase before designing an algorithm (or, equivalently, an automaton, as we will see in the end of this proof) recognizing $HL_k^*(L)$.

According to these remarks we present a non-deterministic algorithm that accepts the iterated k-hairpin lengthening of a regular language L over V. The main idea is quite simple: we choose non-deterministically a factor of the input word, and compute the sets S_1, S_1', S_2, S_2', S, and S' determined by this factor. Then, we just check if these sets form an acceptable 6-tuple or not, and if the chosen factor is in L; if both checks return positive answers, we accept the input word, otherwise we reject it.

This algorithm is made of several different phases, which we present separately. Assume that L is specified by a deterministic automaton accepting it $M = (Q, V, q_o, F, \delta)$. Also, for the first seven phases of the algorithm we assume that the input word w has at least length $2k + 2$, otherwise it belongs to $HL_k^*(L)$ if and only if it lies in L. For simplicity, if S is an ordered list with m elements $(x_1, x_2, \ldots, x_m)$, we denote by $S[\ell]$ the ℓ-th element of that list, namely x_ℓ. By deleting an element $S[\ell]$ from a list as the above we obtain a new list with $m - 1$ elements, $(x_1, x_2, \ldots, x_{\ell-1}, x_{\ell+1}, \ldots, x_m)$. By adding a new element x to the end of the list S above we obtain a new list with $m + 1$ elements $(x_1, x_2, \ldots, x_m, x)$.

In the first phase we initialize the lists we use in the rest of the computation.

In the second phase of the algorithm we just read the first k symbols of the input word, and store them in the word $last_k$.

The third phase of the algorithm is used to compute the lists S_1 and S_1'. We also decide non-deterministically when we have read the first k symbols of a

---

Algorithm 3.1. Phase 1: Initialization

Initialize the ordered lists $S_1, S_2, S'_1, S'_2, S, S'$ as the void lists;
Initialize the words $last_k = \lambda$ and $last_{k+1} = \lambda$;
Set the variables $phase = 2$, $i = 1$, $f = 1$, and $count = 0$; set the state $q = q_0$;

Algorithm 3.2. Phase 2

---

while $phase = 2$ **do**
    Add the symbol $w[i]$ at the end of $last_k$;
    Increase i by 1;
    if $i = k + 1$ **then**
        Set $phase = 3$;
    end if
end while

---

factor of w, that is tested if it is in L, and used to test whether w is in $HL_k^*(L)$ or not.

At the beginning of the fourth phase of the algorithm we have already analyzed the first k symbols of the chosen factor. In this phase we start computing the set S. We also count the number of symbols read after the first k symbols of the factor, in order to detect the precise moment when we should start adding words to S' as well. As soon as we reach this moment or when we decide non-deterministically that we reached the last k symbols of the factor, we move on to one of the following phases. In the fifth phase we finish computing the set

---

Algorithm 3.3. Phase 3

---

while $phase = 3$ **do**
    Add the symbol $w[i]$ at the end of $last_k$, set $last_{k+1} = last_k$, and delete the first symbol of $last_k$; Let t be the number of elements in S_1;
    if $count < k$ **then**
        Increase $count$ by 1;
    else
        $count = k$;
    end if
    if there exists $\ell < t$ such that $S_1[\ell] = last_{k+1}$ **then**
        Delete $S_1[\ell]$ from S_1, set $S'_1[\ell + 1] = \min(k, S'_1[\ell + 1] + S'_1[\ell])$, delete $S'_1[\ell]$ from S'_1, set $count = 1$;
    else
        if $S_1[t] = last_{k+1}$ **then**
            Delete $S_1[t]$ from S_1, delete $S'_1[t]$ from S'_1;
        end if
    end if
    Insert $last_{k+1}$ at the end of S_1, insert $count$ at the end of S'_1;
    Increase i by 1.
    Choose non-deterministically between $phase = 3$ or setting $phase = 4$;
end while

Algorithm 3.4. Phase 4

1: Set $t = 1$; let $q = \delta(q_0, last_k)$;
2: **while** $phase = 4$ **do**
3:     Set $q = \delta(q, w[i])$;
4:     Add the symbol $w[i]$ at the end of $last_k$, set $last_{k+1} = last_k$, and delete the first symbol of $last_k$;
5:     Insert $last_{k+1}$ in S, if it was not in this list already; Increase t by 1;
6:     Increase i by 1;
7:     **if** $t = k + 1$ **then**
8:         Set $phase = 5$;
9:     **else**
10:         Choose non-deterministically between $phase = 4$ or setting $phase = 6$;
11:     **end if**
12: **end while**

S' and start computing the set S. In this phase, as in the previous one, we can also decide non-deterministically that we have reached the last k symbols of the chosen factor, case in which we move to Phase 6.

Algorithm 3.5. Phase 5

1: **while** $phase = 5$ **do**
2:     Set $q = \delta(q, w[i])$;
3:     Add the symbol $w[i]$ at the end of $last_k$, set $last_{k+1} = last_k$, and delete the first symbol of $last_k$;
4:     Insert $last_{k+1}$ in S, if it is not already in this list; Insert $last_{k+1}$ in S', if it was not in this list already;
5:     Increase i by 1;
6:     Choose non-deterministically between $phase = 5$ or setting $phase = 6$;
7: **end while**

In the sixth phase we finish reading the chosen factor as well as we finish computing S. After this phase q is in F if and only if the non-deterministically chosen factor is in L.

In the seventh phase we compute the lists S_2 and S'_2. The computation described in this phase stops at the moment we have reached the end of the input word. Then we move along to phase 8, where the decision is made.

Finally, once we have computed all the needed lists, we use the eighth phase to decide whether the input word was obtained by hairpin lengthening from the non-deterministically chosen factor. The decision taken in this phase relies on the preprocessing that we have already mentioned: we assume that we know which 6-tuples of lists are acceptable and which are not.

By the remarks made prior to the detailed description of the eight phases it is clear that our algorithm decides exactly $HL_k^*(L)$.

Algorithm 3.6. Phase 6

1: Set $t = 1$;
2: **while** $phase = 6$ **do**
3:     Set $q = \delta(q, w[i])$;
4:     Add the symbol $w[i]$ at the end of $last_k$, set $last_{k+1} = last_k$, and delete the first symbol of $last_k$;
5:     Insert $last_{k+1}$ in S', if it was not in this list already;
6:     Increase i by 1;
7:     Increase t by 1;
8:     **if** $t = k + 1$ **then**
9:         Set $phase = 7$;
10:     **end if**
11: **end while**

Algorithm 3.7. Phase 7

1: Set $count = 0$.
2: **while** $phase = 7$ **do**
3:     Add the symbol $w[i]$ at the end of $last_k$, set $last_{k+1} = last_k$, and delete the first symbol of $last_k$;
4:     **if** $last_{k+1}$ appears in S_2 **then**
5:         **if** $count < k$ **then**
6:             Increase $count$ by 1;
7:         **else**
8:             $count = k$;
9:         **end if**
10:     **else**
11:         Insert $last_{k+1}$ at the end of S_2;
12:         Add $count$ at the end of S'_2 and reset $count = 1$ when S_2 contains at least two elements;
13:     **end if**
14:     Increase i by 1.
15:     **if** $i = n + 1$ **then**
16:         Add $count$ at the end of S'_2; set $phase = 8$;
17:     **end if**
18: **end while**

Algorithm 3.8. Phase 8: Decision

1: **if** $|w| < 2k + 2$ and $w \in L$ **then**
2:     ACCEPT; HALT;
3: **else**
4:     **if** $|w| < 2k + 2$ **then**
5:         REJECT; HALT;
6:     **end if**
7: **end if**
8: **if** $|w| \geq 2k + 2$ and $(S_1, S'_1, S_2, S'_2, S, S')$ is acceptable and $q \in F$ **then**
9:     ACCEPT; HALT;
10: **else**
11:     REJECT;
12: **end if**

Further, one can easily see that this approach can be implemented on a finite automaton. The states of this automaton store the six lists we use, and all the other variables used in the algorithm, except for i (clearly, only a constant memory is needed to do this). The transitions are defined according to the processing described in the phases above. Clearly, the variable i should not be memorized, since this variable is used only to read the symbols of the input word one by one, and this can be done by a finite automaton. Thus, if L is regular, then $HL_k^*(L)$ is also regular.

In conclusion, the class of regular languages is closed under iterated k-hairpin lengthening.   □

4   Hairpin Shortening of Regular Languages

We get the following result:

Theorem 3. *The class of regular languages is closed under k-hairpin shortening for any $k \geq 1$.*

Proof. Let us take a regular language L, and denote by $A = (Q, V, q_0, F, \delta)$ the deterministic finite automaton accepting L. We assume that there exists no state $q \in Q$ such that $q_0 = \delta(q, a)$, for some letter $a \in V$. For a fixed integer $k \geq 1$, we show that HSS_k is regular. We define the non-deterministic automaton $A' = (Q', V, q_0, F', \delta')$, as follows:

$$Q' = Q \cup Q \times Q \cup Q \times Q \times \{x \cup [x] \cup (x) | x \in V^i, \text{ where } 0 \leq i \leq k\},$$
$$F = \{(q, q, (\lambda)) | q \in Q\},$$

$$\delta'(q_1, a) = \delta(q_1, a) \cup \{(\delta(q_1, a), q_2) | f = \delta(q_2, \overline{a}) \text{ and } f \in F\},$$
$$\delta'((q_1, q_2), a) = \{(\delta(q_1, a), q_2') | q_2 = \delta(q_2', \overline{a})\} \cup$$
$$\{(\delta(q_1, a), q_2', \lambda) | q_2 = \delta(q_2', \overline{a})\},$$
$$\delta'((q_1, q_2, x), a) = \{(\delta(q_1, a), q_2, ax) | |x| < k\} \cup \{(\delta(q_1, a), q_2, [x]) | |x| = k\},$$
$$\delta'((q_1, q_2, [x]), a) = \{(\delta(q_1, a), q_2, [x])\} \cup \{(\delta(q_1, a), q_2, (x))\},$$
$$\delta'((q_1, q_2, (xa)), \overline{a}) = \{(q_1, q_2', (x)) | q_2 = \delta(q_2', \overline{a})\}.$$

Our automaton accepts all words $\gamma\delta\alpha\beta\overline{\alpha}$ where $\gamma\delta\alpha\beta\overline{\alpha}\overline{\gamma} \in L$. In order to see how it works, please note that at the beginning the automaton just reads the prefix of γ, possibly empty that precedes δ. After that, with a non-deterministic choice we ensure that reading δ, we would have a word that has $\overline{\delta^R}$ as a suffix in L. Next we again non-deterministically start to read α, and remember the letters for a later comparison with the end of the word. Please note that since the number of factors of length k is finite, the automaton remains finite. Finally, we non-deterministically start to read β and guess when β ends; then we compare the rest of the word with the complement image of the reverse of α, and accept in the case we end up in the same state.

It is easy to see that A' accepts exactly $HSS_k(L)$. Due to the closure of regular languages on the reverse operation, the result follows.   □

A direct consequence is:

Corollary 1. *The class of regular languages is closed under finitely iterated k-hairpin shortening for any $k \geq 1$.*

Let us finally look at the iterated case of hairpin shortening of a regular language. Clearly, given a language $L \subseteq V^*$ and a word $w \in V^*$, if $w \in HS_k^*(L)$, then $HL_k^*(w) \cap L \neq \emptyset$ must hold.

Theorem 4. *The class of regular languages is closed under iterated k-hairpin shortening for all $k \geq 1$.*

Proof. The proof is to some extend similar to that of Theorem 2. Let $A = (Q, V, q_0, F, \delta)$ be a deterministic finite automaton with the transition function δ totally defined. We now consider the following nondeterministic algorithm consisting of three phases: *initialization, update, decision.*

Algorithm 4.1. Phase 1: Initialization

1: Choose a pair of states $(q, r) \in Q \times Q$ such that $\delta(q, w) = r$;
2: $P = w[1..k]$; $P_{2k+1} = w[1..2k+1]$;
3: $S = w[|w| - k + 1..|w|]$; $S_{2k+1} = w[|w| - 2k..|w|]$;
4: $L = \{x \in V^* \mid w = uxv, |x| = k + 1, u, v \in V^*, |v| \geq k + 1\}$;
5: $R = \{x \in V^* \mid w = uxv, |x| = k + 1, u, v \in V^*, |u| \geq k + 1\}$;
6: Set *phase* := 2;

The input of this algorithm is the automaton A, an integer $k \geq 1$, and a word $w \in V^*$ with $|w| \geq 2k + 2$. Clearly, any word shorter than $2k + 2$ is in $HS_k^*(L)$ if and only if it belongs to L. The first phase can be accomplished by scanning once from left to right the input word. The second phase is devoted to the update of all variables initialized in the first phase.

Phases 2 and 3 are executed alternatively until a decision is made in Phase 3.

It is not hard to see that the algorithm decides any input, as in the first phase we make a number of steps proportional to the length of the input, while in the second and third phase we make only a finite number of steps, depending only on k. Note that in these latter phases the algorithm does not read any input symbol, so, basically, the only part of the algorithm where we need to have access to the input is Phase 1. The proof is complete as soon as we note that this algorithm can be implemented on a finite automaton.   □

Algorithm 4.2. Phase 2: Update

1: Choose only one of the two situations:
2: **if** $\overline{P^R a} \in R$ for some $a \in V$ **then**
3:    $R = R \cup \{P_{2k+1}[k+1..2k+1]\}$
4:    $L = L \cup \{aP\}$;
5:    $P = aP[1..k-1]$; $P_{2k+1} = aP_{2k}[1..2k]$;
6:    $q = q'$ such that $\delta(q', a) = q$;
7: **end if**
8: **if** $\overline{aS^R} \in L$ for some $a \in V$ **then**
9:    $L = L \cup \{S_{2k+1}[1..k+1]\}$
10:    $R = R \cup \{Sa\}$;
11:    $S = S[2..k]a$; $S_{2k+1} = S_{2k+1}[2..2k+1]a$;
12:    $r = \delta(r, a)$;
13: **end if**
14: *phase* := 3;

Algorithm 4.3. Phase 3: Decision

1: **if** $q = q_0$ and $r \in F$ **then**
2:    ACCEPT;
3: **else**
4:    **if** (q, r, L, R, P, S) has been considered already in Phase 2 **then**
5:       REJECT;
6:    **else**
7:       *phase* := 2;
8:    **end if**
9: **end if**

Acknowledgments. We would like to express our gratitude to the *Alexander von Humboldt Foundation* who has given us the great opportunity of having Jürgen Dassow for a while walking beside us and helping us along our journey of life. We thank Professor Jürgen Dassow for his tremendous and invaluable assistance, support, and guidance.

The authors would also like to thank to the anonymous referees for their comments and suggestions that led to a better presentation of this paper.

Florin Manea's work is currently supported by the *DFG* grant 582014. The work of Robert Mercas is currently supported by the *Alexander von Humboldt Foundation*.

References

1. Bottoni, P., Labella, A., Manca, V., Mitrana, V.: Superposition based on watson-crick-like complementarity. Theory of Computing Systems 39(4), 503–524 (2006)
2. Castellanos, J., Mitrana, V.: Some remarks on hairpin and loop languages. In: Words, Semigroups, and Transductions 2001, pp. 47–58 (2001)
3. Cheptea, D., Martín-Vide, C., Mitrana, V.: A new operation on words suggested by dna biochemistry: Hairpin completion. Transgressive Computing, 216–228 (2006)

4. Deaton, R., Murphy, R., Garzon, M., Franceschetti, D., Stevens Jr., S.: Good encodings for DNA-based solutions to combinatorial problems. In: Proceedings of the Second Annual Meeting on DNA Based Computer. DIMACS, vol. 44, pp. 247–259 (1996)
5. Diekert, V., Kopecki, S.: Complexity Results and the Growths of Hairpin Completions of Regular Languages (Extended Abstract). In: Domaratzki, M., Salomaa, K. (eds.) CIAA 2010. LNCS, vol. 6482, pp. 105–114. Springer, Heidelberg (2011)
6. Diekert, V., Kopecki, S.: Language theoretical properties of hairpin formations. Theoretical Computer Science 429, 65–73 (2012)
7. Diekert, V., Kopecki, S., Mitrana, V.: On the Hairpin Completion of Regular Languages. In: Leucker, M., Morgan, C. (eds.) ICTAC 2009. LNCS, vol. 5684, pp. 170–184. Springer, Heidelberg (2009)
8. Garzon, M., Deaton, R., Nino, L., Stevens, E., Wittner, M.: Encoding genomes for DNA computing. In: Koza, J., Banzhaf, W., Chellapilla, K., Deb, K., Dorigo, M., Fogel, D., Garzon, M., Goldberg, D., Iba, H., Riolo, R. (eds.) Genetic Programming 1998: Proceedings of the Third Annual Conference, pp. 684–690. Morgan Kaufmann, Madison (1998)
9. Ito, M., Leupold, P., Manea, F., Mitrana, V.: Bounded hairpin completion. Information and Computation 209(3), 471–485 (2011)
10. Kari, L., Konstantinidis, S., Sosík, P., Thierrin, G.: On Hairpin-Free Words and Languages. In: De Felice, C., Restivo, A. (eds.) DLT 2005. LNCS, vol. 3572, pp. 296–307. Springer, Heidelberg (2005)
11. Kari, L., Losseva, E., Konstantinidis, S., Sosík, P., Thierrin, G.: A formal language analysis of DNA hairpin structures. Fundam. Inform. 71(4), 453–475 (2006)
12. Kopecki, S.: On iterated hairpin completion. Theoretical Computer Science 412(29), 3629–3638 (2011)
13. Manea, F.: A series of algorithmic results related to the iterated hairpin completion. Theoretical Computer Science 411(48), 4162–4178 (2010)
14. Manea, F., Martín-Vide, C., Mitrana, V.: On some algorithmic problems regarding the hairpin completion. Discrete Applied Mathematics 157(9), 2143–2152 (2009)
15. Manea, F., Martín-Vide, C., Mitrana, V.: Hairpin Lengthening. In: Ferreira, F., Löwe, B., Mayordomo, E., Mendes Gomes, L. (eds.) CiE 2010. LNCS, vol. 6158, pp. 296–306. Springer, Heidelberg (2010)
16. Manea, F., Mitrana, V.: Hairpin Completion Versus Hairpin Reduction. In: Cooper, S.B., Löwe, B., Sorbi, A. (eds.) CiE 2007. LNCS, vol. 4497, pp. 532–541. Springer, Heidelberg (2007)
17. Manea, F., Mitrana, V., Yokomori, T.: Two complementary operations inspired by the DNA hairpin formation: Completion and reduction. Theoretical Computer Science 410(4-5), 417–425 (2009)
18. Manea, F., Martín-Vide, C., Mitrana, V.: Hairpin lengthening: Language theoretic and algorithmic results. Journal of Logic and Compuation (to appear, 2012)
19. Păun, G., Rozenberg, G., Yokomori, T.: Hairpin languages. International Journal of Foundations of Computer Science 12, 837–847 (2001)
20. Rozenberg, G., Salomaa, A. (eds.): Handbook of Formal Languages. Springer, Heidelberg (1997)
21. Sakamoto, K., Gouzu, H., Komiya, K., Kiga, D., Yokoyama, S., Yokomori, T., Hagiya, M.: Molecular computation by dna hairpin formation. Science 288(5469), 1223–1226 (2000)
22. Williams, J.G.K., Kubelik, A.R., Livak, K.J., Rafalski, J.A., Tingey, S.V.: DNA polymorphisms amplified by arbitrary primers are useful as genetic markers. Nucleic Acids Research 18(22), 6531–6535 (1990)

One-Sided Random Context Grammars
with Leftmost Derivations

Alexander Meduna and Petr Zemek

Brno University of Technology, Faculty of Information Technology,
IT4Innovations Centre of Excellence,
Božetěchova 1/2, 612 66 Brno, Czech Republic
{meduna,izemek}@fit.vutbr.cz

Abstract. In this paper, we study the generative power of one-sided random context grammars working in a leftmost way. More specifically, by analogy with the three well-known types of leftmost derivations in regulated grammars, we introduce three types of leftmost derivations to one-sided random context grammars and prove the following three results. (I) One-sided random context grammars with type-1 leftmost derivations characterize the family of context-free languages. (II) One-sided random context grammars with type-2 and type-3 leftmost derivations characterize the family of recursively enumerable languages. (III) Propagating one-sided random context grammars with type-2 and type-3 leftmost derivations characterize the family of context-sensitive languages. In the conclusion, the generative power of random context grammars and one-sided random context grammars with leftmost derivations is compared.

Keywords: formal languages, regulated rewriting, one-sided random context grammars, leftmost derivations, generative power.

1   Introduction

The investigation of grammars that perform leftmost derivations is central to formal language theory as a whole. Indeed, from a practical viewpoint, leftmost derivations fulfill a crucial role in parsing, which represents a key application area of formal grammars (see [1,2,7,21]). From a theoretical viewpoint, an effect of leftmost derivation restrictions to the power of grammars restricted in this way represents an intensively investigated area of this theory as clearly indicated by many studies on the subject. More specifically, [3,4,17,18,32] contain fundamental results concerning leftmost derivations in classical Chomsky grammars, [6,14,19,30,33] and Section 5.3 in [9] give an overview of the results concerning leftmost derivations in regulated grammars published until late 1980's, and [8,10,11,20,23,25] together with Section 7.3 in [24] present several follow-up results. In addition, [15,16,31] cover language-defining devices introduced with some kind of leftmost derivations, and [5] discusses the recognition complexity of derivation languages of various regulated grammars with leftmost derivations. Finally, [16,22,28] study grammar systems working under the leftmost derivation restriction, and [12,13,29] investigates leftmost derivations in terms of P systems.

H. Bordihn, M. Kutrib, and B. Truthe (Eds.): Dassow Festschrift 2012, LNCS 7300, pp. 160–173, 2012.
© Springer-Verlag Berlin Heidelberg 2012

The present paper approaches this topic in terms of one-sided random context grammars. Recall that a *one-sided random context grammar* (see [26,27]) represents a variant of a random context grammar (see [9] and Chapter 3 in the second volume of [32]). In this variant, a set of *permitting symbols* and a set of *forbidding symbols* are attached to every rule, and its set of rules is divided into the set of *left random context rules* and the set of *right random context rules*. A left random context rule can rewrite a nonterminal if each of its permitting symbols occurs to the left of the rewritten symbol in the current sentential form while each of its forbidding symbols does not occur there. A right random context rule is applied analogically except that the symbols are examined to the right of the rewritten symbol.

Specifically, this paper introduces three types of leftmost derivation restrictions placed upon one-sided random context grammars. In the *type-1 derivation restriction*, during every derivation step, the leftmost occurrence of a nonterminal has to be rewritten. In the *type-2 derivation restriction*, during every derivation step, the leftmost occurrence of a nonterminal which can be rewritten has to be rewritten. In the *type-3 derivation restriction*, during every derivation step, a rule is chosen, and the leftmost occurrence of its left-hand side is rewritten.

The paper demonstrates the following three results. (I) One-sided random context grammars with type-1 leftmost derivations characterize the family of context-free languages. (II) One-sided random context grammars with type-2 and type-3 leftmost derivations characterize the family of recursively enumerable languages. (III) Propagating one-sided random context grammars with type-2 and type-3 leftmost derivations characterize the family of context-sensitive languages.

The paper is organized as follows. First, Section 2 gives all the necessary terminology. Then, Section 3 rigorously establishes the results mentioned above. In the conclusion, Section 4 compares the generative power of random context grammars and that of one-sided random context grammars with leftmost derivations.

2   Preliminaries and Definitions

We assume that the reader is familiar with formal language theory (see [32]). For a set Q, 2^Q denotes the power set of Q. For an alphabet (finite nonempty set) V, V^* represents the free monoid generated by V under the operation of concatenation. The unit of V^* is denoted by ε. Set $V^+ = V^* - \{\varepsilon\}$; algebraically, V^+ is thus the free semigroup generated by V under the operation of concatenation. For $x \in V^*$, $|x|$ denotes the length of x, and $\mathrm{alph}(x)$ denotes the set of symbols occurring in x.

A *context-free grammar* is a quadruple, $G = (N, T, P, S)$, where N and T are two disjoint alphabets, $S \in N$, and $P \subseteq N \times (N \cup T)^*$ is a finite relation. Set $V = N \cup T$. The components V, N, T, P, and S are called the *total alphabet*, the alphabet of *nonterminals*, the alphabet of *terminals*, the set of *rules*, and the *start symbol*, respectively. Each $(A, x) \in P$ is written as $A \to x$ throughout this paper. If $A \rightarrow x \in P$ implies that $|x| \geq 1$, then G is *propagating*. The *direct derivation relation* over V^*, symbolically denoted by $\Rightarrow$, is defined as follows: $uAv \Rightarrow uxv$ in G if and only if $u, v \in V^*$ and $A \rightarrow x \in P$. Let $\Rightarrow^n$ and $\Rightarrow^*$ denote the nth power of $\Rightarrow$, for some $n \geq 0$, and the reflexive-transitive closure of $\Rightarrow$, respectively. The *language of G* is denoted by $L(G)$ and defined as $L(G) = \{w \in T^* \mid S \Rightarrow^* w\}$.

A *one-sided random context grammar* (see [27]) is a quintuple, $G = (N, T, P_L, P_R, S)$, where N and T are two disjoint alphabets, $S \in N$, and $P_L, P_R \subseteq N \times (N \cup T)^* \times 2^N \times 2^N$ are two finite relations. Set $V = N \cup T$. The components V, N, T, P_L, P_R and S are called the *total alphabet*, the alphabet of *nonterminals*, the alphabet of *terminals*, the set of *left random context rules*, the set of *right random context rules*, and the *start symbol*, respectively. Each $(A, x, U, W) \in P_L \cup P_R$ is written as $\lfloor A \rightarrow x, U, W \rfloor$ throughout this paper. For $\lfloor A \rightarrow x, U, W \rfloor \in P_L$, U and W are called the *left permitting context* and the *left forbidding context*, respectively. For $\lfloor A \rightarrow x, U, W \rfloor \in P_R$, U and W are called the *right permitting context* and the *right forbidding context*, respectively. If $\lfloor A \rightarrow x, U, W \rfloor \in P_L \cup P_R$ implies that $|x| \geq 1$, then G is *propagating*. The *direct derivation relation* over V^*, symbolically denoted by $\Rightarrow$, is defined as follows. Let $u, v \in V^*$ and $\lfloor A \rightarrow x, U, W \rfloor \in P_L \cup P_R$. Then, $uAv \Rightarrow uxv$ in G if and only if

$$\lfloor A \rightarrow x, U, W \rfloor \in P_L, U \subseteq \mathrm{alph}(u) \text{ and } W \cap \mathrm{alph}(u) = \emptyset$$

or

$$\lfloor A \rightarrow x, U, W \rfloor \in P_R, U \subseteq \mathrm{alph}(v) \text{ and } W \cap \mathrm{alph}(v) = \emptyset$$

Let $\Rightarrow^n$ and $\Rightarrow^*$ denote the nth power of $\Rightarrow$, for some $n \geq 0$, and the reflexive-transitive closure of $\Rightarrow$, respectively. The *language of G* is denoted by $L(G)$ and defined as $L(G) = \{w \in T^* \mid S \Rightarrow^* w\}$.

2.1   Leftmost Derivations

By analogy with the discussion of leftmost derivations in [9], we next place three types of leftmost derivation restrictions on one-sided random context grammars.

In the first derivation restriction type, during every derivation step, the leftmost occurrence of a nonterminal has to be rewritten. This type of leftmost derivations corresponds to the well-known leftmost derivations in context-free grammars.

Definition 1. Let $G = (N, T, P_L, P_R, S)$ be a one-sided random context grammar. The *type-1 direct leftmost derivation relation* over V^*, symbolically denoted by $_{\mathrm{lm}}\overset{1}{\Rightarrow}$, is defined as follows. Let $u \in T^*$, $A \in N$ and $x, v \in V^*$. Then, $uAv \; _{\mathrm{lm}}\overset{1}{\Rightarrow} uxv$ in G if and only if $uAv \Rightarrow uxv$ in G.

Let $_{\mathrm{lm}}\overset{1}{\Rightarrow}{}^n$ and $_{\mathrm{lm}}\overset{1}{\Rightarrow}{}^*$ denote the nth power of $_{\mathrm{lm}}\overset{1}{\Rightarrow}$, for some $n \geq 0$, and the reflexive-transitive closure of $_{\mathrm{lm}}\overset{1}{\Rightarrow}$, respectively. The $_{\mathrm{lm}}\overset{1}{\Rightarrow}$-*language of G* is denoted by $L(G, {}_{\mathrm{lm}}\overset{1}{\Rightarrow})$ and defined as $L(G, {}_{\mathrm{lm}}\overset{1}{\Rightarrow}) = \{w \in T^* \mid S \; _{\mathrm{lm}}\overset{1}{\Rightarrow}{}^* w\}$.   □

Notice that if the leftmost occurrence of a nonterminal cannot be rewritten by any rule, then the derivation is blocked.

In the second derivation restriction type, during every derivation step, the leftmost occurrence of a nonterminal that can be rewritten has to be rewritten.

Definition 2. Let $G = (N, T, P_L, P_R, S)$ be a one-sided random context grammar. The *type-2 direct leftmost derivation relation* over V^*, symbolically denoted by $_{\text{lm}}{\overset{2}{\Rightarrow}}$, is defined as follows. Let $u, x, v \in V^*$ and $A \in N$. Then, $uAv \,_{\text{lm}}{\overset{2}{\Rightarrow}} uxv$ in G if and only if $uAv \Rightarrow uxv$ in G and there is no $B \in N$ and $y \in V^*$ such that $u = u_1 B u_2$ and $u_1 B u_2 Av \Rightarrow u_1 y u_2 Av$ in G.

Let $_{\text{lm}}{\overset{2}{\Rightarrow}}{}^n$ and $_{\text{lm}}{\overset{2}{\Rightarrow}}{}^*$ denote the nth power of $_{\text{lm}}{\overset{2}{\Rightarrow}}$, for some $n \geq 0$, and the reflexive-transitive closure of $_{\text{lm}}{\overset{2}{\Rightarrow}}$, respectively. The $_{\text{lm}}{\overset{2}{\Rightarrow}}$-*language of G* is denoted by $L(G, _{\text{lm}}{\overset{2}{\Rightarrow}})$ and defined as $L(G, _{\text{lm}}{\overset{2}{\Rightarrow}}) = \{w \in T^* \mid S \,_{\text{lm}}{\overset{2}{\Rightarrow}}{}^* w\}$. □

In the third derivation restriction type, during every derivation step, a rule is chosen, and the leftmost occurrence of its left-hand side is rewritten.

Definition 3. Let $G = (N, T, P_L, P_R, S)$ be a one-sided random context grammar. The *type-3 direct leftmost derivation relation* over V^*, symbolically denoted by $_{\text{lm}}{\overset{3}{\Rightarrow}}$, is defined as follows. Let $u, x, v \in V^*$ and $A \in N$. Then, $uAv \,_{\text{lm}}{\overset{3}{\Rightarrow}} uxv$ in G if and only if $uAv \Rightarrow uxv$ in G and $\text{alph}(u) \cap \{A\} = \emptyset$.

Let $_{\text{lm}}{\overset{3}{\Rightarrow}}{}^n$ and $_{\text{lm}}{\overset{3}{\Rightarrow}}{}^*$ denote the nth power of $_{\text{lm}}{\overset{3}{\Rightarrow}}$, for some $n \geq 0$, and the reflexive-transitive closure of $_{\text{lm}}{\overset{3}{\Rightarrow}}$, respectively. The $_{\text{lm}}{\overset{3}{\Rightarrow}}$-*language of G* is denoted by $L(G, _{\text{lm}}{\overset{3}{\Rightarrow}})$ and defined as $L(G, _{\text{lm}}{\overset{3}{\Rightarrow}}) = \{w \in T^* \mid S \,_{\text{lm}}{\overset{3}{\Rightarrow}}{}^* w\}$. □

Notice the following difference between the second and the third type. In the former, a leftmost occurrence of a rewritable nonterminal is chosen first, and then, a choice of a rule with this nonterminal on its let-hand side is made. In the latter, a rule is chosen first, and then, the leftmost occurrence of its left-hand side is rewritten.

2.2 Denotation of Language Families

Throughout the rest of this paper, the language families under discussion are denoted in the following way. The families of context-free languages, context-sensitive languages, and recursively enumerable languages are denoted by $\mathscr{L}_{\text{CF}}^\varepsilon$, $\mathscr{L}_{\text{CS}}$, and $\mathscr{L}_{\text{RE}}^\varepsilon$, respectively.

The language family generated by one-sided random context grammars is denoted by $\mathscr{L}_{\text{ORC}}^\varepsilon$. The language families generated by one-sided random context grammars with type-1 leftmost derivations, one-sided random context grammars with type-2 leftmost derivations, and one-sided random context grammars with type-3 leftmost derivations are denoted by $\mathscr{L}_{\text{ORC}}^\varepsilon(_{\text{lm}}{\overset{1}{\Rightarrow}})$, $\mathscr{L}_{\text{ORC}}^\varepsilon(_{\text{lm}}{\overset{2}{\Rightarrow}})$, and $\mathscr{L}_{\text{ORC}}^\varepsilon(_{\text{lm}}{\overset{3}{\Rightarrow}})$, respectively.

The notation without ε stands for the corresponding propagating family. For example, $\mathscr{L}_{\text{ORC}}$ denotes the language family generated by propagating one-sided random context grammars.

3 Results

In this section, we prove results I through III, given next.

I. One-sided random context grammars with type-1 leftmost derivations characterize $\mathscr{L}_{CF}^{\varepsilon}$ (Theorem 1). An analogical result holds for propagating one-
sided random context grammars (Theorem 2).

II. One-sided random context grammars with type-2 leftmost derivations characterize $\mathscr{L}_{RE}^{\varepsilon}$ (Theorem 3). Propagating one-sided random context grammars
with type-2 leftmost derivations characterize $\mathscr{L}_{CS}$ (Theorem 4).

III. One-sided random context grammars with type-3 leftmost derivations characterize $\mathscr{L}_{RE}^{\varepsilon}$ (Theorem 5). Propagating one-sided random context grammars
with type-3 leftmost derivations characterize $\mathscr{L}_{CS}$ (Theorem 6).

3.1   Type-1 Leftmost Derivations

First, we consider one-sided random context grammars with type-1 leftmost
derivations.

Lemma 1. *For every context-free grammar G, there is a one-sided random context grammar H such that $L(H, {}_{lm}^{1}{\Rightarrow}) = L(G)$. Furthermore, if G is propagating,
then so is H.*

Proof. Let $G = (N, T, P, S)$ be a context-free grammar. Construct the one-sided
random context grammar $H = (N, T, P', P', S)$, where

$$P' = \big\{ \lfloor A \to x, \emptyset, \emptyset \rfloor \mid A \to x \in P \big\}$$

As the rules in P' have their permitting and forbidding contexts empty, any
successful type-1 leftmost derivation in H is also a successful derivation in G,
so the inclusion $L(H, {}_{lm}^{1}{\Rightarrow}) \subseteq L(G)$ holds. On the other hand, let $w \in L(G)$ be
a string successfully generated by G. Then, it is well known that there exists a
successful leftmost derivation of w in G. Observe that such a leftmost derivation
is also possible in H. Thus, the other inclusion $L(G) \subseteq L(H, {}_{lm}^{1}{\Rightarrow})$ holds as well.
Finally, notice that whenever G is propagating, then so is H. Hence, the theorem
holds. $\square$

Lemma 2. *For every one-sided random context grammar G, there is a context-
free grammar H such that $L(H) = L(G, {}_{lm}^{1}{\Rightarrow})$. Furthermore, if G is propagating,
then so is H.*

Proof. Let $G = (N, T, P_L, P_R, S)$ be a one-sided random context grammar.
In what follows, symbols $\langle$ and $\rangle$ are used to clearly unite more symbols into a
single compound symbol. Construct the context-free grammar

$$H = \big(N', T, P, \langle S, \emptyset \rangle\big)$$

in the following way. Initially, set $N' = \{\langle A, Q \rangle \mid A \in N, Q \subseteq N\}$ and $P = \emptyset$
(without any loss of generality, we assume that $N' \cap V = \emptyset$). Perform (1) and (2),
given next:

(1) for each $\lfloor A \rightarrow y_0 Y_1 y_1 Y_2 y_2 \cdots Y_h y_h, U, W \rfloor \in P_R$, where $y_i \in T^*$, $Y_j \in N$, for all i and j, $0 \leq i \leq h$, $1 \leq j \leq h$, for some $h \geq 0$, and for each $\langle A, Q \rangle \in N'$ such that $U \subseteq Q$ and $W \cap Q = \emptyset$, add the following rule to P:

$$\langle A, Q \rangle \rightarrow y_0 \langle Y_1, Q \cup \{Y_2, Y_3, \ldots, Y_h\} \rangle y_1$$
$$\langle Y_2, Q \cup \{Y_3, \ldots, Y_h\} \rangle y_2$$
$$\vdots$$
$$\langle Y_h, Q \rangle y_h$$

(2) for each $\lfloor A \rightarrow y_0 Y_1 y_1 Y_2 y_2 \cdots Y_h y_h, \emptyset, W \rfloor \in P_L$, where $y_i \in T^*$, $Y_j \in N$, for all i and j, $0 \leq i \leq h$, $1 \leq j \leq h$, for some $h \geq 0$, and for each $\langle A, Q \rangle \in N'$, add the following rule to P:

$$\langle A, Q \rangle \rightarrow y_0 \langle Y_1, Q \cup \{Y_2, Y_3, \ldots, Y_h\} \rangle y_1$$
$$\langle Y_2, Q \cup \{Y_3, \ldots, Y_h\} \rangle y_2$$
$$\vdots$$
$$\langle Y_h, Q \rangle y_h$$

Before proving that $L(H) = L(G, {}_{\mathrm{lm}}^{1}\!\!\Rightarrow)$, let us give an insight into the construction. As G always rewrites the leftmost occurrence of a nonterminal, we use compound nonterminals of the form $\langle A, Q \rangle$ in H, where A is a nonterminal, and Q is a set of nonterminals that appear to the right of this occurrence of A. When simulating rules from P_R, the check for the presence and absence of symbols is accomplished by using Q. Also, when rewriting A in $\langle A, Q \rangle$ to some y, the compound nonterminals from N' are generated instead of nonterminals from N.

Rules from P_L are simulated analogously; however, notice that if the permitting set of such a rule is nonempty, it is never applicable in G. Therefore, such rules are not introduced to P'. Furthermore, since there are no nonterminals to the left of the leftmost occurrence of a nonterminal, no check for their absence is done.

Clearly, $L(G, {}_{\mathrm{lm}}^{1}\!\!\Rightarrow) \subseteq L(H)$. The opposite inclusion, $L(H) \subseteq L(G, {}_{\mathrm{lm}}^{1}\!\!\Rightarrow)$, can be proved by analogy with the proof of Lemma 1 by simulating the leftmost derivation of every $w \in L(H)$ by G. Observe that since the check for the presence and absence of symbols in H is done in the second components of the compound nonterminals, each rule introduced to P in (1) and (2) can be simulated by a rule from P_R and P_L from which it is created.

Finally, notice that whenever G is propagating, then so is H. Hence, the theorem holds. $\square$

Theorem 1. $\mathscr{L}_{\mathrm{ORC}}^{\varepsilon}({}_{\mathrm{lm}}^{1}\!\!\Rightarrow) = \mathscr{L}_{\mathrm{CF}}^{\varepsilon}$

Proof. By Lemma 1, $\mathscr{L}_{\mathrm{CF}}^{\varepsilon} \subseteq \mathscr{L}_{\mathrm{ORC}}^{\varepsilon}({}_{\mathrm{lm}}^{1}\!\!\Rightarrow)$. By Lemma 2, $\mathscr{L}_{\mathrm{ORC}}^{\varepsilon}({}_{\mathrm{lm}}^{1}\!\!\Rightarrow) \subseteq \mathscr{L}_{\mathrm{CF}}^{\varepsilon}$. Consequently, $\mathscr{L}_{\mathrm{ORC}}^{\varepsilon}({}_{\mathrm{lm}}^{1}\!\!\Rightarrow) = \mathscr{L}_{\mathrm{CF}}^{\varepsilon}$, so the theorem holds. $\square$

Theorem 2. $\mathscr{L}_{\mathrm{ORC}}({}_{\mathrm{lm}}^{1}\!\!\Rightarrow) = \mathscr{L}_{\mathrm{CF}}$

Proof. Since it is well-known that any context-free grammar that does not generate the empty string can be converted to an equivalent propagating context-free grammar, this theorem follows from Lemmas 1 and 2. $\square$

3.2   Type-2 Leftmost Derivations

Next, we turn our attention to one-sided random context grammars with type-2 leftmost derivations.

Lemma 3. *For every one-sided random context grammar G, there is a one-sided random context grammar H such that $L(H, {}_{\mathrm{lm}}^2\!\!\Rightarrow) = L(G)$. Furthermore, if G is propagating, then so is H.*

Proof. Let $G = (N, T, P_L, P_R, S)$ be a one-sided random context grammar. We construct the one-sided random context grammar H in such a way that always allows it to rewrite an arbitrary occurrence of a nonterminal. Construct

$$H = \big(N', T, P'_L, P'_R, S\big)$$

as follows. Initially, set $\bar{N} = \{\bar{A} \mid A \in N\}$, $\hat{N} = \{\hat{A} \mid A \in N\}$, $N' = N \cup \bar{N} \cup \hat{N}$, and $P'_L = P'_R = \emptyset$ (without any loss of generality, we assume that N, $\bar{N}$, and $\hat{N}$ are pairwise disjoint). Define the function ψ from 2^N to $2^{\bar{N}}$ as $\psi(\emptyset) = \emptyset$ and

$$\psi(\{A_1, A_2, \ldots, A_n\}) = \{\bar{A}_1, \bar{A}_2, \ldots, \bar{A}_n\}$$

Perform (1) through (3), given next:

(1) for each $A \in N$,
   (1.1) add $\lfloor A \to \bar{A}, \emptyset, N \cup \hat{N} \rfloor$ to P'_L,
   (1.2) add $\lfloor \bar{A} \to \hat{A}, \emptyset, N \cup \bar{N} \rfloor$ to P'_R,
   (1.3) add $\lfloor \hat{A} \to A, \emptyset, \bar{N} \cup \hat{N} \rfloor$ to P'_R;
(2) for each $\lfloor A \to y, U, W \rfloor \in P_R$, add $\lfloor A \to y, U, W \rfloor$ to P'_R;
(3) for each $\lfloor A \to y, U, W \rfloor \in P_L$, add $\lfloor A \to y, \psi(U), \psi(W) \cup N \cup \hat{N} \rfloor$ to P'_L.

Before proving that $L(H) = L(G)$, let us informally explain (1) through (3). Rules from (2) and (3) simulate the corresponding rules from P_R and P_L, respectively. Rules from (1) allow H to rewrite any occurrence of a nonterminal.

Consider a sentential form $x_1 A x_2$, where $x_1, x_2 \in (N \cup T)^*$ and $A \in N$. To rewrite A in H using type-2 leftmost derivations, all occurrences of nonterminals in x_1 are first rewritten to their barred versions by rules from (1.1). Then, A can be rewritten by a rule from (2) or (3). By rules from (1.1), every occurrence of a nonterminal in the current sentential form is then rewritten to its barred version. Rules from (1.2) then start rewriting barred nonterminals to hatted nonterminals. This is done from the right to the left. Finally, hatted nonterminals are rewritten to their original versions by rules from (1.3). This is also done from the right to the left.

To establish $L(H, {}_{\mathrm{lm}}^2\!\!\Rightarrow) = L(G)$, we prove two claims. First, Claim 1 shows how derivations of G are simulated by H. Then, Claim 2 demonstrates the converse—that is, it shows how derivations of H are simulated by G.

Claim 1. If $S \Rightarrow^n x$ in G, where $x \in V^*$, for some $n \geq 0$, then $S \, {}_{\mathrm{lm}}^2\!\!\Rightarrow^* x$ in H.

Proof. This claim is established by induction on $n \geq 0$.

Basis. For $n = 0$, this claim obviously holds.

Induction Hypothesis. Suppose that there exists $n \geq 0$ such that the claim holds for all derivations of length ℓ, where $0 \leq \ell \leq n$.

Induction Step. Consider any derivation of the form $S \Rightarrow^{n+1} w$ in G, where $w \in V^*$. Since $n + 1 \geq 1$, this derivation can be expressed as $S \Rightarrow^n x \Rightarrow w$, for some $x \in V^+$. By the induction hypothesis, $S \underset{lm}{\overset{2}{\Rightarrow}}^* x$ in H. Next, we consider all possible forms of $x \Rightarrow w$ in G, covered by the following two cases—(i) and (ii).

(i) *Application of* $\lfloor A \rightarrow y, U, W \rfloor \in P_R$. Let $x = x_1 A x_2$ and $r = \lfloor A \rightarrow y, U, W \rfloor \in P_R$, where $x_1, x_2 \in V^*$ such that $U \subseteq \text{alph}(x_2)$ and $W \cap \text{alph}(x_2) = \emptyset$, so $x_1 A x_2 \Rightarrow x_1 y x_2$ in G. If $x_1 \in T^*$, then $x_1 A x_2 \underset{lm}{\overset{2}{\Rightarrow}} x_1 y x_2$ in H by the corresponding rule introduced in (2), and the induction step is completed for (i). Therefore, assume that $\text{alph}(x_1) \cap N \neq \emptyset$. Let $x_1 = z_0 Z_1 z_1 Z_2 z_2 \cdots Z_h z_h$, where $z_i \in T^*$ and $Z_j \in N$, for all i and j, $0 \leq i \leq h$, $1 \leq j \leq h$, for some $h \geq 1$. By rules introduced in (1.1),

$$z_0 Z_1 z_1 Z_2 z_2 \cdots Z_h z_h A x_2 \underset{lm}{\overset{2}{\Rightarrow}}^* z_0 \bar{Z}_1 z_1 \bar{Z}_2 z_2 \cdots \bar{Z}_h z_h A x_2 \text{ in } H$$

By the corresponding rule to r introduced in (2),

$$z_0 \bar{Z}_1 z_1 \bar{Z}_2 z_2 \cdots \bar{Z}_h z_h A x_2 \underset{lm}{\overset{2}{\Rightarrow}} z_0 \bar{Z}_1 z_1 \bar{Z}_2 z_2 \cdots \bar{Z}_h z_h y x_2 \text{ in } H$$

By rules introduced in (1.1) through (1.3),

$$z_0 \bar{Z}_1 z_1 \bar{Z}_2 z_2 \cdots \bar{Z}_h z_h y x_2 \underset{lm}{\overset{2}{\Rightarrow}}^* z_0 Z_1 z_1 Z_2 z_2 \cdots Z_h z_h y x_2 \text{ in } H$$

which completes the induction step for (i).

(ii) *Application of* $\lfloor A \rightarrow y, U, W \rfloor \in P_L$. Let $x = x_1 A x_2$ and $r = \lfloor A \rightarrow y, U, W \rfloor \in P_L$, where $x_1, x_2 \in V^*$ such that $U \subseteq \text{alph}(x_1)$ and $W \cap \text{alph}(x_1) = \emptyset$, so $x_1 A x_2 \Rightarrow x_1 y x_2$ in G. To complete the induction step for (ii), proceed by analogy with (i), but use a rule from (3) instead of a rule from (2).

Observe that cases (i) and (ii) cover all possible forms of $x \Rightarrow w$ in G. Thus, the claim holds. □

Set $V = N \cup T$ and $V' = N' \cup T$. Define the homomorphism τ from V'^* to V^* as $\tau(A) = \tau(\bar{A}) = \tau(\hat{A}) = A$, for all $A \in N$, and $\tau(a) = a$, for all $a \in T$.

Claim 2. If $S \underset{lm}{\overset{2}{\Rightarrow}}^n x$ in H, where $x \in V'^*$, for some $n \geq 0$, then $S \Rightarrow^* \tau(x)$ in G, and either $x \in (\bar{N} \cup T)^* V^*$, $x \in (\bar{N} \cup T)^* (\hat{N} \cup T)^*$, or $x \in (\hat{N} \cup T)^* V^*$.

Proof. This claim is established by induction on $n \geq 0$.

Basis. For $n = 0$, this claim obviously holds.

Induction Hypothesis. Suppose that there exists $n \geq 0$ such that the claim holds for all derivations of length ℓ, where $0 \leq \ell \leq n$.

Induction Step. Consider any derivation of the form $S \underset{lm}{\overset{2}{\Rightarrow}}{}^{n+1} w$ in H, where $w \in V'^*$. Since $n + 1 \geq 1$, this derivation can be expressed as $S \underset{lm}{\overset{2}{\Rightarrow}}{}^n x \underset{lm}{\overset{2}{\Rightarrow}} w$, for some $x \in V'^+$. By the induction hypothesis, $S \Rightarrow^* \tau(x)$ in G, and either $x \in (\bar{N} \cup T)^* V^*$, $x \in (\bar{N} \cup T)^* (\hat{N} \cup T)^*$, or $x \in (\hat{N} \cup T)^* V^*$. Next, we consider all possible forms of $x \underset{lm}{\overset{2}{\Rightarrow}} w$ in H, covered by the following five cases— (i) through (v).

(i) *Application of a rule introduced in (1.1).* Let $\lfloor A \to \bar{A}, \emptyset, N \cup \hat{N} \rfloor \in P'_L$ be a rule introduced in (1.1). Observe that this rule is applicable only if $x = x_1 A x_2$, where $x_1 \in (\bar{N} \cup T)^*$ and $x_2 \in V^*$. Then,

$$x_1 A x_2 \underset{lm}{\overset{2}{\Rightarrow}} x_1 \bar{A} x_2 \text{ in } H$$

Since $\tau(x_1 \bar{A} x_2) = \tau(x_1 A x_2)$ and $x_1 \bar{A} x_2 \in (\bar{N} \cup T)^* V^*$, the induction step is completed for (i).

(ii) *Application of a rule introduced in (1.2).* Let $\lfloor \bar{A} \to \hat{A}, \emptyset, N \cup \hat{N} \rfloor \in P'_R$ be a rule introduced in (1.2). Observe that this rule is applicable only if $x = x_1 \bar{A} x_2$, where $x_1 \in (\bar{N} \cup T)^*$ and $x_2 \in (\hat{N} \cup T)^*$. Then,

$$x_1 \bar{A} x_2 \underset{lm}{\overset{2}{\Rightarrow}} x_1 \hat{A} x_2 \text{ in } H$$

Since $\tau(x_1 \hat{A} x_2) = \tau(x_1 \bar{A} x_2)$ and $x_1 \hat{A} x_2 \in (\bar{N} \cup T)^* (\hat{N} \cup T)^*$, the induction step is completed for (ii).

(iii) *Application of a rule introduced in (1.3).* Let $\lfloor \hat{A} \to A, \emptyset, \bar{N} \cup \hat{N} \rfloor \in P'_R$ be a rule introduced in (1.3). Observe that this rule is applicable only if $x = x_1 \hat{A} x_2$, where $x_1 \in (\hat{N} \cup T)^*$ and $x_2 \in V^*$. Then,

$$x_1 \hat{A} x_2 \underset{lm}{\overset{2}{\Rightarrow}} x_1 A x_2 \text{ in } H$$

Since $\tau(x_1 A x_2) = \tau(x_1 \hat{A} x_2)$ and $x_1 A x_2 \in (\hat{N} \cup T)^* V^*$, the induction step is completed for (iii).

(iv) *Application of a rule introduced in (2).* Let $\lfloor A \to y, U, W \rfloor \in P'_R$ be a rule introduced in (2) from $\lfloor A \to y, U, W \rfloor \in P_R$, and let $x = x_1 A x_2$ such that $U \subseteq \text{alph}(x_2)$ and $W \cap \text{alph}(x_2) = \emptyset$. Then,

$$x_1 A x_2 \underset{lm}{\overset{2}{\Rightarrow}} x_1 y x_2 \text{ in } H$$

and

$$\tau(x_1) A \tau(x_2) \Rightarrow \tau(x_1) y \tau(x_2) \text{ in } G$$

Clearly, $x_1 y x_2$ is of the required form, so the induction step is completed for (iv).

(v) *Application of a rule introduced in (3).* Let $\lfloor A \to y, \psi(U), \psi(W) \cup N \cup \hat{N} \rfloor \in P'_L$ be a rule introduced in (3) from $\lfloor A \to y, U, W \rfloor \in P_L$, and let $x = x_1 A x_2$ such that $\psi(U) \subseteq \text{alph}(x_1)$ and $(\psi(W) \cup N \cup \hat{N}) \cap \text{alph}(x_1) = \emptyset$. Then,

$$x_1 A x_2 \underset{lm}{\overset{2}{\Rightarrow}} x_1 y x_2 \text{ in } H$$

and

$$\tau(x_1) A \tau(x_2) \Rightarrow \tau(x_1) y \tau(x_2) \text{ in } G$$

Clearly, $x_1 y x_2$ is of the required form, so the induction step is completed for (v).

Observe that cases (i) through (v) cover all possible forms of $x \, _{\mathrm{lm}}^2\!\!\Rightarrow w$ in H. Thus, the claim holds. □

We now prove that $L(H, _{\mathrm{lm}}^2\!\!\Rightarrow) = L(G)$. Consider Claim 1 with $x \in T^*$. Then, $S \Rightarrow^* x$ in G implies that $S \, _{\mathrm{lm}}^2\!\!\Rightarrow^* x$ in H, so $L(G) \subseteq L(H, _{\mathrm{lm}}^2\!\!\Rightarrow)$. Consider Claim 2 with $x \in T^*$. Then, $S \, _{\mathrm{lm}}^2\!\!\Rightarrow^* x$ in H implies that $S \Rightarrow^* x$ in G, so $L(H, _{\mathrm{lm}}^2\!\!\Rightarrow) \subseteq L(G)$. Consequently, $L(H, _{\mathrm{lm}}^2\!\!\Rightarrow) = L(G)$.

Finally, notice that whenever G is propagating, then so is H. Hence, the theorem holds. □

Lemma 4. $\mathscr{L}_{\mathrm{ORC}}^\varepsilon(_{\mathrm{lm}}^2\!\!\Rightarrow) \subseteq \mathscr{L}_{\mathrm{RE}}^\varepsilon$

Proof. This inclusion can be obtained by standard simulations, so we leave the proof to the reader. □

Theorem 3. $\mathscr{L}_{\mathrm{ORC}}^\varepsilon(_{\mathrm{lm}}^2\!\!\Rightarrow) = \mathscr{L}_{\mathrm{RE}}^\varepsilon$

Proof. Since $\mathscr{L}_{\mathrm{ORC}}^\varepsilon = \mathscr{L}_{\mathrm{RE}}^\varepsilon$ (see Theorem 2 in [27]), Lemma 3 implies that $\mathscr{L}_{\mathrm{RE}}^\varepsilon \subseteq \mathscr{L}_{\mathrm{ORC}}^\varepsilon(_{\mathrm{lm}}^2\!\!\Rightarrow)$. By Lemma 4, $\mathscr{L}_{\mathrm{ORC}}^\varepsilon(_{\mathrm{lm}}^2\!\!\Rightarrow) \subseteq \mathscr{L}_{\mathrm{RE}}^\varepsilon$. Consequently, we have that $\mathscr{L}_{\mathrm{ORC}}^\varepsilon(_{\mathrm{lm}}^2\!\!\Rightarrow) = \mathscr{L}_{\mathrm{RE}}^\varepsilon$, so the theorem holds. □

Lemma 5. $\mathscr{L}_{\mathrm{ORC}}(_{\mathrm{lm}}^2\!\!\Rightarrow) \subseteq \mathscr{L}_{\mathrm{CS}}$

Proof. Since the length of sentential forms in derivations of propagating one-sided random context grammars is nondecreasing, propagating one-sided random context grammars can be simulated by linear bounded automata. A rigorous proof of this lemma is left to the reader. □

Theorem 4. $\mathscr{L}_{\mathrm{ORC}}(_{\mathrm{lm}}^2\!\!\Rightarrow) = \mathscr{L}_{\mathrm{CS}}$

Proof. Since $\mathscr{L}_{\mathrm{ORC}} = \mathscr{L}_{\mathrm{CS}}$ (see Theorem 1 in [27]), Lemma 3 implies that $\mathscr{L}_{\mathrm{CS}} \subseteq \mathscr{L}_{\mathrm{ORC}}(_{\mathrm{lm}}^2\!\!\Rightarrow)$. By Lemma 5, $\mathscr{L}_{\mathrm{ORC}}(_{\mathrm{lm}}^2\!\!\Rightarrow) \subseteq \mathscr{L}_{\mathrm{CS}}$. Consequently, we have that $\mathscr{L}_{\mathrm{ORC}}(_{\mathrm{lm}}^2\!\!\Rightarrow) = \mathscr{L}_{\mathrm{CS}}$, so the theorem holds. □

3.3 Type-3 Leftmost Derivations

Finally, we consider one-sided random context grammars with type-3 leftmost derivations.

Lemma 6. *For every one-sided random context grammar G, there is a one-sided random context grammar H such that $L(H, _{\mathrm{lm}}^3\!\!\Rightarrow) = L(G)$. Furthermore, if G is propagating, then so is H.*

Proof. Let $G = (N, T, P_L, P_R, S)$ be a one-sided random context grammar. We prove this lemma by analogy with the proof of Lemma 3. That is, we construct the one-sided random context grammar H in such a way that always allows it to rewrite an arbitrary occurrence of a nonterminal. Construct

$$H = \left(N', T, P_L', P_R', S\right)$$

as follows. Initially, set $\bar{N} = \{\bar{A} \mid A \in N\}$, $N' = N \cup \bar{N}$, and $P'_L = P'_R = \emptyset$ (without any loss of generality, we assume that $N \cap \bar{N} = \emptyset$). Define the function ψ from 2^N to $2^{\bar{N}}$ as $\psi(\emptyset) = \emptyset$ and

$$\psi(\{A_1, A_2, \ldots, A_n\}) = \{\bar{A}_1, \bar{A}_2, \ldots, \bar{A}_n\}$$

Perform (1) through (3), given next:

(1) for each $A \in N$,
   (1.1) add $\lfloor A \rightarrow \bar{A}, \emptyset, N \rfloor$ to P'_L;
   (1.2) add $\lfloor \bar{A} \rightarrow A, \emptyset, \bar{N} \rfloor$ to P'_R;
(2) for each $\lfloor A \rightarrow y, U, W \rfloor \in P_R$, add $\lfloor A \rightarrow y, U, W \rfloor$ to P'_R;
(3) for each $\lfloor A \rightarrow y, U, W \rfloor \in P_L$, let $U = \{X_1, X_2, \ldots, X_k\}$, and for each

$$U' \in \{\{Y_1, Y_2, \ldots, Y_k\} \mid Y_i \in \{X_i, \bar{X}_i\}, 1 \leq i \leq k\}$$

add $\lfloor A \rightarrow y, U', W \cup \Psi(W) \rfloor$ to P'_L ($U' = \emptyset$ if and only if $U = \emptyset$).

Before proving that $L(G) = L(H, {}_{\mathrm{lm}}{\overset{3}{\Rightarrow}})$, let us give an insight into the construction. Rules introduced in (1) allow H to rewrite an arbitrary occurrence of a nonterminal. Rules from (2) and (3) simulate the corresponding rules from P_R and P_L, respectively.

Consider a sentential form $x_1 A x_2$, where $x_1, x_2 \in (N \cup T)^*$ and $A \in N$, and a rule, $r = \lfloor A \rightarrow y, U, W \rfloor \in P'_L \cup P'_R$, introduced in (2) or (3). If $A \in \mathrm{alph}(x_1)$, all occurrences of nonterminals in x_1 are rewritten to their barred versions by rules from (1). Then, r is applied, and all barred nonterminals are rewritten back to their non-barred versions. Since not all occurrences of nonterminals in x_1 need to be rewritten to their barred versions before r is applied, all combinations of barred and non-barred nonterminals in the left permitting contexts of the resulting rules in (3) are considered.

The identity $L(H, {}_{\mathrm{lm}}{\overset{3}{\Rightarrow}}) = L(G)$ can be established by analogy with the proof given in Lemma 3, and we leave its proof to the reader. Finally, notice that whenever G is propagating, then so is H. Hence, the theorem holds.   □

Lemma 7. $\mathscr{L}^\varepsilon_{\mathrm{ORC}}({}_{\mathrm{lm}}{\overset{3}{\Rightarrow}}) \subseteq \mathscr{L}^\varepsilon_{\mathrm{RE}}$

Proof. This inclusion can be obtained by standard simulations, so we leave the proof to the reader.   □

Theorem 5. $\mathscr{L}^\varepsilon_{\mathrm{ORC}}({}_{\mathrm{lm}}{\overset{3}{\Rightarrow}}) = \mathscr{L}^\varepsilon_{\mathrm{RE}}$

Proof. Since $\mathscr{L}^\varepsilon_{\mathrm{ORC}} = \mathscr{L}^\varepsilon_{\mathrm{RE}}$ (see Theorem 2 in [27]), Lemma 6 implies that $\mathscr{L}^\varepsilon_{\mathrm{RE}} \subseteq \mathscr{L}^\varepsilon_{\mathrm{ORC}}({}_{\mathrm{lm}}{\overset{3}{\Rightarrow}})$. By Lemma 7, $\mathscr{L}^\varepsilon_{\mathrm{ORC}}({}_{\mathrm{lm}}{\overset{3}{\Rightarrow}}) \subseteq \mathscr{L}^\varepsilon_{\mathrm{RE}}$. Consequently, we have that $\mathscr{L}^\varepsilon_{\mathrm{ORC}}({}_{\mathrm{lm}}{\overset{3}{\Rightarrow}}) = \mathscr{L}^\varepsilon_{\mathrm{RE}}$, so the theorem holds.   □

Lemma 8. $\mathscr{L}_{\mathrm{ORC}}({}_{\mathrm{lm}}{\overset{3}{\Rightarrow}}) \subseteq \mathscr{L}_{\mathrm{CS}}$

Proof. This lemma can be established by analogy with the proof of Lemma 5.

Theorem 6. $\mathscr{L}_{\mathrm{ORC}}({}_{\mathrm{lm}}{\overset{3}{\Rightarrow}}) = \mathscr{L}_{\mathrm{CS}}$

Proof. Since $\mathscr{L}_{\mathrm{ORC}} = \mathscr{L}_{\mathrm{CS}}$ (see Theorem 1 in [27]), Lemma 6 implies that $\mathscr{L}_{\mathrm{CS}} \subseteq \mathscr{L}_{\mathrm{ORC}}({}_{\mathrm{lm}}{\overset{3}{\Rightarrow}})$. By Lemma 8, $\mathscr{L}_{\mathrm{ORC}}({}_{\mathrm{lm}}{\overset{3}{\Rightarrow}}) \subseteq \mathscr{L}_{\mathrm{CS}}$. Consequently, we have that $\mathscr{L}_{\mathrm{ORC}}({}_{\mathrm{lm}}{\overset{3}{\Rightarrow}}) = \mathscr{L}_{\mathrm{CS}}$, so the theorem holds.   □

4   Concluding Remarks

In this final section, we compare the results achieved in the previous section with some well-known results of formal language theory. More specifically, we relate the language families generated by one-sided random context grammars with leftmost derivations to the language families generated by random context grammars with leftmost derivations (in what follows, by random context grammars, we always mean random context grammars with both permitting and forbidding contexts, see [9] for the details).

The language families generated by random context grammars, random context grammars with type-1 leftmost derivations, random context grammars with type-2 leftmost derivations, and random context grammars with type-3 leftmost derivations are denoted by $\mathscr{L}_{RC}^{\varepsilon}$, $\mathscr{L}_{RC}^{\varepsilon}(_{lm}^{1}{\Rightarrow})$, $\mathscr{L}_{RC}^{\varepsilon}(_{lm}^{2}{\Rightarrow})$, and $\mathscr{L}_{RC}^{\varepsilon}(_{lm}^{3}{\Rightarrow})$, respectively (see [9] for the definitions of all these families). The notation without ε stands for the corresponding propagating family. For example, $\mathscr{L}_{RC}$ denotes the language family generated by propagating random context grammars.

The fundamental relationships between these families are summarized next.

Corollary 1. $\mathscr{L}_{CF}^{\varepsilon} \subset \mathscr{L}_{RC} \subset \mathscr{L}_{ORC} = \mathscr{L}_{CS} \subset \mathscr{L}_{ORC}^{\varepsilon} = \mathscr{L}_{RC}^{\varepsilon} = \mathscr{L}_{RE}^{\varepsilon}$

Proof. This corollary follows from Theorems 1 and 2 in [27] and from Theorems 1.2.4 and 1.2.5 in [9].   □

Considering type-1 leftmost derivations, we significantly decrease the power of both one-sided random context grammars and random context grammars.

Corollary 2. $\mathscr{L}_{ORC}^{\varepsilon}(_{lm}^{1}{\Rightarrow}) = \mathscr{L}_{RC}^{\varepsilon}(_{lm}^{1}{\Rightarrow}) = \mathscr{L}_{CF}^{\varepsilon}$

Proof. This corollary follows from Theorem 1 in the previous section and from Theorem 1.4.1 in [9].   □

Type-2 leftmost derivations increase the generative power of propagating random context grammars, but the generative power of random context grammars remains unchanged.

Corollary 3

(i) $\mathscr{L}_{ORC}(_{lm}^{2}{\Rightarrow}) = \mathscr{L}_{RC}(_{lm}^{2}{\Rightarrow}) = \mathscr{L}_{CS}$

(ii) $\mathscr{L}_{ORC}^{\varepsilon}(_{lm}^{2}{\Rightarrow}) = \mathscr{L}_{RC}^{\varepsilon}(_{lm}^{2}{\Rightarrow}) = \mathscr{L}_{RE}^{\varepsilon}$

Proof. This corollary follows from Theorems 3 and 4 in the previous section and from Theorem 1.4.4 in [9].   □

Finally, type-3 leftmost derivations are not enough for propagating random context grammars to generate the family of context-sensitive languages.

Corollary 4

(i) $\mathscr{L}_{RC}(_{lm}^{3}{\Rightarrow}) \subset \mathscr{L}_{ORC}(_{lm}^{3}{\Rightarrow}) = \mathscr{L}_{CS}$

(ii) $\mathscr{L}_{ORC}^{\varepsilon}(_{lm}^{3}{\Rightarrow}) = \mathscr{L}_{RC}^{\varepsilon}(_{lm}^{3}{\Rightarrow}) = \mathscr{L}_{RE}^{\varepsilon}$

Proof. This corollary follows from Theorems 5 and 6 in the previous section, from Theorem 1.4.5 in [9], and from Remarks 5.11 in [10].   □

Acknowledgments. This work was supported by the following grants: FRVŠ MŠMT FR271/2012/G1, BUT FIT-S-11-2, EU CZ 1.05/1.1.00/02.0070, and CEZ MŠMT MSM0021630528.

References

1. Aho, A.V., Lam, M.S., Sethi, R., Ullman, J.D.: Compilers: Principles, Techniques, and Tools, 2nd edn. Addison-Wesley, Boston (2006)
2. Aho, A.V., Ullman, J.D.: The Theory of Parsing, Translation and Compiling. Parsing, vol. I. Prentice-Hall, New Jersey (1972)
3. Baker, B.S.: Non-context-free grammars generating context-free languages. Information and Control 24(3), 231–246 (1974)
4. Cannon, R.L.: Phrase structure grammars generating context-free languages. Information and Control 29(3), 252–267 (1975)
5. Cojocaru, L., Mäkinen, E.: On the complexity of Szilard languages of regulated grammars. Tech. rep., Department of Computer Sciences, University of Tampere, Tampere, Finland (2010)
6. Cremers, A.B., Maurer, H.A., Mayer, O.: A note on leftmost restricted random context grammars. Information Processing Letters 2(2), 31–33 (1973)
7. Cytron, R., Fischer, C., LeBlanc, R.: Crafting a Compiler. Addison-Wesley, Boston (2009)
8. Dassow, J., Fernau, H., Păun, G.: On the leftmost derivation in matrix grammars. International Journal of Foundations of Computer Science 10(1), 61–80 (1999)
9. Dassow, J., Păun, G.: Regulated Rewriting in Formal Language Theory. Springer, New York (1989)
10. Fernau, H.: Regulated grammars under leftmost derivation. Grammars 3(1), 37–62 (2000)
11. Fernau, H.: Nonterminal complexity of programmed grammars. Theoretical Computer Science 296(2), 225–251 (2003)
12. Ferretti, C., Mauri, G., Păun, G., Zandron, C.: On three variants of rewriting P systems. Theoretical Computer Science 301(1-3), 201–215 (2003)
13. Freund, R., Oswald, M.: P Systems with Activated/Prohibited Membrane Channels. In: Păun, G., Rozenberg, G., Salomaa, A., Zandron, C. (eds.) WMC 2002. LNCS, vol. 2597, pp. 261–269. Springer, Heidelberg (2003)
14. Ginsburg, S., Spanier, E.H.: Control sets on grammars. Theory of Computing Systems 2(2), 159–177 (1968)
15. Kasai, T.: An hierarchy between context-free and context-sensitive languages. Journal of Computer and System Sciences 4, 492–508 (1970)
16. Lukáš, R., Meduna, A.: Multigenerative grammar systems. Schedae Informaticae 2006(15), 175–188 (2006)
17. Luker, M.: A generalization of leftmost derivations. Theory of Computing Systems 11(1), 317–325 (1977)
18. Matthews, G.H.: A note on asymmetry in phrase structure grammars. Information and Control 7, 360–365 (1964)
19. Maurer, H.: Simple matrix languages with a leftmost restriction. Information and Control 23(2), 128–139 (1973)
20. Meduna, A.: On the Number of Nonterminals in Matrix Grammars with Leftmost Derivations. In: Păun, G., Salomaa, A. (eds.) New Trends in Formal Languages. LNCS, vol. 1218, pp. 27–38. Springer, Heidelberg (1997)

21. Meduna, A.: Elements of Compiler Design. Auerbach Publications, Boston (2007)
22. Meduna, A., Goldefus, F.: Weak leftmost derivations in cooperative distributed grammar systems. In: MEMICS 2009: 5th Doctoral Workshop on Mathematical and Engineering Methods in Computer Science, pp. 144–151. Brno University of Technology, Brno (2009)
23. Meduna, A., Techet, J.: Canonical scattered context generators of sentences with their parses. Theoretical Computer Science 2007(389), 73–81 (2007)
24. Meduna, A., Techet, J.: Scattered Context Grammars and their Applications. WIT Press, Southampton (2010)
25. Meduna, A., Škrkal, O.: Combined leftmost derivations in matrix grammars. In: ISIM 2004: Proceedings of 7th International Conference on Information Systems Implementation and Modelling, Ostrava, CZ, pp. 127–132 (2004)
26. Meduna, A., Zemek, P.: Nonterminal Complexity of One-Sided Random Context Grammars. Acta Informatica 49(2), 55–68 (2012)
27. Meduna, A., Zemek, P.: One-sided random context grammars. Acta Informatica 48(3), 149–163 (2011)
28. Mihalache, V.: Matrix grammars versus parallel communicating grammar systems. In: Mathematical Aspects of Natural and Formal Languages, pp. 293–318. World Scientific Publishing, River Edge (1994)
29. Mutyam, M., Krithivasan, K.: Tissue P systems with leftmost derivation. Ramanujan Mathematical Society Lecture Notes Series 3, 187–196 (2007)
30. Păun, G.: On leftmost derivation restriction in regulated rewriting. Romanian Journal of Pure and Applied Mathematics 30(9), 751–758 (1985)
31. Rosenkrantz, D.J.: Programmed grammars and classes of formal languages. Journal of the ACM 16(1), 107–131 (1969)
32. Rozenberg, G., Salomaa, A. (eds.): Handbook of Formal Languages, vol. 1 through 3. Springer, Berlin (1997)
33. Salomaa, A.: Matrix grammars with a leftmost restriction. Information and Control 20(2), 143–149 (1972)

Earley's Parsing Algorithm
and k-Petri Net Controlled Grammars

Taishin Y. Nishida

Department of Information Sciences, Toyama Prefectural University,
Imizu, 939-0398 Toyama, Japan
nishida@pu-toyama.ac.jp

Abstract. In this paper we modify Earley's parsing algorithm to parse words generated by Petri net controlled grammars. Adding a vector which corresponds to a marking of a Petri net to Earley's algorithm, it is shown that languages generated by a subclass of k-Petri net controlled grammars (introduced by J. Dassow and S. Turaev) are parsed in polynomial time of the length of a word.

1 Introduction

Petri net controlled grammars have been introduced by M. ter Beek and J. Kleijn [1] and then have been extensively studied by J. Dassow and S. Turaev [3–6, 13]. For a context-free grammar, there is a Petri net whose places correspond to the nonterminals of the grammar and whose transitions correspond to the rules of the grammar such that a transition occurs (fires) if and only if the corresponding rule is applied in a derivation of the grammar. That is, the Petri net, called a cf Petri net, represents a sequence of rules which are used in a derivation of the grammar. Adding new places to a cf Petri net, a Petri net controls derivation of a context-free grammar to generate a non-context-free language. Thus Petri net controlled grammars have appeared quite naturally.

In this paper we focus our attention on the membership problem of Petri net controlled grammars. It has been shown that languages generated by most variants of Petri net controlled grammars are included in the class of matrix languages [3, 4, 13]. Thus the membership problem of Petri net controlled grammars might be reduced to that of matrix grammars. But we want to solve the membership problem directly, that is, to parse a word in order to construct a derivation tree. If a variant of Petri net controlled grammars has a fast parsing algorithm, then the grammars will be as easily and frequently used as context-free (without control) grammars, or even will replace context-free grammars, in practical application.

There are two famous fast parsing algorithms, CKY algorithm [10, 14] and Earley's algorithm [8], for context-free grammars. CKY algorithm assumes a grammar in Chomsky normal form. But, since modification of production rules in a Petri net controlled grammar changes the structure of the Petri net, it is not clear that a language generated by a variant of Petri net controlled grammars

H. Bordihn, M. Kutrib, and B. Truthe (Eds.): Dassow Festschrift 2012, LNCS 7300, pp. 174–185, 2012.

is generated by a grammar in the same variant and in Chomsky normal form. This is why CKY algorithm is inappropriate for our purpose. On the other hand, Earley's algorithm has no restriction on context-free grammars. Thus we start to develop a parsing algorithm for Petri net controlled grammars from Earley's algorithm.

In Section 2 basic notions and notations about context-free grammars, Petri nets, and Earley's algorithm are described. A variant of Petri net controlled grammars, k-Petri net controlled grammars (k-PN controlled grammars for short), is introduced in Section 3. Earley's algorithm is extended in Section 4 to parse words generated by k-PN controlled grammars. The algorithm parses a word of length n in time $O(n^3)$ for an unambiguous k-PN controlled grammar and in time $O(n^{k+4})$ for an ambiguous grammar. Section 5 is a conclusion.

2   Preliminaries

We assume that the reader is familiar with rudiments of context-free grammars, regulated grammars, and Petri nets. For notions and notations which are not described in this section, we refer to [2, 7, 9, 11, 12].

2.1   Context-Free Grammars

A context-free grammar is a construct $G = (V, \Sigma, S, R)$ where V and Σ are *nonterminal* and *terminal* alphabets, respectively, with $V \cap \Sigma = \emptyset$, $S \in V$ is the *start symbol*, and $R \subseteq V \times (V \cup \Sigma)^*$ is a finite set of (*production*) *rules*. A rule (A, x) is written as $A \to x$. A word $x \in (V \cup \Sigma)^+$ directly derives $y \in (V \cup \Sigma)^*$, written as $x \overset{r}{\Rightarrow}_G y$, if and only if there is a rule $r : A \to \alpha \in R$ such that $x = x_1 A x_2$ and $y = x_1 \alpha x_2$. We write $x \overset{r}{\Rightarrow} y$ if G is understood and write $x \Rightarrow y$ if we are not interested in the rule r. The reflexive and transitive closure of $\Rightarrow$ is denoted by $\Rightarrow^*$. If there are a sequence of rules $r_1, r_2, \ldots, r_n$ and a sequence of words $w_0, w_1, \ldots, w_n$ such that $w_{i-1} \overset{r_i}{\Rightarrow} w_i$ for every $1 \leq i \leq n$, then we write $w_0 \xrightarrow{r_1 r_2 \cdots r_n} w_n$. We call the sequence of rules $r_1, r_2, \ldots, r_n$ *derivation process* from w_0 to w_n. The language generated by G is defined by $L(G) = \{w \in \Sigma^* \mid S \Rightarrow^*_G w\}$.

Let $G = (V, \Sigma, S, R)$ be a context-free grammar. A rule of the form $A \to \lambda$ is called a λ-*rule*, where λ is the empty word. A rule $A \to \alpha$ is said to be a *chain rule* if $\alpha \in V$. Let σ be a derivation process from S to $w \in \Sigma^*$ in G. Then σ determines a derivation tree, which is denoted by $t(\sigma)$. Let U be the set of nodes of $t(\sigma)$ and let $\nu : U \to V \cup \Sigma \cup \{\lambda\}$ be the labelling function of $t(\sigma)$. We make a *derivation tree in rules*, denoted by $t_R(\sigma)$, by:

(1) Removing every node u which satisfies $\nu(u) \in \Sigma \cup \{\lambda\}$, i.e., a node which is labelled by a terminal or λ.
(2) Replacing every label of nonterminal with the rule which rewrites the nonterminal in the derivation process σ. That is, define a new labelling function $\nu' : U \to R$ by $\nu'(u) = r$ where rule r rewrites the nonterminal $\nu(u)$.

2.2 Petri Net

A Petri net is a quadruple $N = (P, T, F, \phi)$ where P and T are disjoint finite sets of places and transitions, respectively, $F \subseteq (P \times T) \cup (T \times P)$ is the set of directed arcs, $\phi : (P \times T) \cup (T \times P) \to \mathbb{N}$ is a weight function with $\phi(x, y) = 0$ for every $(x, y) \notin F$, where $\mathbb{N}$ is the set of nonnegative integers. A Petri net can be represented by a bipartite directed graph with the node set $P \cup T$ where places are drawn as circles, transitions are rectangles, and arcs as arrows with labels $\phi(p, t)$ or $\phi(t, p)$. If $\phi(p, t) = 1$ or $\phi(t, p) = 1$, then the label is omitted.

A place contains a number of tokens. Each number of tokens in every place is expressed by a mapping $\mu : P \to \mathbb{N}$, which is called a *marking*. For every place $p \in P$, $\mu(p)$ denotes the number of tokens in p. Graphically, tokens are drawn as small solid dots inside circles.

A transition $t \in T$ is enabled by a marking μ if and only if $\mu(p) \geq \phi(p, t)$ for every $p \in P$. In this case t can *occur* (*fire*). An occurrence of a transition t transforms the marking μ into a new marking μ' which is defined by $\mu'(p) = \mu(p) - \phi(p, t) + \phi(t, p)$ for every $p \in P$. More than one transition may be enabled by a marking. In this case one transition is nondeterministically selected and fires. If a transition t occurs in a marking μ and the marking changes to μ', then we write $\mu \xrightarrow{t} \mu'$. A finite sequence $t_1 t_2 \cdots t_k$ of transitions is called an *occurrence sequence* enabled at a marking μ if there are markings $\mu_1, \mu_2, \ldots, \mu_k$ such that $\mu \xrightarrow{t_1} \mu_1 \xrightarrow{t_2} \cdots \xrightarrow{t_k} \mu_k$. In short this sequence can be written as $\mu \xrightarrow{t_1 t_2 \cdots t_k} \mu_k$ or $\mu \xrightarrow{\nu} \mu_k$ where $\nu = t_1 t_2 \cdots t_k$. For each $1 \leq i \leq k$, the marking μ_i is called *reachable* from the marking μ.

A *marked* Petri net is a system $N = (P, T, F, \phi, \iota)$ where (P, T, F, ϕ) is a Petri net, ι is the *initial marking*. Let M be a set of marking. An occurrence sequence ν of transitions is called *successful* for M if it is enabled at the initial marking ι and finished at a final marking τ of M. Thus M is called a set of final markings. If M is understood from the context, we say that ν is a successful occurrence sequence.

Let $N = (P, T, F, \phi)$ be a Petri net. For an arc $e = (u, v)$ in F (note that $(u \in P$ and $v \in T)$ or $(u \in T$ and $v \in P))$, we use the notations $^{\bullet}e = u$ and $e^{\bullet} = v$. A sequence of arcs $e_1, e_2, \ldots, e_n$ is said to be a *path* in N if $e_i^{\bullet} = {}^{\bullet}e_{i+1}$ for every $i \in \{1, \ldots, n-1\}$. A path is a *cycle* if $e_n^{\bullet} = {}^{\bullet}e_1$.

2.3 Earley's Algorithm

Here we introduce Earley's algorithm [8]. Let $G = (V, \Sigma, S, R)$ be a context-free grammar. Let $G' = (V \cup \{S'\}, \Sigma, S', R' = R \cup \{S' \to S\})$ be a new context-free grammar where S' is a new nonterminal. For a rule $r : A \to \alpha$ in P', an item $[A \to \beta \cdot \gamma]$ is said to be an *Earley's state* where $\beta\gamma = \alpha$. Let $S' \Rightarrow^* w = a_1 a_2 \cdots a_m \in \Sigma^*$ be a derivation in G'. If a rule $A \to \alpha\beta$ which is used in the derivation

$$S' \Rightarrow^* \gamma A \delta \Rightarrow \gamma\alpha\beta\delta \Rightarrow^* a_1 \cdots a_k a_{k+1} \cdots a_i \beta \delta$$

satisfies

$$\gamma \Rightarrow^* a_1 \cdots a_k \quad \text{and} \quad \alpha \Rightarrow^* a_{k+1} \cdots a_i,$$

then the state $[A \to \alpha \cdot \beta]$ belongs to a set $E_w(k, i)$. The sets $E_w(k, i)$ of such states $(0 \le k \le i \le m)$ are called *sets of Earley's state*. The next property directly follows from the definition.

Property 1. Let G, G', and w be grammars and a word described in the above paragraph.

(1) If a state of the form $[A \to \cdot\alpha]$ appears in $E_w(k, i)$, then $k = i$. Thus $[S' \to \cdot S] \in E_w(0, 0)$.
(2) $[A \to \alpha\cdot]$ is in $E_w(k, i)$ if and only if $A \Rightarrow^* a_{k+1} \cdots a_i$.
(3) $[S' \to S\cdot]$ is in $E_w(0, m)$ if and only if $S \Rightarrow^* a_1 \cdots a_m$, that is, w is in $L(G)$.
(4) If $[A \to \alpha \cdot B\beta] \in E_w(k, i)$ with $B \in V$, then $[B \to \cdot\gamma] \in E_w(i, i)$ for every B-rule $B \to \gamma$.
(5) If $[A \to \alpha \cdot B\beta] \in E_w(k, i)$ and $[B \to \gamma\cdot] \in E_w(i, j)$, then $[A \to \alpha B \cdot \beta] \in E_w(k, j)$.
(6) If $[A \to \alpha \cdot a\beta] \in E_w(k, i - 1)$ and $a_i = a$, then $[A \to \alpha a \cdot \beta] \in E_w(k, i)$.

The above property shows that an algorithm which constructs every set of Earley's states solves the membership problem for context-free languages. Algorithm 1 constructs sets of Earley's states.

Algorithm 1. An algorithm which constructs sets of Earley's states

---

input: a word $w = X_1 X_2 \cdots X_m$ $(X_i \in \Sigma)$
output: the sets of Earley's states $E_w(k, i)$ $(0 \le k \le i \le m)$
1:   for every k, i with $((k, i) \ne (0, 0))$
2:       $E_w(k, i) = \emptyset$
3:       $E_w(0, 0) = \{[S' \to \cdot S]\}$
4:   do
5:       if $[A \to \alpha \cdot B\beta] \in E_w(0, 0)$ and $[B \to \gamma\cdot] \in E_w(0, 0)$, then
6:           insert $[A \to \alpha B \cdot \beta]$ to $E_w(0, 0)$
7:       if $[A \to \alpha \cdot B\beta] \in E_w(0, 0)$ with $B \in V$, then
8:           insert $[B \to \cdot\gamma]$ to $E_w(0, 0)$ for every B-rule $B \to \gamma$
9:   while $E_w(0, 0)$ is changed
10: do
11:      if $[A \to \alpha \cdot a\beta] \in E_w(k, i - 1)$ and $X_i = a$, then
12:          insert $[A \to \alpha a \cdot \beta]$ in $E_w(k, i)$
13:      if $[A \to \alpha \cdot B\beta] \in E_w(k, i)$ and $[B \to \gamma\cdot] \in E_w(i, j)$, then
14:          insert $[A \to \alpha B \cdot \beta]$ to $E_w(k, j)$
15:      if $[A \to \alpha \cdot B\beta] \in E_w(k, i)$ with $B \in V$, then
16:          insert $[B \to \cdot\gamma]$ to $E_w(i, i)$ for every B-rule $B \to \gamma$
17: while some $E_w(k, i)$ is changed

For an Earley's state $[A \to \alpha\cdot] \in E_w(k, i)$ a set of derivations from A to $a_k \cdots a_i$ can be constructed inductively.

(1) If $\alpha \in \Sigma^*$, then $A \Rightarrow \alpha$ is a derivation.

(2) If $\alpha = u_0 B_1 u_1 \cdots u_{l-1} B_l u_l$ for some $B_1, \ldots, B_l \in V$ and $u_0 u_1 \cdots u_l \in \Sigma^*$ and a set of derivations from B_i to β_i has been constructed for every $i \in \{1, \ldots, l\}$, then the set of derivations from A to $a_k \cdots a_i$ contains all derivations of the form

$$A \Rightarrow u_0 B_1 u_1 \cdots u_{l-1} B_l u_l \Rightarrow^* u_0 \beta_1 u_1 \cdots u_{l-1} B_l u_l \Rightarrow^* \cdots$$

$$\Rightarrow^* u_0 \beta_1 u_1 \cdots u_{l-1} \beta_l u_l = a_k \cdots a_i.$$

If a start symbol S of a grammar does not appear in a right-hand side of any rule, then we do not introduce new start symbol S' and we use states of the form $[S \rightarrow \alpha \cdot]$ to decide whether a word is generated or not. Whenever we discuss Earley's states, a new start symbol S' is implicitly assumed if necessarily.

3 k-Petri Net Controlled Grammars

In this section we define k-Petri net controlled grammars according to [6].

Let $G = (V, \Sigma, S, R)$ be a context-free grammar. A marked Petri net $N = (P, T, F, \phi, \iota)$ is a *cf Petri net* with respect to G under labelling functions (β, γ) if N and (β, γ) satisfy:

(1) $\beta : P \rightarrow V$ and $\gamma : T \rightarrow R$ are bijections.
(2) F and ϕ satisfy:
  - $(p, t) \in F$ if and only if $\gamma(t) = A \rightarrow \alpha$ and $\beta(p) = A$, in this case $\phi(p, t) = 1$.
  - $(t, p) \in F$ if and only if $\gamma(t) = A \rightarrow \alpha$ and $\beta(p) = x$ where $|\alpha|_x \geq 1$, in this case $\phi(t, p) = |\alpha|_x$.
(3) $\iota(p) = 1$ if $\beta(p) = S$ and $\iota(p) = 0$ for every $p \in P - \{\beta^{-1}(S)\}$.

We note that a cf Petri net is uniquely determined from a combination of a context-free grammar G and a pair of labelling functions (β, γ). Therefore, a cf Petri net with respect to G under (β, γ) can be denoted by $PN[G, (\beta, \gamma)]$.

Definition 1. *Let $G_0 = (V, \Sigma, S, R)$ be a context-free grammar and let $N = PN[G_0, (\beta, \gamma)] = (P, T, F, \phi, \iota)$ be a cf Petri net with respect to G_0. A k-Petri net controlled grammar (k-PN controlled grammar) is a quintuple $G = (V, \Sigma, S, R, N_k)$ where V, Σ, S, R are the components from the grammar G_0 and $N_k = (P', T', F', \phi', \iota')$ is a k-Petri net which satisfies:*

(1) *$P' = P \cup Q$ where $Q = \{q_1, \ldots, q_k\}$ is a set of new places.*
(2) *$T' = T$.*
(3) *$F' = F \cup E$ where $E \subseteq (T \times Q) \cup (Q \times T)$ is a set of new arcs. Every arc in E satisfies the following condition;*
  - *for every $t \in T$, $(t, q_i) \in E$ and $(t, q_j) \in E$ imply $i = j$ and $(q_i, t) \in E$ and $(q_j, t) \in E$ imply $i = j$.*
  - *for every $1 \leq i < j \leq k$, there exists no $t \in T$ such that $(t, q_i) \in E$ and $(q_j, t) \in E$.*

- *for every $1 \leq i \leq k$, $(q_i, t) \in E$ for some $t \in T$ if and only if $(t', q_i) \in E$ for some $t' \in T$.*

(4) $\phi'(x, y) = \phi(x, y)$ if $(x, y) \in F$ and $\phi'(x, y) = 1$ if $(x, y) \in E$.

(5) $\iota'(p) = 1$ if $\beta(p) = S$ and $\iota'(p) = 0$ for every $p \in (P - \{\beta^{-1}(S)\}) \cup Q$, i,e., $\iota'(p) = \iota(p)$ if $p \in P$ and $\iota'(p) = 0$ if $p \in Q$.

We call G_0 the underlying grammar *of G.*

In [6], condition (3) is described differently:

- $E = \{(t, q_i) \,|\, t \in T_1^i, 1 \leq i \leq k\} \cup \{(q_i, t) \,|\, t \in T_2^i, 1 \leq i \leq k\}$ such that $T_1^i \subset T$ and $T_2^i \subset T$, $1 \leq i \leq k$ where $T_l^i \cap T_l^j = \emptyset$ for $1 \leq l \leq 2$, $T_1^i \cap T_2^j = \emptyset$ for $1 \leq i < j \leq k$ and $T_1^i = \emptyset$ if and only if $T_2^i = \emptyset$ for any $1 \leq i \leq k$.

It is clear that the two conditions say the same thing. The most important point of condition (3) is that a k-Petri net does not have any cycle of arcs in E. We call it *cycle-free* condition.

Let τ be the marking $\tau(p) = 0$ for every $p \in P \cup Q$. Next we define the derivation in a k-PN controlled grammar G and the language generated by G.

Definition 2. *Let $G = (V, \Sigma, S, R, N_k)$ be a k-PN controlled grammar. A word $\alpha \in (V \cup \Sigma)^*$ is derived in G if $S \stackrel{r_1 r_2 \cdots r_n}{\Longrightarrow} \alpha$ such that $t_1 t_2 \cdots t_n = \gamma^{-1}(r_1 r_2 \cdots r_n) \in T^*$ is an occurrence sequence of the transitions of N_k enabled at the initial marking ι. A derivation $S \stackrel{r_1 r_2 \cdots r_n}{\Longrightarrow} w \in \Sigma^*$ successfully generates a terminal word if $t_1 t_2 \cdots t_n = \gamma^{-1}(r_1 r_2 \cdots r_n) \in T^*$ is an occurrence sequence of the transitions of N_k enabled at the initial marking ι and finished at the final marking τ. The* language generated by G, *denoted by $L(G)$, consists of all words which are successfully generated in G^1.*

In k-PN controlled grammars, new places control sequence of rules in a derivation, which is shown in the next example.

Example 1 (Example 7 of [3]). Let $G = (\{S, A, B\}, \{a, b, c\}, S, R, N_1)$ be a 1-PN controlled grammar where R consists of the following rules

$$r_0 : S \to AB, \; r_1 : A \to aAb, \; r_2 : A \to ab, \; r_3 : B \to cB, \; r_4 : B \to c$$

---

[1] A k-PN controlled grammar can be viewed a kind of *positive* valence grammar [6]. A (context-free) valence grammar G on $\mathbf{z}^k$ for some positive integer k is a construct $G = (V, \Sigma, S, R, \nu)$ where $\mathbf{z}$ is the set of integers, $(V, \Sigma, S, R) = G_0$ is a context-free grammar (the underlying grammar), and $\nu : R \to \mathbf{z}^k$ is the valence function. Let $\sigma = (r_1, r_2, \ldots, r_n)$ be a derivation process from S to $w \in \Sigma^*$ in G_0. The word w is generated by G if and only if $\sum_{j=1}^{n} \nu(r_j) = \mathbf{0}$ and for every initial segment $(r_1, r_2, \ldots, r_i)$ of σ $\sum_{j=1}^{i} \nu(r_j) \geq \mathbf{0}$ where $\mathbf{a} \geq \mathbf{b}$ if and only if $a_j \geq b_j$ for every jth component. A k-PN controlled grammar (V, Σ, S, R, N_k) is a positive valence grammar (V, Σ, S, R, ν) where $\nu : R \to \mathbf{z}^k$ is given by

$$\nu(r) = (\phi(\gamma^{-1}(r), q_1) - \phi(q_1, \gamma^{-1}(r)), \ldots, (\phi(\gamma^{-1}(r), q_k) - \phi(q_k, \gamma^{-1}(r)).$$

Clearly, every component of $\nu(r)$ is one of 1, 0, or -1. It should be noted that the notion of positive valence grammars is different from that of valence grammars. The latter permits $\sum_{j=1}^{i} \nu(r_j) \not\geq \mathbf{0}$ for some initial segment $(r_1, \cdots, r_i)$.

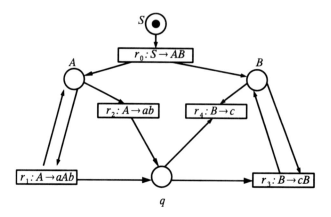

Fig. 1. 1-Petri net controlled grammar generating $\{a^n b^n c^n \mid n > 0\}$

and N_1 is illustrated in Fig. 1, in which rules are drawn in the rectangles of the corresponding transitions. The grammar G generates the language $\{a^n b^n c^n \mid n > 0\}$. □

4   Parsing for k-Petri Net Controlled Grammars

Our aim of this section is to develop a parsing algorithm for k-PN controlled grammars.

First we must consider chain rules and λ-rules. Every context-free grammar can be converted, without changing the generated language, to a grammar with no chain rules and no λ-rules. But it is not obvious whether a k-PN controlled grammar has an equivalent grammar (generating the same language) with no chain rules and no λ-rules since a chain rule or a λ-rule may make a token in Q. Indeed these rules cause a subtle problem (see Section 5). On the other hand, a grammar with chain rules or λ-rules may have infinite derivation trees for a word, which makes parsing very complex. So, in the remaining sections in this paper, we assume that an underlying context-free grammar does not have any chain rules nor λ-rules (but $S \to \lambda$ for the start symbol S with the condition that S does not appear in a right-hand side of any rule).

A parsing algorithm for k-PN controlled grammars is obtained by associating a vector which corresponds to a marking of a k-Petri net to each Earley's state. Let $G = (V, \Sigma, S, R, N_k)$ be a k-PN controlled grammar and let $G_0 = (V, \Sigma, S, R)$ be the underlying grammar of G. For a word $w \in L(G_0)$ we define an Earley's state with a *token counter*, which is a vector from $\mathbf{Z}^k$ with set of integers $\mathbf{Z}$, as follows:

(1) For a rule $A \to \alpha$, the state $[A \to \cdot \alpha]$ has a token counter $(v_1, \ldots, v_k)$ where $v_i = 1$ and $v_j = 0$ for $j \neq i$ if $(\gamma^{-1}(A \to \alpha), q_i) \in E$, $v_i = -1$ and $v_j = 0$ for $j \neq i$ if $(q_i, \gamma^{-1}(A \to \alpha)) \in E$, or $v_i = 0$ for every $1 \leq i \leq k$ otherwise.

(2) If a state $[A \to \alpha B \cdot \beta]$ is obtained from $[A \to \alpha \cdot B\beta]\boldsymbol{v}_1$ and $[B \to \gamma\cdot]\boldsymbol{v}_2$ where $\boldsymbol{v}_1$ and $\boldsymbol{v}_2$ are token counters associated to the states, then $[A \to \alpha B \cdot \beta]$ has the token counter $\boldsymbol{v}_1 + \boldsymbol{v}_2$ in which the addition is the normal component-wise vector addition.

If the underlying grammar is ambiguous, then there may be states $[A \to \alpha \cdot B\beta] \in E_w(i,j)$, $[B \to \gamma\cdot] \in E_w(j,l)$, $[A \to \alpha \cdot B\beta] \in E_w(i,j')$, and $[B \to \gamma'\cdot] \in E_w(j',l)$. In this case a state $[A \to \alpha B \cdot \beta] \in E_w(i,l)$ is constructed more than one way. The state may have different token counters, that is, the situation $[A \to \alpha B \cdot \beta]\boldsymbol{v} \in E_w(i,l)$ and $[A \to \alpha B \cdot \beta]\boldsymbol{v}' \in E_w(i,l)$ with $\boldsymbol{v} \neq \boldsymbol{v}'$ is possible. States with different token counters should be treated differently.

Example 2. In Example 1, the sets of Earley's states $E_w(k,i)$ with token counters for the word $w = a^2b^2c^2$ are shown in the next table.

	$i=0$	$i=1$	$i=2$	$i=3$	$i=4$	$i=5$	$i=6$
	$[S \to \cdot AB](0)$ $[A \to \cdot aAb](0)$ $[A \to \cdot ab](0)$	$[A \to a \cdot Ab](1)$ $[A \to a \cdot b](1)$		$[A \to aA \cdot b](2)$	$[S \to A \cdot B](2)$ $[A \to aAb \cdot](2)$	$[S \to AB \cdot](1)$	$[S \to AB \cdot](0)$
$k=1$		$[A \to \cdot aAb](1)$ $[A \to \cdot ab](1)$	$[A \to a \cdot Ab](1)$ $[A \to a \cdot b](1)$	$[A \to ab \cdot](1)$			
$k=2$			$[A \to \cdot aAb](1)$ $[A \to \cdot ab](1)$				
$k=3$							
$k=4$					$[B \to \cdot cB](-1)$ $[B \to \cdot c](-1)$	$[B \to c \cdot B](-1)$ $[B \to c \cdot](-1)$	$[B \to cB \cdot](-2)$
$k=5$						$[B \to \cdot cB](-1)$ $[B \to \cdot c](-1)$	$[B \to c \cdot B](-1)$ $[B \to c \cdot](-1)$

$\square$

It is obvious that a successful derivation $S \Rightarrow^* w \in \Sigma^*$ in G (under control) implies $[S' \to S \cdot]\boldsymbol{0} \in E_w(0,n)$ where $\boldsymbol{0}$ is the zero vector and $n = |w|$. But the converse is not always the case. Let us consider the next example.

Example 3. Let $G = (\{S, A, B, C\}, \{a, b, c\}, S, R, N_1)$ be a 1-PN controlled grammar where R and N_1 are illustrated in Fig. 2.

The word $bbabc$ is generated by G while the word $bacc$ cannot be generated by G. But, as seen in the next tables, both token counters attached to Earley's states for the words become the zero vector.

Earley's states for the word $bbabc$.

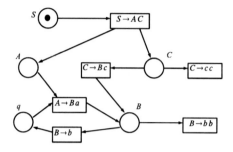

Fig. 2. Petri net of Example 3

$i = 0$	$i = 1$	$i = 2$	$i = 3$	$i = 4$	$i = 5$
$[S \to \cdot AC](0)$			$[S \to A \cdot C](-1)$		$[S \to AC\cdot](0)$
$[A \to \cdot Ba](-1)$	$[A \to B \cdot a](0)$	$[A \to B \cdot a](-1)$	$[A \to Ba\cdot](-1)$		
$[B \to \cdot b](1)$	$[B \to b\cdot](1)$				
$[B \to \cdot bb](0)$	$[B \to b \cdot b](0)$	$[B \to bb\cdot](0)$			
$k = 1$					
	$k = 2$				
		$k = 3$	$[C \to \cdot Bc](0)$	$[C \to B \cdot c](1)$	$[C \to Bc\cdot](1)$
			$[C \to \cdot cc](0)$		
			$[B \to \cdot b](1)$	$[B \to b\cdot](1)$	
			$[B \to \cdot bb](0)$	$[B \to b \cdot b](0)$	
			$k = 4$		
				$k = 5$	

Earley's states for the word $bacc$.

$i = 0$	$i = 1$	$i = 2$	$i = 3$	$i = 4$
$[S \to \cdot AC](0)$		$[S \to A \cdot C](0)$		$[S \to AC\cdot](0)$
$[A \to \cdot Ba](-1)$	$[A \to B \cdot a](0)$	$[A \to Ba\cdot](0)$		
$[B \to \cdot b](1)$	$[B \to b\cdot](1)$			
$[B \to \cdot bb](0)$	$[B \to b \cdot b](0)$			
$k = 1$				
	$k = 2$	$[C \to \cdot Bc](0)$	$[C \to c \cdot c](0)$	$[C \to cc\cdot](0)$
		$[C \to \cdot cc](0)$		
		$[B \to \cdot b](1)$		
		$[B \to \cdot bb](0)$		
		$k = 3$		
			$k = 4$	

□

If a k-Petri net $N_k = (P, T, F, \phi, \iota)$ satisfies "every cycle in N_k does not contain any places in Q", then we can prove equivalence between derivations $S \Rightarrow^* w \in \Sigma^*$ under control of N_k and $[S' \Rightarrow S\cdot]\mathbf{0} \in E_w(0, n)$. We call the condition *strict cycle-free condition*.

Lemma 1. *Let $G = (V, \Sigma, S, R, N_k)$ be a k-PN controlled grammar with strict cycle-free Petri net N_k and let $G_0 = (V, \Sigma, S, R)$ be the underlying grammar of G. For every $w \in L(G_0)$ with $|w| = n$, if $[S' \to S\cdot]\mathbf{0} \in E_w(0, n)$, then $w \in L(G)$.*

Proof. Let $S' \Rightarrow^* w$ be a derivation which is constructed from $[S' \to S\cdot]\mathbf{0} \in E_w(0, n)$. Let $\alpha A \beta$ be a sentential form in the derivation such that there is an arc $(q_l, \gamma^{-1}(A \to \delta)) \in E$ where $A \to \delta$ is the rule in the derivation. If the l-th component of the token counter increases in the derivation $A \Rightarrow \delta \Rightarrow^*$

$u \in \Sigma^*$, then there is a path from $\gamma^{-1}(A \to \delta)$ to q_l in N_k. The path and the arc $(q_l, \gamma^{-1}(A \to \delta))$ forms a cycle, which contradicts to the strict cycle-free condition. Thus the l-th component of the token counter cannot increase in the derivation $A \Rightarrow^* u$.

Since the l-th component of the token counter becomes to 0 after derivation, some positive values in the l-th component must appear in one of the derivations: $S' \Rightarrow^* \alpha A\beta$, $\alpha \Rightarrow^* x$, or $\beta \Rightarrow^* y$ where $xuy = w$. Hence there is a derivation in which an occurrence of a positive value in the l-th component is prior to use the rule $A \to \delta$, that is, $A \to \delta$ can be applied under control of N_k. Therefore, the derivation $S \Rightarrow^* w$ is possible in G. $\qquad\square$

Since to associate a token counter to an Earley's state can be done in a constant time, parsing for a strict cycle-free k-PN controlled grammar is performed in time proportional to multiplication of n and the number of Earley's states with token counters. Number of Earley's states for a context-free grammar is $O(n^2)$. Now we enumerate different token counters. At most n rules are used to generate a word of length n because the grammar has no chain rules and no λ-rules. Then at most one token is generated or consumed when a rule is used. For a token counter $\boldsymbol{v} = (v_1, \ldots, v_k)$ sum of all components of the vector $\boldsymbol{v}' = (|v_1|, \ldots, |v_k|)$ is at most n. There are

$$\binom{n+k}{k} = O(n^k)$$

different vectors in $\mathbb{N}^k$ that sum of all components is n. Considering cases that total tokens are less than n, there are $O(n^{k+1})$ different token counters. Therefore time complexity of Earley's algorithm with token counters is $O(n^{k+4})$. We note that if the underlying grammar is unambiguous, then the time complexity is $O(n^3)$.

The strict cycle-free condition, however, is so strict that some k-PN controlled grammars are excluded from the algorithm, an example is illustrated in Example 3. It is an open problem whether every language generated by a k-PN controlled grammar is generated by a strict cycle-free k-PN controlled grammar or not.

5   Conclusion

We have developed a parsing algorithm for k-PN controlled grammars. If an underlying grammar of a k-PN controlled grammar is unambiguous, then the algorithm is effective, that is, the time complexity is $O(n^3)$ where n is the length of a word. For ambiguous grammars, the algorithm is less effective, that is, the time complexity becomes $O(n^{k+4})$ for k-PN controlled grammars.

There are some restrictions on the algorithms investigated in this paper. The condition of no chain rules and no λ-rules is necessary to avoid infinitely many possibilities in derivation. Let us consider the next 1-PN controlled grammar shown in Fig. 3.

In the underlying grammar, there are infinite derivation trees for every word of the form $a(bb)^+$. For a fixed word in $a(bb)^+$, only one derivation is possible

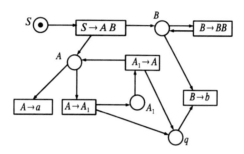

Fig. 3. A 1-PN controlled grammar with chain rules

under the control of the 1-Petri net. A parsing algorithm must select one possible derivation (tree) from infinite candidates. There may be two methods to resolve this problem: converting every k-PN controlled grammar into a grammar without chain rules and λ-rules or developing other algorithms for k-PN controlled grammars with chain rules or λ-rules. These will be done in future.

References

1. ter Beek, M., Kleijn, J.: Petri Net Control for Grammar Systems. In: Brauer, W., Ehrig, H., Karhumäki, J., Salomaa, A. (eds.) Formal and Natural Computing. LNCS, vol. 2300, pp. 220–243. Springer, Heidelberg (2002)
2. Dassow, J., Păun, G.: Regulated rewriting in formal language theory. Springer, Berlin (1989)
3. Dassow, J., Turaev, S.: k-Petri Net Controlled Grammars. In: Martín-Vide, C., Otto, F., Fernau, H. (eds.) LATA 2008. LNCS, vol. 5196, pp. 209–220. Springer, Heidelberg (2008)
4. Dassow, J., Turaev, S.: Petri net controlled grammars: the case of special Petri nets. Journal of Universal Computer Science 14, 2808–2835 (2009)
5. Dassow, J., Turaev, S.: Petri net controlled grammars: the power of labeling and final markings. Romanian Journal of Information Science and Technology 12, 191–207 (2009)
6. Dassow, J., Turaev, S.: Petri net controlled grammars with a bounded number of additional places. Acta Cybernetica 19, 609–634 (2010)
7. David, R., Alla, H.: Petri nets and grafcet: tool for modelling discrete event systems. Prentice Hall, Hertfordshire (1992)
8. Earley, J.: An efficient context-free parsing algorithm. Communications of the ACM 13, 94–102 (1970)
9. Hopcroft, J.H., Ullman, J.: Introduction to automata theory, languages, and computation. Addison-Wesley, Reading (1979)
10. Kasami, T.: An efficient recognition and syntax algorithm for context-free languages, Scientific Report AFCRL-65-758, Air force Cambridge Research Lab., Bedford, Mass (1965)

11. Reisig, W., Rozenberg, G. (eds.): APN 1998. LNCS, vol. 1491. Springer, Heidelberg (1998)
12. Rozenberg, G., Salomaa, A.: Handbook of Formal Languages, vol. 1-3. Springer, Berlin (1997)
13. Turaev, S.: Petri net controlled grammars. In: Proc. 3rd Doctoral Workshop on MEMICS 2007, Znojmo, Czech Republic, pp. 233–240 (2007)
14. Younger, D.H.: Recognition and parsing of context-free languages in time n^3. Information and Control 10, 189–208 (1967)

Descriptional Complexity
of Input-Driven Pushdown Automata

Alexander Okhotin[1,*], Xiaoxue Piao[2], and Kai Salomaa[2]

[1] Department of Mathematics, University of Turku,
20014 Turku, Finland
`alexander.okhotin@utu.fi`
[2] School of Computing, Queen's University,
Kingston, Ontario K7L 3N6, Canada
{`piao,ksalomaa`}`@cs.queensu.ca`

Abstract. It is known that a nondeterministic input-driven push-down automaton (IDPDA) can be determinized. Alur and Madhusudan ("Adding nesting structure to words", J.ACM 56(3), 2009) showed that a deterministic IDPDA simulating a nondeterministic IDPDA with n states and stack symbols may need, in the worst case, $2^{\Omega(n^2)}$ states. In their construction, the equivalent deterministic IDPDA does, in fact, not need to use the stack. This paper considers the size blow-up of determinization in more detail, and gives a lower bound construction, that is tight within a multiplicative constant, with respect to the size of the non-deterministic automaton both for the number of states and the number of stack symbols. The paper also surveys the recent results on operational state complexity of IDPDAs, and on the cost of converting a non-deterministic automaton to an unambiguous one, and an unambiguous automaton to a deterministic one.

Keywords: descriptional complexity, state complexity, nondeterminism, input-driven pushdown.

1 Introduction

In an input-driven pushdown automaton computation, the current input symbol determines whether the automaton performs a push operation, a pop operation, or does not touch the stack. *Input-driven pushdown automata* were originally introduced by Mehlhorn [37] in 1980, who showed that the languages recognized by such automata, called *input-driven languages* in the following, have space complexity $O\left(\frac{\log^2 n}{\log \log n}\right)$. This bound was further improved to $O(\log n)$ by von Braunmühl and Verbeek [8,9], and later Rytter [49] obtained a different algorithm with the same space requirements. Von Braunmühl and Verbeek [9] also considered the nondeterministic variant of the model, and proved it to be equal in power to the deterministic case. Input-driven languages were shown to be in NC^1 by Dymond [15].

---

* Supported by the Academy of Finland under grant 134860.

H. Bordihn, M. Kutrib, and B. Truthe (Eds.): Dassow Festschrift 2012, LNCS 7300, pp. 186–206, 2012.
© Springer-Verlag Berlin Heidelberg 2012

The model was reintroduced by Alur and Madhusudan [2] in 2004 under the name of *visibly pushdown automata*, and there has been considerably more work on the model since then [1,3,11,43,45]. In particular, Alur and Madhusudan established bounds for the descriptional complexity of determinizing a nondeterministic automaton of this kind [3], and showed that the corresponding class of languages has strong closure properties. Furthermore, part of the recent literature considers an equivalent model called *nested word automaton* [3,4,46,50]. Nested words provide a natural data model for applications like XML document processing, where the data has a dual linear-hierarchical structure. In particular, the $2^{\Omega(n^2)}$ lower bound for the descriptional complexity of determinizing nondeterministic automata was established using the model of nested word automata [3].

This paper sticks to the original name of *input-driven pushdown automata* (IDPDA) for this machine model. Though the name "visibly pushdown automaton" has been more widespread in the recent literature, the authors believe that the original name better describes the operation of the machine model[1]. In the following, when referring to work on nested word automata we, without separate mention, use terminology associated with IDPDAs, that is, talk about *states* and *stack symbols* of an IDPDA, instead of linear and hierarchical states of a nested word automaton.

Input-driven pushdown automata were shown by Gauwin et al. [17] to be equivalent to the *pushdown forest automata* of Neumann and Seidl [39]. A pushdown forest automaton can be viewed, roughly speaking, as a tree automaton that traverses a tree in depth-first left-to-right order and is equipped with a synchronized pushdown. That is, the machine pushes the stack when going down to the leftmost child and pops the stack when returning from the rightmost child. The class of tree languages recognized by pushdown forest automata coincides with the regular tree languages; however, a pushdown forest automaton may be considerably more concise than an ordinary bottom-up tree automaton: Alur and Madhusudan [3] establish this succinctness gap by comparing nested word automata and bottom-up tree automata.

A tree-walking transducer model with a synchronized pushdown, albeit working on ranked trees and with an arbitrary traversal strategy, was considered already by Engelfriet et al. [16]. Kamimura and Slutzki [26] have proved in 1981, that the nondeterministic and deterministic variants of such graph-walking automata equipped with a synchronized pushdown are equivalent. Instead of trees, the automata of Kamimura and Slutzki [26] operate on directed acyclic graphs.

The subject of this paper is **descriptional complexity** of input-driven languages, which is measured in terms of the number of states and pushdown symbols in several kinds of IDPDAs. This work further describes the capabilities of IDPDAs, and also contributes to the ongoing studies of the state complexity of other important language families. Descriptional complexity of regular

---

[1] For example, Bollig [7] actually uses multiple times the term "input-driven" to describe the operation of visibly pushdown automata, in spite of being unaware of this name from Mehlhorn [37].

languages has been fruitfully studied over half a century, and particularly intensively in the last two decades, see a recent survey by Holzer and Kutrib [24]. The size cost of determinizing finite automata was determined in the early papers of Rabin and Scott [47] and Lupanov [34], the costs of unambiguous nondeterminism were settled by Leung [33], and the tradeoffs between two-way and one-way finite automata were determined by Kapoutsis [27]. Similar questions in the case of a one-letter alphabet were addressed by Chrobak [12], Mereghetti and Pighizzini [38], Okhotin [41,42] and Kunc and Okhotin [28]. State complexity of basic operations on DFAs was determined by Maslov [36] and by Yu et al. [54], and the subject was exhaustively studied in numerous follow-up papers; Holzer and Kutrib [22] similarily settled the state complexity of operations on NFAs, while Jirásková and Okhotin [25] and Kunc and Okhotin [29] obtained the first results on the complexity of operations on two-way finite automata. Turning to context-free languages, their descriptional complexity is known to involve undecidability problems and non-recursive tradeoffs, brought to light by Hartmanis [21]; their succinctness with respect to the number of nonterminal symbols in a grammar was studied in the works of Gruska [19], Domaratzki et al. [14] and Dassow and Stiebe [13]. As the input-driven languages are strictly contained between the regular languages and the context-free languages, and possess good closure properties, the study of their descriptional complexity belongs to the same research area.

The first result on the succinctness of input-driven pushdown automata was obtained by Alur and Madhusudan [2,3], who determined the costs of their determinization in terms of the number of states. This line of research was continued by the authors [45], who refined this result by investigating the succinctness of an intermediate class with unambiguous nondeterminism (UIDPDA). The complexity of basic operations on input-driven languages was gradually determined in a series of papers by Han and Salomaa [20], Piao and Salomaa [46], Salomaa [50] and Okhotin and Salomaa [44]. In this paper, the above results shall be reviewed, and, in particular, the size blow-up of determinization shall be considered in more detail. We give a lower bound for the size of a deterministic IDPDA equivalent to a given nondeterministic IDPDA that is tight within a multiplicative constant on the size of the original automaton, both with respect to the number of states and the number of stack symbols. The results of Alur and Madhusudan [2,3] give the same lower bound for the number of states of the equivalent deterministic IDPDA; however, in their construction, the deterministic IDPDA does not need to use the stack at all, and could be replaced by a deterministic finite automaton (DFA).

The IDPDA model has been intensively studied, because it allows determinization, and because of its strong closure and decidability properties [3]. Input-driven pushdown automata with multiple stacks have been introduced and shown to retain many of the desirable closure and decidablity properties [30]. Another promising extension of the input-driven pushdown automaton model are the height-deterministic pushdown automata of Nowotka and Srba [40]. In the last section, we discuss possible extensions of the IDPDA model, as well as, the realization of such automata as graph automata by Madhusudan and Parlato [35] that extends the correspondence between IDPDAs and nested word automata.

2   Preliminaries

We assume the reader to be familiar with basic notions of formal languages [48,52]. Good references on descriptional complexity of finite automata include surveys by Holzer and Kutrib [23,24] and a handbook article by Yu [53].

The cardinality of a finite set F is $|F|$ and the set of subsets of F is 2^F. Let $\mathbb{N}$ denote the set of positive integers and for $n \in \mathbb{N}$ denote $[n] = \{1, \ldots, n\}$. The domain of a binary relation $R \subseteq X \times Y$ is $\mathrm{dom}(R) = \{x \in X \mid (\exists y \in Y)(x, y) \in R\}$. The image of R is $\mathrm{im}(R) = \{y \in Y \mid (\exists x \in X)(x, y) \in R\}$. The relation R is said to be *left-total* if $\mathrm{dom}(R) = X$ and R is *surjective* if $\mathrm{im}(R) = Y$. A left-total and surjective binary relation is called a *correspondence*. For every integer $n \geqslant 1$, the number of correspondences $R \subseteq [n] \times [n]$ is denoted by $\mathrm{corr}(n)$. It would be possible to give a recursive formula for the number of correspondences on $[n] \times [n]$, however, for our purposes it is sufficient to have a lower bound $2^{\Omega(n^2)}$. Consider that

$$\mathrm{corr}(n) \geqslant (2^n - 1)^{n-1}. \tag{1}$$

The lower bound of (1) follows from the observation that if f is a function $[n - 1] \to (2^{[n]} - \{\varnothing\})$, then the relation

$$\{ (x, y) \mid x \in [n - 1], y \in f(x) \} \cup (\{n\} \times [n])$$

is a correspondence on $[n] \times [n]$.

In the following, Σ denotes a finite input alphabet. The set of strings over Σ is Σ^*. The length of a string $w \in \Sigma^*$ is denoted by $|w|$, and the empty string ε is the unique string of length 0. Let $\Sigma^+ = \Sigma^* \setminus \{\varepsilon\}$ be the set of nonempty strings, let Σ^m with $m \geqslant 0$ be the set of strings of length exactly m, and let $\Sigma^{\leqslant m}$ denote the set of strings of length at most m. For any subalphabet $\Gamma \subseteq \Sigma$, the projection $\pi_\Gamma \colon \Sigma^* \to \Gamma^*$ is the homomorphism that maps all symbols in Γ to themselves and erases the symbols from $\Sigma \setminus \Gamma$.

Next, we recall and introduce definitions and notation concerning input-driven pushdown automata (IDPDA) [37], also known under the name of visibly pushdown automata. The reader is referred to Alur and Madhusudan [2,3] for a more detailed presentation and examples.

An IDPDA is a pushdown automaton, in which the type of the current input symbol determines whether the next operation pushes onto the stack, pops from the stack, or does not touch the stack. In the formal definition, the input alphabet is an *action alphabet*, defined as a triple $\widetilde{\Sigma} = (\Sigma_{+1}, \Sigma_{-1}, \Sigma_0)$, in which the components Σ_{+1}, Σ_{-1} and Σ_0 are finite disjoint sets. In the following, unless otherwise mentioned, Σ_{+1}, Σ_{-1} and Σ_0 always refer to components of an action alphabet, and their union is denoted by Σ. A string over $\widetilde{\Sigma}$ is an ordinary string over Σ, where each symbol is assigned a "type" depending on the component it belongs to.

Definition 2.1. *A nondeterministic input-driven pushdown automaton, NIDPDA is a tuple*

$$A = (\widetilde{\Sigma}, \Gamma, Q, q_0, F, \delta_i, \delta_{\mathrm{push}}, \delta_{\mathrm{pop}}) \tag{2}$$

where $\Sigma = \Sigma_{+1} \cup \Sigma_{-1} \cup \Sigma_0$ is the input alphabet, Γ is the finite set of stack symbols, Q is the finite set of internal states, $q_0 \in Q$ is the start state, $F \subseteq Q$ is the set of final states, $\delta_i : Q \times \Sigma_0 \to 2^Q$ is the internal transition function, and $\delta_{\mathrm{push}} : Q \times \Sigma_{+1} \to 2^{Q \times \Gamma}$ and $\delta_{\mathrm{pop}} : Q \times (\Gamma \cup \{\bot\}) \times \Sigma_{-1} \to 2^Q$ are the transition functions determining the push and pop operations, respectively. The symbol $\bot \notin \Gamma$ is used to denote the empty stack.

A *configuration of A* is a tuple (q, w, u), where $q \in Q$ is the state, $w \in \Sigma^*$ is the remaining input and $u \in \Gamma^*$ is the stack contents[2]. The *height* of the stack of the configuration (q, w, u) is $|u|$ (and hence the height of the empty stack is zero). The set of configurations of A is $\mathcal{C}(A)$ and we define the single step computation relation $\vdash_A \subseteq \mathcal{C}(A) \times \mathcal{C}(A)$ as follows.

Operation not changing the stack: $(q, aw, u) \vdash_A (q', w, u)$, for all $a \in \Sigma_0$, $q' \in \delta_i(q, a)$, $w \in \Sigma^*$ and $u \in \Gamma^*$.

Push operation: $(q, aw, u) \vdash_A (q', w, \gamma u)$, for all $a \in \Sigma_{+1}$, $(q', \gamma) \in \delta_{\mathrm{push}}(q, a)$, $\gamma \in \Gamma$, $w \in \Sigma^*$ and $u \in \Gamma^*$.

Pop operation: $(q, aw, \gamma u) \vdash_A (q', w, u)$ for all $a \in \Sigma_{-1}$, $q' \in \delta_{\mathrm{pop}}(q, \gamma, a)$, $\gamma \in \Gamma$, $w \in \Sigma^*$ and $u \in \Gamma^*$; furthermore, $(q, aw, \varepsilon) \vdash_A (q', w, \varepsilon)$, for all $a \in \Sigma_{-1}$, $q' \in \delta_{\mathrm{pop}}(q, \bot, a)$ and $w \in \Sigma^*$.

According to the last case, when the automaton A encounters an element $a \in \Sigma_{-1}$ with an empty stack, A can make a state transition chosen from $\delta_{\mathrm{pop}}(q_1, \bot, a)$ (where $q_1 \in Q$ is the current state) and the stack remains empty. When reading a symbol from $\Sigma_0 \cup \Sigma_{+1}$, the next state does not depend on the top stack symbol. Naturally, given an IDPDA A, it would be possible to construct an IDPDA B that simulates the computation of A and keeps track (in its state) of the topmost stack symbol in the corresponding computation of A. However, this transformation would need to increase the number of states and the number of stack symbols of A. The descriptional complexity results on IDPDAs assume that the transition relations are defined as in Definition 2.1.

The initial configuration of A on input w is $\mathcal{C}_A^{\mathrm{init}}(w) = (q_0, w, \varepsilon)$. The language recognized by A is defined as

$$L(A) = \{w \in \Sigma^* \mid \mathcal{C}_A^{\mathrm{init}}(w) \vdash_A^* (q, \varepsilon, u) \text{ for some } q \in F, u \in \Gamma^*\}.$$

The distinguishing property of input-driven pushdown automata is that the type of the stack operation is always determined by the input symbol and, in particular, the height of the stack that any NIDPDA reaches in an arbitrary computation at the end of an input w is uniquely determined by w.

---

[2] A stack ε is, for the purposes of the transition relation δ_{pop}, interpreted to contain the bottom of stack symbol $\bot$.

A configuration (q, ε, u), with $q \in Q$, $u \in \Gamma^*$ and with the remaining input empty, is called a *terminal configuration*. If $C_A^{\text{init}}(w) \vdash_A^* C$, where C is a terminal configuration, we say that the *stack height* of w is the height of the stack of C. As observed above, the stack height of a string w is a property of w that does not depend on the nondeterministic computation of A reaching the terminal configuration C.

An NIDPDA A as in (2) is *deterministic*, a DIDPDA, if its transition functions δ_i, δ_{push} and δ_{pop} give at most one action in each configuration, that is, are defined as partial functions $\delta_i \colon Q \times \Sigma_0 \to Q$, $\delta_{\text{push}} \colon Q \times \Sigma_{+1} \to Q \times \Gamma$ and $\delta_{\text{pop}} \colon Q \times (\Gamma \cup \{\bot\}) \times \Sigma_{-1} \to Q$.

The computations of an NIDPDA accept by final state only. There is a natural correspondence between NIDPDAs and nested word automata [3], where the stack contents of the NIDPDA consists of the sequence of hierarchical states assigned to currently pending call symbols in the corresponding nested word automaton computation. The above definition of acceptance corresponds to a so called *linearly accepting* nested word automaton [3] that does not care about the hierarchical states assigned to pending calls.

In the following sections, when citing results from papers using the nested word automaton formalism, often without separately mentioning it, we translate them to the notation of input-driven pushdown automata.

3   Size Explosion of Determinization

An extension of the subset construction allows a deterministic simulation of an NIDPDA. The following upper bound and lower bounds for the size blow-up of determinization were established in two papers by Alur and Madhusudan [2,3], respectively.

By the *size* of an NIDPDA A we mean the sum of the number of states of A and the number of stack symbols of A.

Theorem 3.1 (Alur and Madhusudan [2,3])). *An NIDPDA with h states and k stack symbols has an equivalent DIDPDA with $2^{k \cdot h}$ states and $2^{k \cdot k}$ stack symbols. Conversely, there exists a family of languages*

$$L_n = \{<u_1\$v_1\#u_2\$v_2\# \cdots \#u_m\$v_m\%v>u \mid m \geqslant 1,\ u_i, v_i \in \{0,1\}^+ :$$
$$\exists t \in \{1, \ldots, m\} :\ u = u_t,\ v = v_t,\ |u| = |v| = \lfloor \log_2 n \rfloor\},$$

with $n \geqslant 1$, such that L_n has an NIDPDA of size $O(n)$ and any DIDPDA recognizing L_n needs $2^{\Omega(n^2)}$ states.

A small NIDPDA for the witness language L_n begins its computation on the left bracket $<$ by guessing the string u and pushing it to the stack, as well as remembering it in the internal states. Then it scans over the pairs $u_i\$v_i$, until it nondeterministically chooses one of them, and verifies that the first component of the chosen pair is the string u stored in the internal state. Next, it remembers the second component of the next pair as v, scans until the marker $\%$, and verifies that the string after the marker % is exactly the string v remembered in the state. It remains to recall the value of u from the stack upon reading the right bracket $>$, and check that it matches the string given in the end of the input.

This language is designed to enable efficient use of nondeterminism when guessing stack symbols to be pushed. At the same time, a deterministic computation cannot utilize the stack in any useful way. This means that a minimal equivalent DIDPDA is essentially an ordinary deterministic finite automaton, and does not need the stack.

Here we consider the size exposion of determinization in a little more detail. We construct a family of languages $L_{k,h}$, for $k, h \in \mathbb{N}$ with $k \leqslant h$, such that each $L_{k,h}$ has an NIDPDA with $c \cdot h$ states and k stack symbols, where c is a constant independent of k and h, and any DIDPDA for the language $L_{k,h}$ needs $2^{k \cdot h}$ states and $2^{\Omega(k^2)}$ stack symbols.

First we recall from [45,46] some techniques for establishing lower bounds for DIDPDAs. Lemmas 3.1 and 3.2 are inspired by the fooling set methods used for lower bounds for NFAs [5,18,52]. When dealing with input-driven pushdown automata we need to rely on fooling set type lower bound methods also in the case of deterministic automata and, in general, the methods do not always establish optimal lower bounds [50]. Further variants of the lower bound methods can be found in the literature [20,45,46] where they have been used to establish tight lower bounds for operational state complexity of deterministic and nondeterministic IDPDAs.

Definition 3.1. *Let $\widetilde{\Sigma} = (\Sigma_{+1}, \Sigma_{-1}, \Sigma_0)$ be an action alphabet and $L \subseteq \Sigma^*$.*

(a) *A finite set $S \subseteq \Sigma^*$ is a k-separator set for L, $k \geqslant 0$, if*
   (i) *each string of S has stack height k,*
   (ii) *each string of S is a prefix of some string of L, and,*
   (iii) *for any $u, v \in S$, $u \neq v$, there exists $x \in \Sigma^*$ such that $ux \in L$ if and only if $vx \notin L$.*

(b) *A finite set $S \subseteq \Sigma^*$ is a separator set (for L) if S is a k-separator set (for L) for some $k \geqslant 0$.*
   A separator set $\{u_1, \ldots, u_n\}$ is a linear separator set if there exists $v \in \Sigma^*$ such that $u_i = vu'_i$, $u'_i \in \Sigma_0^*$, $i = 1, \ldots, n$.

Any two strings of a linear separator set contain only symbols of Σ_0 after their longest common prefix.

The lemma below is from Piao and Salomaa [46] and Lemma 3.2 is from Okhotin and Salomaa [45]. For completeness we include a short proof for the lemmas.

Lemma 3.1 (Piao and Salomaa [46]). *Let A be a DIDPDA with set of states Q and set of stack symbols Γ. If S is a k-separator set for $L(A)$, then*

$$|\Gamma|^k \cdot |Q| \geqslant |S|.$$

Proof. Consider any distinct strings $u_1, u_2 \in S$, $u_1 \neq u_2$. Since u_i is a prefix of some string of $L(A)$, there exists a unique terminal configuration (q_i, ε, v_i), $q_i \in Q$, $v_i \in \Gamma^*$, such that $C_A^{\text{init}} \vdash_A^* (q_i, \varepsilon, v_i)$, $i = 1, 2$. The stack height of each element of S is k and, hence, $v_i \in \Gamma^k$, $i = 1, 2$. Since (q_i, ε, v_i) uniquely determines the computation of A on a suffix following u_i, condition (a-iii) of Definition 3.1 implies that $q_1 \neq q_2$ or $v_1 \neq v_2$, and the claim follows. □

Lemma 3.2 (Okhotin and Salomaa [45]). *If a language L has a linear separator set S, then any DIDPDA for L has at least $|S|$ states.*

Proof. Let $S = \{u_1, \ldots, u_m\}$ and let A be an arbitrary DIDPDA recognizing L. Since u_i is a prefix of some string of L, the computation of A on u_i is defined and ends in a configuration C_i with a state q_i with $1 \leqslant i \leqslant m$. For the sake of contradiction, assume that A has fewer than $|S|$ states; then $q_i = q_j$ for some $i \neq j$. Since after their common prefix, u_i and u_j contain only symbols of Σ_0 and A is deterministic, we know that the stack contents of C_i and C_j are the same. Now, as in the proof of the previous lemma, we get a contradiction from the fact that S is a separator set. □

A limitation of Lemma 3.1 is the requirement that all strings of a separator set need to have the same stack height. It seems not easy to develop analogous conditions without this requirement. On the other hand, as will be done below, by using more than one k-separator set (with different values of k) for the same language and the linear separator sets of Lemma 3.2 we can obtain improved lower bounds for the size of DIDPDAs.

Next we define the languages used for our size lower bound for converting an NIDPDA to a deterministic automaton.

Choose $\widetilde{\Sigma} = (\Sigma_{+1}, \Sigma_{-1}, \Sigma_0)$, where $\Sigma_0 = \{0, \$, \#, \%\}$, $\Sigma_{+1} = \{<\}$ and $\Sigma_{-1} = \{>\}$, and let $k, h \in \mathbb{N}$. For $1 \leqslant r \leqslant k$ and $1 \leqslant s \leqslant h$ we define the language

$$M_{r,s,0}^{k,h} = \{ <u_1\$v_1\#u_2\$v_2\# \cdots \#u_m\$v_m\%v_t>u_t \mid u_i, v_i \in 0^+, \tag{3}$$
$$i = 1, \ldots, m, \; m \geqslant 1, \; |u_t| = r, |v_t| = s, \; t \in \{1, \ldots, m\} \}$$

The union of all the sets $M_{r,s,0}^{k,h}$ with $1 \leqslant r \leqslant k$ and $1 \leqslant s \leqslant h$ would be roughly analogous to the language used by Alur and Madhusudan [3] to establish the lower bound for the size blow-up of the number of states in determinization.

Define a family of languages $M_{x,i}^{k,h}$, $1 \leqslant x \leqslant k$, inductively on i as follows:

$$M_{x,0}^{k,h} = \bigcup_{1 \leqslant s \leqslant h} M_{x,s,0}^{k,h}, \tag{4}$$

$$M_{x,i+1}^{k,h} = \{<u_1\$v_1\#u_2\$v_2\# \cdots u_m\$v_m \, \alpha_i>u_t \mid u_j, v_j \in 0^+, \tag{5}$$
$$j = 1, \ldots, m, \; m \geqslant 1, \; (\exists t \in \{1, \ldots, m\}) \; \text{such that}$$
$$|u_t| = x, |v_t| = r, \; \text{and} \; \alpha_i \in M_{r,i}^{k,h}, 1 \leqslant r \leqslant k\}.$$

Any computation of an NIDPDA on any $w_j \in M_{x,j}^{k,h}$ with $j \geqslant 1$ must perform $j + 1$ push-operations on the stack followed by $j + 1$ pop-operations. According to definition (5), if $i \geqslant 1$, then a string w_{i+1} of $M_{x,i+1}^{k,h}$ always contains a unique string of $M_{r,i}^{k,h}$, for some $1 \leqslant r \leqslant k$, as a substring. If $i = 0$, then $M_{x,i+1}^{k,h}$ has a unique substring from $M_{r,s,0}^{k,h}$, for some $1 \leqslant r \leqslant k$, $1 \leqslant s \leqslant h$. In the notations of (5) this is α_i, and α_i is called the *directly enclosed substring of* w_{i+1}.

In the notation of (5), the choice of the substring $u_t \$ v_t$ with $1 \leqslant t \leqslant m$ need not be unique, but the directly enclosed substring of $w_{i+1} \in M_{x,i+1}^{k,h}$ is determined as the unique substring beginning with the second occurrence of the symbol $< \in \Sigma_{+1}$ in w_{i+1}, and ending with the symbol preceding the last occurrence of the symbol $> \in \Sigma_{-1}$.

Now we define

$$L_{k,h} = \bigcup_{\substack{1 \leqslant x \leqslant k, \\ i \geqslant 0}} M_{x,i}^{k,h}.$$

Lemma 3.3. *There exists a constant* $c \in \mathbb{N}$ *such that for* $k, h \geqslant 1$, *the language* $L_{k,h}$ *is recognized by an NIDPDA with* $c \cdot \max\{k, h\}$ *states and* k *stack symbols.*

Proof. We describe the construction of an NIDPDA A recognizing $L_{k,h}$. The below discussion assumes that the input does not contain any left brackets $<$ after the first occurrence of a right bracket $>$ (that is, the projection of the input to $\{<, >\}$ belongs to $<^* >^*$). Clearly, this property can be checked by doubling the number of states.

The set of stack symbols of A is $\{\gamma_1, \ldots, \gamma_k\}$. On the first input symbol $<$, A pushes a nondeterministically chosen symbol γ_j, with $1 \leqslant j \leqslant k$, to the stack. In general, when A on input symbol $<$ pushes γ_j, it "remembers" j in the state and nondeterministically chooses to operate as in (i) or (ii) below. As explained in the following, except for the first input symbol $<$, the stack symbol γ_j is determined by the state of the computation at $<$.

(i) A nondeterministically guesses that it has read the last occurrence of $<$. The computation of A verifies that the input before the first symbol $>$ is of the form $(0^+\$0^+\#)^*0^+\$0^+\%0^{j'}$, and A nondeterministically "selects" one substring $0^{i_1}\$0^{i_2}$ delimited by the $\#$-markers and verifies that $i_1 = j$ and $i_2 = j'$.

(ii) A nondeterministically guesses that $<$ is not the last occurrence of this symbol. Now the computation verifies that the input before the next $<$ is of the form $(0^+\$0^+\#)^*0^+\$0^+$, and A nondeterministically "selects" one substring $0^{i_1}\$0^{i_2}$ delimited by the $\#$-markers and verifies that $i_1 = j$ and then goes to a state that in the next symbol $<$ pushes γ_{i_2}.

On a symbol $>$, the automaton A pops γ_j from the stack, and the subsequent computation verifies that the substring before the next $>$ is 0^j.

The operation of A between the push and pop operations needs to store and compare strings of the form 0^j, $j \leqslant \max\{k, h\}$ and this can clearly be done with the required number of states. Note that because all occurrences of symbol $<$ precede all occurrences of $>$, at no point in the computation the state of A needs to remember the length of more than one substring 0^j, $1 \leqslant j \leqslant \max\{k, h\}$.

Assuming the input is well-nested, the push and pop operations nondeterministically verify that the input is of the form as specified in the definition of $M_{x,i+1}^{k,h}$ (5), or $M_{r,s,0}^{k,h}$ (3). In the latter case, at the first symbol $<$, according to (i), A guesses that this was the last occurrence of $<$ and pushes γ_r to the stack.

Finally, A can verify that the input is well nested, by storing in the state the information whether or not the stack is empty, and for this it is sufficient to double the number of states.[3]   □

Next we give a lower bound for the size of a DIDPDA recognizing $L_{k,h}$. Recall that $\mathrm{corr}(n)$, $n \geqslant 1$, is the number of correspondences on $[n] \times [n]$.

Lemma 3.4. *Any DIDPDA for the language* $L_{k,h}$, $h \geqslant k \geqslant 1$, *needs at least* $2^{k \cdot h} - 1$ *states and* $\mathrm{corr}(k)$ *stack symbols.*

Proof. For $R \subseteq [k] \times [h]$, denote

$$w_R = <u_1 \# u_2 \# \cdots \# u_{|R|},$$

where the sequence of substrings u_i, $i = 1, \ldots, |R|$, consists of all strings $0^r \$ 0^s$, $(r, s) \in R$, $1 \leqslant r \leqslant k$, $1 \leqslant s \leqslant h$, in a fixed but arbitrary order. Now the set

$$S_1 = \{w_R \mid \varnothing \neq R \subseteq [k] \times [h]\}$$

is a linear separator set for $L_{k,h}$. To see this consider two non-empty relations $R_1 \neq R_2$, and without loss of generality choose $(r, s) \in R_1 - R_2$, $1 \leqslant r \leqslant k$, $1 \leqslant s \leqslant h$. This means that $w_{R_1} \% 0^s > 0^r \in L_{k,h}$ and $w_{R_2} \% 0^s > 0^r \notin L_{k,h}$. All strings of S_1 contain an element of $\Sigma_{+1} \cup \Sigma_{-1}$ only as the first symbol, and hence S_1 is a linear separator set.

Since $|S_1| = 2^{k \cdot h} - 1$, Lemma 3.2 gives the required lower bound for the number of states of an arbitrary DIDPDA for $L_{k,h}$.

To get the lower bound for the number of stack symbols, consider $R \subseteq [k] \times [k]$. By the *string representation of R* we mean a string $w_R = u_1 \# u_2 \# \cdots \# u_{|R|}$ where the substrings u_i, $1 \leqslant i \leqslant |R|$, are all substrings $0^r \$ 0^s$, $1 \leqslant r, s \leqslant k$, in some (fixed but arbitrary) order. Thus for each $R \subseteq [k] \times [k]$ the string w_R is uniquely determined.

For $m \geqslant 2$, define

$$S_m = \{<w_{R_1} <w_{R_2} < \cdots <w_{R_m} \mid$$
$$R_i \subseteq [k] \times [k] \text{ is a correspondence}, i = 1, \ldots, m \}.$$

We show that S_m is an m-separator set for $L_{k,h}$. Consider two distinct strings of S_m,

$$u_1 = <w_{R_1} <w_{R_2} < \cdots <w_{R_m} \quad \text{and} \quad u_2 = <w_{R_1'} <w_{R_2'} < \cdots <w_{R_m'}.$$

Thus, there exists $1 \leqslant i \leqslant m$ such that $R_i \neq R_i'$ and, without loss of generality, we can choose $(r_i^0, s_i^0) \in R_i - R_i'$. Since each R_j, $1 \leqslant j \leqslant m$, is a correspondence,

---

[3] This is the same construction that converts an arbitrary nested string automaton to a linearly accepting one [3].

there exists a chain $(r_j, s_j) \in R_j$, $j = 1, \ldots, m$, such that $s_j = r_{j+1}$, $1 \leqslant j < m$, and $r_i = r_i^0$, $s_i = s_i^0$, that is, the element r_i^0 is in the image of the composed relation $R_1 \circ \cdots \circ R_{i-1}$, and s_i^0 is in the domain of $R_{i+1} \circ \ldots \circ R_m$.

This means that

$$u_1 \% s_m > r_m > r_{m-1} > \cdots > r_2 > r_1 \in L_{k,h}.$$

On the other hand, we note that $(r_i^0, s_i^0) = (r_i, r_{i+1})$ if $i < m$, and $(r_i^0, s_i^0) = (r_m, s_m)$ if $i = m$. Since $(r_i^0, s_i^0) \notin R_i'$, it follows that

$$u_2 \% s_m > r_m > r_{m-1} > \cdots > r_2 > r_1 \notin L_{k,h}.$$

Now Lemma 3.1 implies that if A is an arbitrary DIDPDA for $L_{k,h}$ with set of states Q and set of stack symbols Γ, then

$$|\Gamma|^m \cdot |Q| \geqslant |S_m| \geqslant \mathrm{corr}(k)^m.$$

Since the inequality has to hold for any $m \geqslant 2$, we get $|\Gamma| \geqslant \mathrm{corr}(k)$. □

Using Lemmas 3.3 and 3.4, and the estimation (1) we have:

Theorem 3.2. *For all $k, h \in \mathbb{N}$, $k \leqslant h$, there exists a language $L_{k,h}$ recognized by an NIDPDA with $O(h)$ states and $O(k)$ stack symbols such that any DIDPDA for $L_{k,h}$ needs $\Omega(2^{k \cdot h})$ states and $\Omega(2^{k^2})$ stack symbols.*

Note that the assumption $k \leqslant h$ was used in Lemma 3.3, and it is needed in Theorem 3.2 to guarantee that the number of states of an NIDPDA for $L_{k,h}$ does not depend on k.

4　Unambiguous Nondeterminism and Its Succinctness

Nondeterministic machines of any kind have an important special case, in which every accepted string has a unique accepting computation. In particular, this restriction yields unambiguous finite automata (UFA) and unambiguous complexity classes, such as UL and UP. For UFAs, it is known from Leung [33] that simulating an n-state UFA requires, in the worst case, a DFA with 2^n states, while the NFA–UFA tradeoff is $2^n - 1$. In the case of a one-letter alphabet, the UFA–DFA and NFA–UFA tradeoffs are known from Okhotin [41,42].

Applying the same restriction to IDPDAs leads to the notion of an *unambiguous input-driven pushdown automaton* (UIDPDA). The succinctness tradeoffs between NIDPDAs, UIDPDAs and DIDPDAs, which are the subject of this section, are similar to the case of finite automata: converting a UIDPDA of size n to a DIDPDA requires $2^{\Theta(n^2)}$ states in the worst case, and the same number of states is necessary for converting NIDPDAs to UIDPDAs [44].

The UIDPDA–DIDPDA tradeoff is established using the lower bound method of Lemma 3.1 for DIDPDAs. However, the known lower bound languages used for the NIDPDA–DIDPDA tradeoff, as presented in Theorems 3.1 and 3.2, do not have small UIDPDAs. The next theorem presents an example with a similarly high lower bound on the size of a DIDPDA, which is, however, recognized by a small UIDPDA.

For every binary string $w \in \{0,1\}^*$, let $(w)_2$ denote the nonnegative integer with the binary notation w.

Theorem 4.1 (Okhotin and Salomaa [44]). *For every $n \geqslant 1$, the language*

$$\{<x_0\#x_1\#\ldots\#x_\ell\$v>u \mid x_i \in \{0,1\}^*, 1 \leqslant i \leqslant \ell,$$
$$u, v \in \{0,1\}^{\lceil \log n \rceil}, \text{bit number } (v)_2 \text{ in } x_{(u)_2} \text{ is } 1\}$$

is recognized by an UIDPDA with $O(n)$ states, but every DIDPDA for this language needs 2^{n^2} states.

The small automaton recognizing this language can be unambiguous, because all the nondeterministic guesses it has to make are explicitly written out in the end of the string as $v>u$. Then, any guesses made are eventually verified, and hence at most one possible sequence of guesses may lead to acceptance.

In the beginning of the computation, when reading the left bracket $<$, the automaton guesses a string $u \in \{0,1\}^{\lceil \log n \rceil}$, stores it in the stack, and at the same time preserves it in the current state. Inside the brackets, it looks for the string $x_{(u)_2}$, and then nondeterministically guesses a true bit in this string, remembering its number in the current state. Next, the automaton reads until the marker $\$$, and then checks that the string $v \in \{0,1\}^{\lceil \log n \rceil}$ represents the binary notation of the number of that bit: that is, verifies that v has earlier been guessed correctly. Finally, once the value of u guessed in the beginning is popped from the stack after reading the right bracket $>$, it remains to check that it matches the intended value of u included in the input after the bracket.

Turning to the size tradeoff between NIDPDAs and UIDPDAs, proving that every UIDPDA for a given language must have a certain number of states requires a different lower bound method. Such a method is obtained by generalizing a result by Schmidt [51] for unambiguous finite automata.

Lemma 4.1 (Okhotin and Salomaa [44]; cf. Schmidt [51], Leung [33]). *Let $L \subseteq \Sigma^*$ be a language. Let $F = \{(x_i, y_i) \mid i = 1, \ldots, n\}$ be a paired k-set, and define an integer matrix $M = M(F, L) \in \mathbb{Z}^{n \times n}$ by setting $M_{i,j} = 1$ if $x_iy_j \in L$, and $M_{i,j} = 0$ otherwise. Then, if L is recognized by an UIDPDA A with a set of states Q and a pushdown alphabet Γ, this implies*

$$|\Gamma|^k \cdot |Q| \geqslant \operatorname{rank} M.$$

Using this lower bound method, one can establish the following tradeoff.

Theorem 4.2 (Okhotin and Salomaa [44]). *For every $n \geqslant 1$, the language*

$$L_n = \{ <\#u_1\$v_1\#u_2\$v_2\#\cdots\#u_m\$v_m\%v'_1\$u'_1\#v'_2\$u'_2\#\cdots\#v'_{m'}\$u'_{m'}> \mid \quad (6)$$
$$m, m' \geqslant 1, u_i, v_i, u'_j, v'_j \in a^+, \exists s, t : u_s = u'_t, v_s = v'_t, |u_s|, |v_s| \leqslant n\}.$$

is recognized by an NIDPDA with $O(n)$ states, but every UIDPDA for L_n needs at least $2^{\lfloor \frac{n^2}{2} \rfloor - 1}$ states.

Upon reading the left bracket $<$, the NIDPDA guesses a string u, pushes it to the stack and preserves it in the state, and then detects a pattern of the form $< \ldots u\$v \ldots \% \ldots v\$u \ldots >$ in the input. First, it nondeterministically selects a pair in the left part of the string, and verifies that its first component is the earlier guessed string u. Then it memorizes the second component v of this pair and proceeds to the right part, where another pair is nondeterministically chosen. The first component of this pair should be v, the second component is memorized, and checked against the value of u stored in the stack upon the transition by the right bracket $>$.

The lower bound on the size of every UIDPDA recognizing this language is obtained by using a set of pairs of strings ($<x_i\%$, $y_i>$), where each x_i and y_i encodes a subset of $[1, n] \times [1, n]$ of size $\frac{n^2}{2}$ (assuming n is even). The sets of pairs used in x_i and in y_i form a disjoint partition of $[1, n] \times [1, n]$, and hence each string $<x_i\%y_i>$ is not in L_n, but $<x_i\%y_j> \in L_n$ for all $i \neq j$. Therefore, the corresponding matrix, as defined in Lemma 4.1, has a zero diagonal and ones in the rest of its elements. Hence, the matrix has full rank, and the lemma asserts the desired lower bound.

The results in Theorems 4.1–4.2 refine the NIDPDA–DIDPDA tradeoff, as established by Alur and Madhusudan [2,3]. However, the given bounds apply only to the number of states. Establishing stronger lower bounds that also take into account the number of stack symbols, as in Theorem 3.2, is left as a task for future research.

5   State Complexity of Operations

Alur and Madhusudan [2] established the closure of input-driven languages under basic operations, such as the Boolean operations, concatenation and Kleene star. For each operation, they defined an effective construction of a DIDPDA for the result of the operation, given DIDPDA(s) for its argument(s). For some operations, their constructions were later found to be optimal; for some other operations, they have been superceded with improved constructions.

This section reviews the results on the exact number of states needed to represent basic operations on input-driven pushdown automata with a given number of states. The case of union and intersection is the simplest one, handled by the well-known *direct product construction* in the same way as for finite automata.

Lemma 5.1 (Alur and Madhusudan [2]). *Let A and B be two DIDPDAs over the same alphabet $(\Sigma_{+1}, \Sigma_{-1}, \Sigma_0)$, with the sets of states P and Q, respectively, and with the pushdown alphabets Γ and Ω. Then there exists DIDPDAs C and D, each with the set of states $P \times Q$ and with the pushdown alphabet $\Gamma \times \Omega$, which recognize the languages $L(A) \cup L(B)$ and $L(A) \cap L(B)$, respectively.*

As established by Piao and Salomaa [46], this construction is optimal with respect to both the number of states and the number of stack symbols.

Next, consider the concatenation. The closure of input-driven languages under this operation was established by Alur and Madhusudan [2], who constructed an NIDPDA with $m+n+const$ states for the concatenation of an m-state and an n-state DIDPDAs. Determinizing this NIDPDA yields a DIDPDA with $2^{\Theta((m+n)^2)}$ states for the desired concatenation. The same construction was used to prove the closure of input-driven languages under two concatenation-like operations, Kleene star and reversal, producing $2^{\Theta(n^2)}$-state upper bounds on their state complexity.

This construction can be substantially improved. It turns out, that, informally speaking, the amount of nondeterminism required to simulate these concatenation-like operations is less than in a general NIDPDA. Instead of simulating general nondeterminism, it is sufficient to calculate the following simpler data structure: *the behaviour of a DIDPDA on the last well-nested substring.*

Consider that the computation of a DIDPDA on any well-nested string has a simple form: it finishes processing such a string with the same stack contents as in the beginning, without ever attempting to pop any symbols underneath. Hence, the behaviour of an automaton on a well-nested string w can be characterized by a function $f_w \colon Q \to Q$, which maps the initial state of the automaton to its state after processing the string. The behaviour on a concatenation uv is a function composition $f_{uv} = f_v \circ f_u$.

As shown in the following lemma, the behaviour function can be calculated using only $n^n \cdot |\Sigma_{+1}|$ states.

Lemma 5.2 (Okhotin and Salomaa [45, implicit in Lemma 4]). *For every DIDPDA A with the set of states Q and the pushdown alphabet Γ, there exists a DIDPDA C with the set of states Q^Q and the pushdown alphabet $Q^Q \times \Sigma_{+1}$, that calculates the behaviour of A on the longest well-nested suffix of the read portion of the input.*

Proof (Sketch of a proof.). The construction uses a simplified notation for the transition function, with $\delta_c(q)$ for $c \in \Sigma_0$ (instead of $\delta_i(q,c)$, as per Definition 2.1), $\delta_<(q)$ for $< \in \Sigma_{+1}$ (instead of $\delta_{push}(q,<)$), and $\delta_>(q,\gamma)$ for $> \in \Sigma_{-1}$ (instead of $\delta_{pop}(q,\gamma,>)$).

The initial state of C is $q'_0 = id$, the identity function on Q. This is the behaviour on ε.

On any symbol $c \in \Sigma_0$, the automaton calculates a composition of the behaviour on the previously read string with the behaviour on the next symbol:

$$\delta'_c(f) = \delta_c \circ f.$$

Whenever the automaton goes into the next level of brackets, it pushes the behaviour on the current level into the stack and begins calculating a new behaviour on the inner level. To this end, for $f : Q \to Q$ and $< \in \Sigma_{+1}$, the transitions are as follows:

$$\delta'_<(f) = id,$$
$$\gamma'_<(f) = (f, <).$$

When the automaton eventually returns into the previous level of brackets, it composes the behaviour on the inner level with the behaviour on the previously read prefix on the outer level: for $< \in \Sigma_{+1}$, $> \in \Sigma_{-1}$, and $f, g \colon Q \to Q$, the transitions are:

$$\delta'_>(g, (f, <)) = h \circ f,$$

where $h \colon Q \to Q$ is defined by

$$h(q) = \delta_> \big(g(\delta_<(q)), \gamma_<(q) \big).$$

$\square$

Consider how this calculation of behaviours is applied to constructing a DIDPDA for reversal.

Lemma 5.3 (Okhotin and Salomaa [45]). *For every DIDPDA A over an alphabet $(\Sigma_{+1}, \Sigma_{-1}, \Sigma_0)$ with a set of states Q and a pushdown alphabet Γ, there exists a DIDPDA C over the inverted alphabet $(\Sigma_{-1}, \Sigma_{+1}, \Sigma_0)$ with the set of states $Q^Q \times 2^Q$ and the pushdown alphabet $Q^Q \times 2^Q \times \Sigma_{-1}$ that recognizes the language $L(A)^R$.*

Given a string uv, where u is its longest well-nested prefix, the automaton C is given a string $v^R u^R$ and should simulate the computation of A on uv. The goal of the construction is that C calculates (i) the behaviour of A on u, as the first component of its state, and (ii) the set of states, beginning from which A would accept the string uv, as the second component.

The initial state of C is $q'_0 = (id, F)$, where F is the set of accepting states of A. The transitions of C manipulate the first component of its states basically accordingly in Lemma 5.2; the fact that the input is read in the reverse, changes only the order of function composition. What is done on the second component of the states, mostly resembles the *subset construction* for finite automata, Whenever a well-nested suffix has to be taken into account in the second component, the automaton C uses the behaviour on this component to calculate the pre-image of the set of states at the previous step.

This construction asserts that the reversal is representable using $2^{O(n \log n)}$ states, and, as shown below, this upper bound is tight.

Lemma 5.4 (Piao and Salomaa [46]). *Let $\Sigma_{+1} = \{<\}$, $\Sigma_{-1} = \{>\}$, $\Sigma_0 = \{a, b, c\}$. For $n \geqslant 1$, the language*

$$L_n = \bigcup_{u \in \{a,b\}^{\lceil \log n \rceil}} u < (\{a, b\}^* c)^{n-1} uc(\{a, b\}^* c)^* >$$

has an IDPDA with $O(n)$ states, while any IDPDA for its reversal requires at least $2^{\Omega(n \log n)}$ states.

Theorem 5.1. *The state complexity of reversal of DIDPDAs is $2^{\Theta(n \log n)}$.*

Similar constructions apply to concatenation and Kleene star. The following lemma defines a DIDPDA recognizing a concatenation of two given DIDPDAs.

Lemma 5.5 (Okhotin and Salomaa [45]). *Let A and B be any DIDPDAs over an alphabet $(\Sigma_{+1}, \Sigma_{-1}, \Sigma_0)$, let P and Q be their respective sets of states, let Γ and Ω be their pushdown alphabets. Then there exists a DIDPDA C with the set of states $P \times 2^Q \times 2^Q \times Q^Q$ and the pushdown alphabet $\Gamma \times 2^Q \times 2^Q \times Q^Q \times \Sigma_{+1}$ that recognizes the language $L(A) \cdot L(B)$.*

The first component of the states of C is used for a plain simulation of A. The fourth component is the behaviour of B on the last well-nested suffix, calculated as in Lemma 5.2. The second and the third component implement two subset constructions for B, referring to different partitions of the bracket structure of the input into concatenations $L(A) \cdot L(B)$.

The next lemma presents a lower bound on the state complexity of concatenation.

Lemma 5.6 (Okhotin and Salomaa [45]). *For every $m, n \geqslant 1$, there exists a language K_m recognized by an m-state DFA, and a language L_n recognized by an DIDPDA with $O(n)$ states and n pushdown symbols, such that every DIDPDA for their concatenation $K_m L_n$ requires at least mn^n states.*

These bounds match, up to a constant multiple in the exponent, leading to the following theorem.

Theorem 5.2. *The state complexity of the concatenation operation for DIDPDAs is $m \cdot 2^{\Theta(n \log n)}$.*

The next lemma gives an upper bound on the state complexity of Kleene star.

Lemma 5.7 (Okhotin and Salomaa [45]). *Let B be any DIDPDA over an alphabet $(\Sigma_{+1}, \Sigma_{-1}, \Sigma_0)$, let Q be its set of states, and let Γ be its pushdown alphabet. Then there exists a DIDPDA C with the set of states $2^Q \times 2^Q \times Q^Q$ and the pushdown alphabet $\Gamma \times 2^Q \times 2^Q \times Q^Q \times \Sigma_{+1}$ that recognizes the language $L(A)^*$.*

This construction is very similar to the one for the concatenation, but follows the acceptance decisions of the instances of the simulated automaton B, as encoded in the subset constructions, instead of following a simulated automaton A. A matching lower bound is also known.

Lemma 5.8 (Salomaa [50]). *There exists a sequence of DIDPDAs $\{A_n\}_{n \geqslant 1}$, with $O(n)$ states and stack symbols, such that the total number of states and stack symbols in any DIDPDA recognizing $L(A)^*$ is at least $2^{n \log n}$.*

These results are combined to the following asymptotic estimation.

Theorem 5.3. *The state complexity of Kleene star of DIDPDAs is $2^{\Theta(n \log n)}$.*

One further result on the complexity of operations on DIDPDAs is that a homomorphic image requires $2^{\Theta(n^2)}$ states in the worst case. The upper bound is immediate by constructing a NIDPDA for the homomorphic image, and the lower bound follows by using a variant of the languages L_n from Theorem 4.2, in which the pairs to be "guessed" are annotated by special symbols, and these symbols are erased by a homomorphism, so that any DIDPDA recognizing the homomorphic image has to deal with the unmodified languages L_n.

Theorem 5.4 (Okhotin and Salomaa [44]). *Let h be a homomorphism and A a DIDPDA with n states. The language $h(L(A))$ can be recognized by a DIDPDA with $2^{O(n^2)}$ states.*

Conversely, there exists a homomorphism h and regular languages L'_n, with $n \geqslant 1$, recognized by a DIDPDA with $O(n)$ states, such that any DIDPDA for $h(L'_n)$ needs 2^{n^2} states.

Finally, consider the complexity of operations on NIDPDAs. Most operations can be represented directly, with a minimal overhead: for instance, union of an m-state and an n-state NIDPDAs has an NIDPDA with $m + n + const$ states, and concatenation requires the same number of states; Kleene star and reversal require only $n + const$ states. However, there are two exceptions. First, intersection requires mn states.

Theorem 5.5 (Han and Salomaa [20]). *Let $k, \ell, m, n \geqslant 1$, let m be divisible by k, and let n be divisible by ℓ. Then there exist NIDPDAs A, B with m and n states, respectively, and with k and ℓ pushdown symbols, such that every NIDPDA for $L(A) \cap L(B)$ has at least mn states and $k\ell$ pushdown symbols.*

Secondly, complementing NIDPDAs is hard.

Theorem 5.6 (Okhotin and Salomaa [44]). *For every $n \geqslant 1$, consider the language L_n defined in Theorem 4.2, which has a NIDPDA with $O(n)$ states. Let A be any NIDPDA recognizing its complement $\overline{L_n}$, let Q be its set of states and let Γ be its pushdown alphabet. Then, $|Q| \cdot |\Gamma| \geqslant 2^{n^2}$.*

Accordingly, the worst-case state complexity of complementing an n-state NIDPDA is $2^{\Theta(n^2)}$.

In contrast, every DIDPDA can be complemented simply by inverting its set of accepting states. This situation is similar to the case of finite automata, where every DFA is complemented by changing its accepting states, thus using n states, while complementing NFAs requires up to 2^n states [6].

The known results on the complexity of operations on deterministic and nondeterministic IDPDAs are summarized in the below table, which compares them to similar results for finite automata.

	DFA	NFA	DIDPDA	NIDPDA
$\cup$	mn [36]	$m + n + 1$ [22]	$\Theta(mn)$ [46]	$m + n + O(1)$ [2]
$\cap$	mn [36]	mn [22]	$\Theta(mn)$ [46]	$\Theta(mn)$ [20]
$\sim$	n	2^n [6]	n	$2^{\Theta(n^2)}$ [44]
$\cdot$	$m \cdot 2^n - 2^{n-1}$ [36]	$m + n$ [22]	$m2^{\Theta(n \log n)}$ [45]	$m + n + O(1)$ [2]
$*$	$\frac{3}{4}2^n$ [36]	$n + 1$ [22]	$2^{\Theta(n \log n)}$ [45]	$n + O(1)$ [2]
R	2^n [32]	$n + 1$ [22]	$2^{\Theta(n \log n)}$ [45]	$n + O(1)$ [2]

The main open problem is the complexity of operations for UIDPDAs. In fact, already for unambiguous finite automata (UFA), these questions appear to be quite difficult, with only very modest bounds known up to date [41,42].

6   Extensions and Open Problems

An attractive feature of input-driven pushdown automata is that every nondeterministic automaton has an equivalent deterministic automaton. The *height-deterministic pushdown automata* of Nowotka and Srba [40], where the stack height is a priori fixed at any point of a nondeterministic computation, are a natural model that properly extends IDPDAs. A nondeterministic realtime height-deterministic automaton always has an equivalent deterministic automaton. This is established using a construction analogous to the one used by Alur and Madhusudan [2] for determinizing NIDPDAs.

La Torre, Madhusudan and Parlato [30] have shown that multi-stack input-driven pushdown automata are closed under Boolean operations and have a decidable inclusion/equivalence problem, albeit equivalence is not tractable even in the deterministic case. In a multi-stack input-driven automaton the input letter determines the type of the operation on each of the stacks and, furthermore, any computation of the machine can be split into a fixed number of phases where each phase can pop only one of the stacks. Two-stack input-driven automata are not closed under complement without the restriction that each computation can be split into a fixed number of phases (as described above) [7]. It was claimed by Carotenuto et al. [10] that nondeterministic two-stack input-driven pushdown automata could be determinized; however, this claim does not hold [31].

Analogously to an input-driven pushdown automaton being implemented by an automaton operating on nested words, other types of machines with auxiliary storage can be realized using the graph automata of Madhusudan and Parlato [35] that operate on specialized graphs. Madhusudan and Parlato [35] deduce from the structure of the corresponding graphs the decidability of the emptiness problem for a number of multi-stack automaton models with various restrictions on the allowable computations. Descriptional complexity of such graph automata [35] remains a topic for further study.

References

1. Alur, R., Kumar, V., Madhusudan, P., Viswanathan, M.: Congruences for Visibly Pushdown Languages. In: Caires, L., Italiano, G.F., Monteiro, L., Palamidessi, C., Yung, M. (eds.) ICALP 2005. LNCS, vol. 3580, pp. 1102–1114. Springer, Heidelberg (2005)
2. Alur, R., Madhusudan, P.: Visibly pushdown languages. In: Proceedings of ACM Symposium on Theory of Computing, STOC 2004, Chicago, USA, pp. 202–211 (2004)
3. Alur, R., Madhusudan, P.: Adding nesting structure to words. J. Assoc. Comput. Mach. 56(3) (2009)

4. Arenas, M., Barceló, P., Libkin, L.: Regular Languages of Nested Words: Fixed Points, Automata, and Synchronization. In: Arge, L., Cachin, C., Jurdziński, T., Tarlecki, A. (eds.) ICALP 2007. LNCS, vol. 4596, pp. 888–900. Springer, Heidelberg (2007)

5. Birget, J.-C.: Intersection and union of regular languages and state complexity. Inform. Process. Lett. 43, 185–190 (1992)

6. Birget, J.-C.: Partial orders on words, minimal elements of regular languages, and state complexity. Theoret. Comput. Sci. 119, 267–291 (1993)

7. Bollig, B.: On the expressive power of 2-stack visibly pushdown automata. Logical Methods in Computer Science 4(4), paper 16 (2008)

8. von Braunmühl, B., Verbeek, R.: Input-Driven Languages are Recognized in $\log n$ Space. In: Karpinski, M. (ed.) FCT 1983. LNCS, vol. 158, pp. 40–51. Springer, Heidelberg (1983)

9. von Braunmühl, B., Verbeek, R.: Input driven languages are recognized in $\log n$ space. North-Holland Mathematics Studies 102, 1–19 (1985)

10. Carotenuto, D., Murano, A., Peron, A.: 2-Visibly Pushdown Automata. In: Harju, T., Karhumäki, J., Lepistö, A. (eds.) DLT 2007. LNCS, vol. 4588, pp. 132–144. Springer, Heidelberg (2007)

11. Chervet, P., Walukiewicz, I.: Minimizing Variants of Visibly Pushdown Automata. In: Kučera, L., Kučera, A. (eds.) MFCS 2007. LNCS, vol. 4708, pp. 135–146. Springer, Heidelberg (2007)

12. Chrobak, M.: Finite automata and unary languages. Theoret. Comput. Sci. 47, 149–158 (1986), Errata 302, 497–498 (2003)

13. Dassow, J., Stiebe, R.: Nonterminal complexity of some operations on context-free languages. Fundam. Inform. 83, 35–49 (2008)

14. Domaratzki, M., Pighizzini, G., Shallit, J.: Simulating finite automata with context-free grammars. Inform. Proc. Lett. 84, 339–344 (2002)

15. Dymond, P.: Input-driven languages are in $\log n$ depth. Inform. Process. Lett. 26, 247–250 (1988)

16. Engelfriet, J., Rozenberg, G., Slutzki, G.: Tree transducers, L systems, and two-way machines. J. Comput. System Sci. 20, 150–202 (1980)

17. Gauwin, O., Niehren, J., Roos, Y.: Streaming tree automata. Inform. Proc. Lett. 109, 13–17 (2008)

18. Glaister, I., Shallit, J.: A lower bound technique for the size of nondeterministic finite automata. Inform. Process. Lett. 59, 75–77 (1996)

19. Gruska, J.: Descriptional complexity of context-free languages. In: Mathematical Foundations of Computer Science (MFCS 1973), Strbské Pleso, High Tatras, Czechoslovakia, September 3-8, pp. 71–83 (1973)

20. Han, Y.-S., Salomaa, K.: Nondeterministic state complexity of nested word automata. Theoret. Comput. Sci. 410, 2961–2971 (2009)

21. Hartmanis, J.: On the succinctness of different representations of languages. SIAM J. Comput. 9, 114–120 (1980)

22. Holzer, M., Kutrib, M.: Nondeterministic descriptional complexity of regular languages. Internat. J. Foundations of Comput. Sci. 14, 1087–1102 (2003)

23. Holzer, M., Kutrib, M.: Nondeterministic Finite Automata—Recent Results on the Descriptional and Computational Complexity. In: Ibarra, O.H., Ravikumar, B. (eds.) CIAA 2008. LNCS, vol. 5148, pp. 1–16. Springer, Heidelberg (2008)

24. Holzer, M., Kutrib, M.: Descriptional and computational complexity of finite automata—A survey. Inf. Comput. 209, 456–470 (2011)

25. Jirásková, G., Okhotin, A.: On the State Complexity of Operations on Two-Way Finite Automata. In: Ito, M., Toyama, M. (eds.) DLT 2008. LNCS, vol. 5257, pp. 443–454. Springer, Heidelberg (2008)

26. Kamimura, T., Slutzki, G.: Parallel and two-way automata on directed ordered acyclic graphs. Inform. Control 49, 10–51 (1981)

27. Kapoutsis, C.A.: Removing Bidirectionality from Nondeterministic Finite Automata. In: Jedrzejowicz, J., Szepietowski, A. (eds.) MFCS 2005. LNCS, vol. 3618, pp. 544–555. Springer, Heidelberg (2005)

28. Kunc, M., Okhotin, A.: Describing Periodicity in Two-Way Deterministic Finite Automata Using Transformation Semigroups. In: Mauri, G., Leporati, A. (eds.) DLT 2011. LNCS, vol. 6795, pp. 324–336. Springer, Heidelberg (2011)

29. Kunc, M., Okhotin, A.: State Complexity of Operations on Two-Way Deterministic Finite Automata over a Unary Alphabet. In: Holzer, M. (ed.) DCFS 2011. LNCS, vol. 6808, pp. 222–234. Springer, Heidelberg (2011)

30. La Torre, S., Madhusudan, P., Parlato, G.: A robust class of context-sensitive languages. In: Proceedings of the 22nd IEEE Symposium on Logic in Computer Science, LICS, pp. 161–170. IEEE Computer Society Press (2007)

31. La Torre, S., Madhusudan, P., Parlato, G.: A note posted at, http://www.cs.uicu.edu/~madhu/vpa/wrong-proof-CMP07.html

32. Leiss, E.L.: Succinct representation of regular languages by Boolean automata. Theoret. Comput. Sci. 13, 323–330 (1981)

33. Leung, H.: Descriptional complexity of NFA of different ambiguity. Internat. J. Foundations Comput. Sci. 16(5), 975–984 (2005)

34. Lupanov, O.B.: A comparison of two types of finite automata. Problemy Kibernetiki 9, 321–326 (1963) (in Russian)

35. Madhusudan, P., Parlato, G.: The tree width of auxiliary storage. In: Proc. 38th ACM Symposium on Principles of Programming Languages, POPL 2011, pp. 283–294 (2011)

36. Maslov, A.N.: Estimates of the number of states of finite automata. Soviet Mathematics Doklady 11, 1373–1375 (1970)

37. Mehlhorn, K.: Pebbling Mountain Ranges and its Application to DCFL-Recognition. In: de Bakker, J.W., van Leeuwen, J. (eds.) ICALP 1980. LNCS, vol. 85, pp. 422–435. Springer, Heidelberg (1980)

38. Mereghetti, C., Pighizzini, G.: Optimal simulations between unary automata. SIAM J. Comput. 30(6), 1976–1992 (2001)

39. Neumann, A., Seidl, H.: Locating Matches of Tree Patterns in Forests. In: Arvind, V., Sarukkai, S. (eds.) FST TCS 1998. LNCS, vol. 1530, pp. 134–146. Springer, Heidelberg (1998)

40. Nowotka, D., Srba, J.: Height-Deterministic Pushdown Automata. In: Kučera, L., Kučera, A. (eds.) MFCS 2007. LNCS, vol. 4708, pp. 125–134. Springer, Heidelberg (2007)

41. Okhotin, A.: Unambiguous Finite Automata over a Unary Alphabet. In: Hliněný, P., Kučera, A. (eds.) MFCS 2010. LNCS, vol. 6281, pp. 556–567. Springer, Heidelberg (2010)

42. Okhotin, A.: Unambiguous finite automata over a unary alphabet. Inform. and Computation (to appear)

43. Okhotin, A.: Comparing Linear Conjunctive Languages to Subfamilies of the Context-Free Languages. In: Černá, I., Gyimóthy, T., Hromkovič, J., Jefferey, K., Králović, R., Vukolić, M., Wolf, S. (eds.) SOFSEM 2011. LNCS, vol. 6543, pp. 431–443. Springer, Heidelberg (2011)

44. Okhotin, A., Salomaa, K.: Descriptional Complexity of Unambiguous Nested Word Automata. In: Dediu, A.-H., Inenaga, S., Martín-Vide, C. (eds.) LATA 2011. LNCS, vol. 6638, pp. 414–426. Springer, Heidelberg (2011)
45. Okhotin, A., Salomaa, K.: State Complexity of Operations on Input-Driven Pushdown Automata. In: Murlak, F., Sankowski, P. (eds.) MFCS 2011. LNCS, vol. 6907, pp. 485–496. Springer, Heidelberg (2011)
46. Piao, X., Salomaa, K.: Operational state complexity of nested word automata. Theoret. Comput. Sci. 410, 3290–3302 (2009)
47. Rabin, M.O., Scott, D.: Finite automata and their decision problems. IBM Journal of Research and Development 3, 114–125 (1959)
48. Rozenberg, G., Salomaa, A. (eds.): Handbook of Formal Languages, vol. I–III. Springer (1997)
49. Rytter, W.: An application of Mehlhorn's algorithm for bracket languages to $\log n$ space recognition of input-driven languages. Inform. Process. Lett. 23, 81–84 (1986)
50. Salomaa, K.: Limitations of lower bound methods for deterministic nested word automata. Inform. Comput 209, 580–589 (2011)
51. Schmidt, E.M.: Succinctness of Description of Context-Free, Regular and Unambiguous Languages, Ph. D. thesis. Cornell University (1978)
52. Shallit, J.: A Second Course in Formal Languages and Automata Theory. Cambridge University Press (2009)
53. Yu, S.: Regular languages. In: [48], vol. I, pp. 41–110
54. Yu, S., Zhuang, Q., Salomaa, K.: The state complexity of some basic operations on regular languages. Theoret. Comput. Sci. 125, 315–328 (1994)

Towards "Fypercomputations" (in Membrane Computing)

Gheorghe Păun

Institute of Mathematics of the Romanian Academy,
P.O. Box 1-764, RO-014700 Bucharest, Romania,
and Research Group on Natural Computing,
Department of Computer Science and Artificial Intelligence,
University of Sevilla,
Avda. Reina Mercedes s/n, 41012 Sevilla, Spain
`george.paun@imar.ro, gpaun@us.es`

Abstract. Looking for ideas which would lead to computing devices able to compute "beyond the Turing barrier" is already a well established research area of computing theory; such devices are said to be able of doing *hypercomputations*. It is also a dream and a concern of computability to speed-up computing devices; we propose here a name for the case when this leads to polynomial solutions to problems known to be (at least) **NP**-complete: *fypercomputing* – with the initial **F** coming from "fast".

In short: *fypercomputing means going polynomially beyond* **NP**.

The aim of these notes is to briefly discuss the existing ideas in membrane computing which lead to fypercomputations and to imagine new ones, some of them at the level of speculations, subject for further investigation.

Keywords: Turing computing, Hypercomputing, Membrane computing, Complexity.

1 Foreword: Jürgen the Fast

These pages are dedicated to the 65th birthday of Jürgen Dassow. The age of retirement! I have met him in 1983 (it was my first trip out of Romania), we collaborated a lot, we met periodically in the framework of IMYCS - International Meeting of Young Computer Scientists, at Smolenice Castle, Slovakia. A young computer scientist is now retiring! He is maybe not the first, but I am not aware of others (to whom I was close enough). I feel this as an end of an epoch... The romantic time of grammars and automata, of sequential computing, of theory for the sake of theory, with aesthetic motivation. I remember the time spent in Magdeburg as a Humboldt fellow in 1991-92 and after that as a visiting researcher. Jürgen was, after 1993, the rector of the university. A very dynamical period, after the unification of Germany, with a lot of bureaucratic work – but with Jürgen switching in an amazingly quick way from administrative matters

H. Bordihn, M. Kutrib, and B. Truthe (Eds.): Dassow Festschrift 2012, LNCS 7300, pp. 207–220, 2012.

to scientific matters. I have never seen such a distributed and fast mind – and this fits with the topic of my notes. I cannot imagine Jürgen "doing nothing", so I am wishing to him, young pensioner, a long and active life!

2   Introduction – From Hyper... to Fyper...

The aim of this note is to briefly discuss an important research topic in computer science in general, in membrane computing in particular, namely ways to speed-up computations. The goal is to obtain devices able to significantly improve from this point of view, e.g., to solve intractable problems (usually, **NP**-complete or even harder) in a feasible time (usually, polynomial). Our constant framework is membrane computing (the reader is supposed to be familiar with the domain, or (s)he can consult [17], [19], [21], and the domain website from [30]), but some considerations/speculations are more general. (By the way, Jürgen is a co-author of one of the earliest papers in membrane computing, [7].)

The model we have in mind is that of hypercomputations, already with a large literature (see, e.g., [6], [9], [25], or the recent survey from [27]). The goal is to imagine computing machineries able "to compute the uncomputable", to compute more than Turing machines. More than a dozen of ideas were proposed and proved to reach the above goal: oracles (already considered by Turing), introducing real numbers in the device, accelerating the functioning of machines, using ingredients of an analogical nature and so on.

Also in membrane computing there were reported attempts of this kind. We mention here the papers [4] and [26]. The first one passes beyond Turing by means of acceleration (we come back below to this idea), the latter by constructing "lineages" of P systems, following the model of [28].

It should be noted that there also are people who do not believe in hypercomputation, refuting it as a circular trick (basically, one introduces uncomputable components in a computing device which is then shown to compute more than a Turing machine...); one of the most explicit voices in this respect is Martin Davis – see, e.g., [8].

We are looking here not for the *power* of computing devices, but for their *efficiency*, and we propose the term *fypercomputation* to name the case when a device can solve an intractable problem in a polynomial time. Surprisingly enough, this research vista has not received a similar attention as hypercomputation; the literature of the latter one is much more extended (more explicit), there were organized debates about it, workshops and special sessions in general conferences. Complexity theory discusses many speed-up procedures, but not at the level of "breaking the **NP** barrier", to put it in the style of "breaking the Turing barrier". The time-space trade-off is a common sense in algorithmics, but using an exponential space is not considered a serious candidate for fyper-computations – excepting the recent natural computing area, molecular (DNA) computing in special.

It is true, there are many papers which can be put under the flag of fypercomputing. They exploit ideas from physics, such as [24] (the abstract of the paper is worth recalling: *We propose to "boost" the speed of communication and computation by immersing the computing environment into a medium whose index of refraction is smaller than one, thereby trespassing the speed-of-light barrier*), propose analogical computations, such as [2] and [3]. The area of DNA computing is full of such ideas, its very beginning is of this kind, [1]. The starting point is the fact that the DNA molecules are a very compact support of information, one can operate with bits stored at a molecular level, hence in a small space (in a test tube) we can have a huge number of molecules. "Huge" however, has no precise mathematical meaning; when we need "exponentially many" molecules, what is perfect in theory becomes unfeasible in a lab...

But not this is the position we adopt here: however "unrealistic" (for current knowledge, be it physical, biological or of another nature), we look for "nice" ideas able to lead to fypercomputations. And membrane computing abounds already in such investigations.

3 Ways towards Fyper... in Membrane Computing

The main practical goal of natural computing is to learn computing ideas from nature and to implement/use them in computer science, starting with a better use of the existing computers and ending – the dream of DNA computing and of quantum computing, at least – with the construction of computers of a new kind, maybe using new materials. In all cases, the aim is to be more efficient, especially with respect to the duration of computations.

Membrane computing started with a more theoretical goal in mind: just learn interesting ideas for computability from the structure and the functioning of the living cells.

The basic computing models (called *cell-like P systems*) considered in this area consist of an hierarchical arrangement of *membranes*, delimiting *compartments* (also called *regions*) where multisets of *objects* (symbols from a given alphabet) are placed; these objects evolve according to *evolution rules* also associated to regions.

The membranes are labeled with symbols in a given alphabet (in most cases we use natural numbers). Each region is precisely identified by its upper membrane, hence we can identify regions by means of the labels of the membranes which delimit them. The external membrane, the one separating the cell from the *environment* is called the *skin* membrane; a membrane without any membrane inside is said to be *elementary*. We also call region the environment of the system (and usually we label it with 0).

The rules are of a multiset rewriting type (corresponding to biochemical reactions taking place in a cell) or of other forms, in general, inspired from biology (transport across membranes in the form of symport or aniport, membrane division, membrane creation, etc.). Starting from an initial configuration of the system (with initial multisets of objects associated with the regions) and using the rules in various ways (the most investigated one is the nondeterministic maximally parallel way), one passes from a configuration to another one. Such a sequence of *transitions* among configurations is called a *computation*. With a *halting* computation (one reaching a configuration where no rule can be applied) we can associate a result, for instance, as the number of objects present in a given region. We will give some further details in the next sections, this quick description of a P system and of its functioning should be sufficient for understanding the general discussion which follows.

The class of cell-like P systems is the basic one (and most investigated) in membrane computing. Several other types of similar devices were considered. For instance, instead of symbol objects, one can consider string objects (then, the rules should be string processing rules, such as rewriting, the splicing from DNA computing, etc.), or, instead of an hierarchical arrangement of membranes (hence described by a tree), we can place the membranes in the nodes of an arbitrary graph – one obtains in this way the *tissue-like P systems*. Also P systems inspired from the neurons architecture and functioning were introduced – a class of a particular interest is that of *spiking neural P systems* (SN P systems, for short), where *neurons* (membranes) placed in the nodes of a graph communicate to each other by means of *spikes*, electrical impulses of identical shapes (hence, the system uses only one object, the spike, present in various numbers of copies in different neurons) which are processed by specific *spiking* rules. We refer to [13] and to the corresponding chapter of [21] for details.

Besides the computing power of P systems (in comparison with classic computing devices, such as Turing machines and their restrictions), also the efficiency issue was addressed, and, in order to speed-up computations, the first proposal was to make use of the biological operation of cell division. In this way, P systems with active membranes have appeared, [18], and they proved indeed to speed-up computations enough for solving **NP**-complete problems in a polynomial time. (The typical rules are of the form $[a]_h \to [b]_h[c]_h$, where a, b, c are objects and h is a label; the membrane h, containing the object a and maybe other objects, is divided into two new membranes, with the same label h; in the first copy, a is replaced by b, in the second one it is replaced by c; any objects different from a are replicated in the two new membranes. Clearly, the new membranes with label h use the same set of rules as the former membrane h; otherwise stated, the label of a membrane identifies the set of rules which can be applied to it.)

An important point should be made here. P systems (with rules of the form $a \to u$, where a is an object and u is a multiset of objects) are able to create exponentially many objects in a linear time: using the rule $a \to aa$ in the maximally parallel manner for n times, we get 2^n copies of a. However, such an exponential workspace is not enough in order to essentially speed-up computations: the so-called Milano Theorem from [29] says that a P system (with rules as above, but without membrane division) can be simulated by a Turing machine with a polynomial overhead. Membrane division produces an exponential workspace, but also introduces some *organization* of it, some *localization*. In a specified membrane, specific rules are used, which is not possible without separating the objects among "protected reactors" – like in biology (compartments with specific chemicals evolving according to specific sets of reactions).

This lesson, of localization, with specific "chemistry" taking place in each compartment, is applied also in the case of P systems with membrane creation, [16], where rules of the form $a \rightarrow [b]_h$ are used, transforming an object a into a new membrane, with label h and containing the object b (as above, the label of the membrane identifies the set of rules to apply to its objects). This time, the aspect mentioned above, i.e., the power of the localization, is still more visible: The rules for creating membranes produce only a linear number of membranes, not an exponential one, as in the case of division, but the exponential space is produced by object evolution rules (such as $a \rightarrow aa$).

There are many other important details in this area, for instance, additional ingredients which have (or not) an influence on the efficiency of P systems with membrane division or membrane creation (for instance, membrane polarization, membrane dissolution, division of only elementary membranes or also of non-elementary membranes). In most cases, polynomial solutions to **NP**-complete problems – often, also of **PSPACE**-complete problems – are obtained. Fyper-computing! (The list of computationally hard problems addressed in terms of membrane computing is very large, ranging from decidability problems, such as SAT and QSAT, to numerical problems – e.g., counting various parameters in a graph; a chapter of [21] is devoted to this research direction, and the reader is refereed to it for details.)

Further two similar ideas were investigated in membrane computing, both of them related to the previous two. The first one is based on string replication, [5], for P systems with string objects (the rules are of the form $a \rightarrow u_1 \| u_2$, where a is a symbol, u_1, u_2 are strings; when rewriting a string xay by such a rule we obtain two strings, xu_1y and xu_2y, maybe placed in two different membranes, because the strings u_1, u_2 can also have associated target indications). Replicating a string looks like dividing a membrane: the symbols of each string remain together, hence "localized", like being encapsulated in separated membranes. Fypercomputations are again obtained (in [5] one shows how SAT can be solved in polynomial time in this framework).

A different approach is that of considering arbitrarily large *pre-computed re-sources*. For instance, we can assume as given in advance, for free (from the point of view of the computing time), a spiking neural P system (this is the case of [14], [15]), with arbitrarily many neurons and synapses, but with a small (finite) number of spikes inside and having a regular structure; otherwise stated, the given system is large, but it cannot contain more than a bounded amount of information, not enough, for instance, to encode the solution to the problem we want to solve. Then, we plug in a code of the (instance of the) problem to solve, in the form of a bounded number of spikes (a bounded number of bits), and this activates the arbitrarily large "hardware", which eventually provides the solution in a polynomial time.

This last idea is only briefly investigated in membrane computing. Issues related to the conditions to be imposed to the given pre-computed resources should be further considered.

Anyway, we count already four ideas, all of them having biological motivations (maybe better: all of them suggested by biological facts), which lead to fypercomputations in the membrane computing area.

There is a large bibliography of this direction of research. We refer here only to the survey paper [22], where the basic ideas of computational complexity for P systems can be found, and to the chapter of [21] devoted to this topic, [23].

4   Further Ways towards Fypercomputations in Membrane Computing

In membrane computing there are several versions of the previously mentioned speed-up tools, for instance, operations of separating membranes according to their contents, budding membranes, etc., but they have the same philosophy and the same functions as membrane division, so we do not count them separately here, but we look for ideas essentially different.

(1) The first candidate is the *acceleration*. The idea of an accelerated Turing machine is old in computer science: imagine that the machine is "clever", it learns from its own functioning, in the following way; after performing a step in a time unit, it performs better for the second step, which is completed in half of the time necessary for the first step – and so on, at each step halving the time with respect to the previous step. If the first step takes one time unit, then the second one takes $1/2$ time units, the third one $1/4$ and so on, hence in two time units the computation ends.

Important: we have here two clocks, an internal one, of the machine, and an external one, of the observer. The internal clock is faster and faster, so that the computation ends in two time units *measured by the external clock*, that of the observer/user.

Accelerated Turing machines can solve the halting problem, hence they compute what usual Turing machines cannot. See references in [4], where the idea is extended to P systems: It is known from biology that nature creates membranes also for enhancing the reactions inside (if the reactants are closer, then the probability to collide and react is higher). If "smaller is faster", then take it as in accelerated Turing machines: the reactions taking place in a membrane placed inside another membrane are twice faster than in the parent membrane. Using membrane creation rules, create then membranes inside membranes, in an unbounded hierarchy, which means faster and faster towards the "center" of the membrane structure. Like for accelerated Turing machines, such P systems were proved to "compute the uncomputable".

Well, then why not using this trick also in complexity sense? Natural enough, but Martin Davis is smiling ironically: we accelerate in order to speed-up... In two (external) time units we solve any problem, whatever complex it is (remember that most classes of P systems are Turing complete, they can do whatever a Turing machine can do).

A way to make the things interesting is to accelerate only parts of a P system, thus having several levels of time speed. For instance, imagine a P system with several internal membranes, all of them governed by the same clock, but some, say, elementary membranes being accelerated: for each of them we have a constant k_h (h is the label of the membrane) and, after introducing a "subproblem" in membrane h (e.g., a string x and a language L – by means of a grammar), we get the answer (the fact whether or not $x \in L$) in at most k_h time units (as measured by the global clock). Not all types of P systems are universal when using only one membrane, so, for such cases, the local acceleration can be indeed of interest (i.e., non-trivial). What other restrictions to consider remains as a research topic.

(2) For instance, we can accelerate not elementary membranes, but rules: a given rule takes one time unit for the first application, half for the second application, and so on. If all rules are accelerated, we go back to globally accelerated systems, but if we allow only to a bounded number of rules to get faster in this way, it is possible to obtain interesting results.

(3) Maybe also objects can be accelerated: the descendants of an object react faster than the father object, irrespective which are the rules which act on them and irrespective of the membranes where they are.

(4) The previous ideas suggest a speculation of a *science-fiction* type, but let us place it here, the context is speculative enough. Namely, we mentioned before (at least) two clocks, an external one, of the observer (or of the higher membranes in the structure) and the local clock(s), of the accelerated element, membrane, rule, object. Always, the inner clock is (much) faster than the external one, it performs sometimes an exponential number of steps while the external one only ticks once. We can then imagine that the inner time is orthogonal to the external time, hence the time has a 2D structure! Is this too much with respect to the "classic" time, in general interpreted as linear (or circular, spiral, in various philosophies)? As a speculation, I would not refuse to think about this.

So: what about considering the time as a plane, not as a line, with the observer only sensing one dimension of it, but with the possibility of some "processors" to run along the other dimension, doing this at-no-time for the observer? (After all, the current models of the universe, the so-called membrane theory – having nothing common with membrane computing! –, which has replaced/generalized the string theories, deals with eleven dimensions, enough for "lateral" times...)

Well, passing from this speculation to precise models, to symbols, algorithms and theorems, is another story... (but what else are the oracles than processors working in a "lateral" time – and space?).

5   P Systems with ω-Populations of Objects

(5) Let us come back to the ground, namely to the recently introduced *reaction systems* (we call them *R systems*) – see [10], [11], [12]. An attempt to bridge the two research areas, P systems and R systems, was done in [20]; the present section can be seen as a continuation of this attempt.

Both areas deal with populations of reactants (molecules) which evolve by means of reactions, with several basic differences. Most of these differences are not mentioned here (e.g., the compartmental structure of models in membrane computing versus the lack of membranes in reaction systems; the focus on evolution, not on computation, in reaction systems; the unique form of rules in reaction systems and so on), and we recall the two basic ones, the crucial postulates of R systems, in the formulation from [10]:

The way that we define the result of a set of reactions on a set of elements formalizes the following two assumptions that we made about the chemistry of a cell:

(i) *We assume that we have the "threshold" supply of elements (molecules) – either an element is present and then we have "enough" of it, or an element is not present. Therefore we deal with a qualitative rather than quantitative (e.g., multisets) calculus.*

(ii) *We do not have the "permanence" feature in our model: if nothing happens to an element, then it remains/survives (status quo approach). On the contrary, in our model, an element remains/survives only if there is a reaction sustaining it.*

Bringing the first assumption in membrane computing is a way to obtain fypercomputations! This, however, is not very surprising: if each object is present in "enough" copies, this means already that we have an arbitrarily large workspace. The only problem is how to use this unstructured space for efficient computations.

Because we want to get a little bit more technical here, we introduce some formal prerequisites.

Given an alphabet O of symbols called here *objects*, we denote by O^* the set of all strings over O, the empty string λ included, and by $O^+ = O^* - \{\lambda\}$ the set of all non-empty strings over O. A multiset-rewriting rule (over O; we also say *evolution rule*) is a pair (u, v), written in the form $u \to v$, where u and v are multisets over O (given as strings over O). The rules are classified according to the complexity (of their left hand side). A rule with at least two objects in its left hand side is said to be *cooperative*. The result in Theorem 1 is given for such rules; a restrictive type of rules is that of *non-cooperative ones*, of the form $u \to v$, with u being an object in O.

Now, a *cell-like P system* (of degree m) is a construct

$$\Pi = (O, \mu, w_1, \ldots, w_m, R_1, \ldots, R_m, i_{in}, i_{out}),$$

where O is the alphabet of objects, μ is the membrane structure (with m membranes), given as an expression of labeled parentheses, $w_1, \ldots, w_m$ are (strings over O representing) multisets of objects present in the m regions of μ at the beginning of a computation, $R_1, \ldots, R_m$ are finite sets of evolution rules associated with the regions of μ, and i_{in}, i_{out} are the labels of input and output regions, respectively, $i_{in} \in \{1, 2, \ldots, m\}, i_{out} \in \{0, 1, 2, \ldots, m\}$; if $i_{out} = 0$, this indicates that the output is obtained in the environment. If the system is used in the generative mode, then i_{in} is omitted, and if the system is used in the accepting mode, then i_{out} is omitted. The number m of membranes in μ is called the *degree* of Π.

The rules in sets R_i are of the form $u \to v$, as specified above, with $u \in O^+$, but with the objects in v also having associated *target indications*, i.e., $v \in (O \times \{here, out, in\})^*$. After using a rule $u \to v$, the objects in u are consumed, and those in v are produced; if $(a, here)$ appears in v, then a remains in the same compartment of the system where the rule was used, if (a, out) is in v, then the object a is moved immediately in the region surrounding the compartment where the rule was used (this is the environment if the rule is used in the skin region), and if (a, in) is in v, then a is sent to one of the inner membranes, nondeterministically chosen (if there is no membrane inside the membrane where the rule is meant to be applied, then the use of the rule is forbidden). The indication *here* is omitted, we write a instead of $(a, here)$.

The rules are used in the nondeterministic maximally parallel manner: in each membrane, a multiset of rules is applied such that there is no larger multiset of rules which is applicable in that membrane.

In the generative mode, the result of a computation consists of the number of objects in region i_{out} in the moment when the computation halts, i.e., no rule can be applied in any membrane of the system. In the accepting mode, a number is introduced in the region i_{in}, e.g., in the form of the multiplicity of a given object, and, if the computation halts, then this number is accepted. A P system can also be used in the computing mode, with a number introduced in region i_{in} and the result obtained in region i_{out}, in the moment when the computation halts. P systems can be also used for solving decidability problems: an encoding of a problem (of an instance) is introduced in membrane i_{in}, in the form of a multiset of objects, and the answer, YES or NO, is obtained, for instance, by halting, or by sending out of the system a distinguished object *yes* or *no*.

Note that in the previous definitions multisets play a crucial role, objects not evolving by a rule remain unchanged, and that always successful computations are defined by halting – all these are essential differences with respect to R systems.

Consider now a P system with some of its objects being present in arbitrarily many copies, like in R systems. More exactly, we consider P systems which contain certain distinguished elementary membranes, whose objects are present in arbitrarily many copies (for instance, if an object a is introduced from outside in such a membrane, then inside the membrane it becomes a^ω; it enters as a single copy, and immediately multiplies inside to arbitrarily many, like in reaction systems).

Let us call such a system an ωP system.

The arbitrary multiplicity of objects introduces an important change in the functioning of a usual system. For instance, if we have the objects a, b, c in a distinguished membrane, together with the rules $ab \to d, ac \to e$, then both these rules can be (and should be) applied, because we have *enough* copies of a for both rules; we obtain d, e, with *all* copies of a, b, c being consumed. If also an object f is present together with a, b, c, then it remains unchanged.

We do not adopt here also the second assumption from the definition of R systems, because then all objects from a halting configuration vanish. Please note that the "easy solution" of introducing dummy rules of the form $f \to f$ for all objects which cannot evolve otherwise does not work either: the computation will never stop.

This apparently innocent change in the structure of P systems, i.e., considering objects with ω multiplicity, is able to speed-up a P system to the level of fypercomputations.

Theorem 1. SAT *can be solved (in a uniform way) in a polynomial time by an ωP system.*

Proof. Let us consider the SAT problem for n variables, $x_1, x_2, \ldots, x_n$, and m clauses. We consider x_l and $\neg x_l$, $1 \le l \le n$, as symbols, we denote by Lit their alphabet, and we also denote $Val = \{t_i, f_i \mid 1 \le i \le n\}$. If $\alpha \in Lit$, then $\neg\alpha$ is its negation (e.g., $\neg(\neg x_i) = x_i$).

An instance $\gamma = C_1 \wedge C_2 \wedge \ldots \wedge C_m$ of SAT(n, m), with $C_i = y_{i,1} \vee y_{i,2} \vee \ldots \vee y_{i,k_i}$, for $y_{i,j} \in Lit$, $1 \le j \le k_i$, is encoded as

$$code(\gamma) = y_{1,1}^{(1)} \ldots y_{1,k_1}^{(1)} y_{2,1}^{(2)} \ldots y_{2,k_2}^{(2)} \ldots y_{m,1}^{(m)} \ldots y_{m,k_m}^{(m)}.$$

We now construct the following ωP system (the unique elementary membrane is the distinguished one, the region where the objects are present with ω multiplicity):

$$\Pi = (O, \mu, w_1, w_2, R_1, R_2, 1, 0),$$
$$O = \{\alpha^{(j)} \mid \alpha \in Lit, 1 \le j \le m\} \cup Val$$
$$\cup \{a, \mathsf{yes}, d, d_1, d_2, \ldots, d_{m+1}\} \cup \{\langle aw \rangle \mid w \in Val^*, |w| \le m\},$$
$$\mu = [\ [\ ]_2\ ]_1,$$
$$w_1 = a d_{m+1}, \ w_2 = \lambda,$$
$$R_1 = \{a \to (\langle a \rangle, in), \ d_1 \to (d, in), \ \mathsf{yes} \to (\mathsf{yes}, out)\}$$
$$\cup \{x_i^{(1)} \to (t_i, in), \ \neg x_i^{(1)} \to (f_i, in) \mid 1 \le i \le n\}$$
$$\cup \{x_i^{(j+1)} \to x_i^{(j)}, \ \neg x_i^{(j+1)} \to \neg x_i^{(j)} \mid 1 \le i \le n, 2 \le j \le m\}$$
$$\cup \{d_{j+1} \to d_j \mid 1 \le j \le m\},$$
$$R_2 = \{\langle aw \rangle x_i \to \lambda \mid \neg x_i \in w, w \in Val^+, 1 \le i \le n\}$$
$$\cup \{\langle aw \rangle \neg x_i \to \lambda \mid x_i \in w, w \in Val^+, 1 \le i \le n\}$$
$$\cup \{\langle aw \rangle \alpha \to \langle aw \rangle \mid \alpha \in w, \alpha \in Val, w \in Val^+\}$$
$$\cup \{\langle aw \rangle \alpha \to \langle aw\alpha \rangle \mid \alpha \notin w, \neg\alpha \notin w, \alpha \in Val, w \in Val^*\}$$
$$\cup \{\langle awd \rangle \to (\mathsf{yes}, out) \mid w \in Val^+\}.$$

For an easier understanding, we also present this system in a graphical form, in Figure 1 (besides the membranes and the objects present in the initial configuration, the rules from each regions are specified).

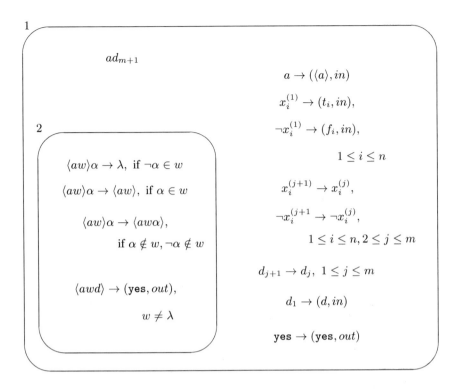

Fig. 1. The ωP system from the proof of Theorem 1

The computation of Π starts after introducing the multiset $code(\gamma)$ in the skin membrane, for a given instance γ of the $\text{SAT}(n, m)$ problem. The "seed" object a enters immediately membrane 2, in the form $\langle a \rangle$, together with the truth values which make true the first clause (the literals in clause C_1 are present in $code(\gamma)$ with superscripts 1). All other objects evolve in the skin membrane, those associated with literals in clauses C_2 to C_m of γ decreasing by one the superscript, d by decreasing by one the subscript.

In the inner membrane we have (at an arbitrary step) several multisets aw in the bracketed form $\langle aw \rangle$ (hence interpreted as symbol objects); initially, $w = \lambda$. Each multiset w contains truths values of variables which make true the clauses of γ, starting with the first clauses in the ordering $C_1, C_2, \ldots$. When the truth values which render true a further clause enter membrane 2, they are checked against the existing truth values, those which made true the previous clauses. It is important to note that *each of the existing sets of truth values "react" with each newly introduced truth value*, in the style of reaction systems, because we have arbitrarily many copies of each object present in membrane 2. If a previous set of truth values w falsifies the new clause, then the object $\langle aw \rangle$ disappears. If it contains a truth value which renders true the new clause, then it survives. When neither of these situations holds, then we add to w a truth value which makes true the new clause (without falsifying the previous clauses).

The process continues for each of the m clauses. In the end, the "checker" d enters membrane 2. If it finds here at least one object $\langle aw \rangle$ with $w \neq \lambda$, this means that at least a truth assignment satisfies all clauses, hence the formula is satisfiable. The signal object yes is produced and sent out of the system. The computation halts. (If the formula is not satisfiable, it halts after $m + 1$ steps.)

Therefore, if the system halts without sending out the object yes, then the formula was unsatisfiable; if the system sends out the object yes, in step $m + 3$, then the formula is satisfiable.

Note that the initial configuration of the system, the rules of each compartment included, is of a polynomial size with respect to n and m; the total alphabet of the system is exponential in size, but the objects are constructed during the computation, not while constructing the system itself.

The construction is uniform (it starts from the problem itself, not from a given instance of the problem – in membrane computing, the weaker case when we start from an instance of the decision problem, is called semi-uniform), and this concludes the proof.   □

The previous system has been presented with two membranes, in order to make clear the ω component of it, the inner membrane, with a "standard" membrane around it. The construction can be easily modified in order to obtain a system with only one membrane (decreasing superscripts and subscripts can be done in the same way for objects supposed to appear in arbitrarily many copies, because each object is modified by only one rule and all copies of an object evolve at the same time, in the same way). Thus, the previous theorem can be also considered as a fypercomputation result for R systems instead of P systems.

We do not know whether the previous result can be improved by imposing the restriction to use only non-cooperative rules.

6   Final Remarks

Following the model of hypercomputation, we have introduced the notion and we considered the issue of "fypercomputation" – of speeding-up computations (of P systems) to such an extent to obtain polynomial time solutions to intractable (typically, **NP**-complete) problems. After briefly mentioning the existing speed-up ideas investigated in membrane computing (membrane division, membrane creation, string replication, and using pre-computed resources), we propose further ideas, such as (local) acceleration, two-dimensional time, using arbitrary populations of objects (like in reaction systems, i.e., assuming for each object that it is either absent or present in arbitrarily many copies). An example of fypercomputation is given for the last idea.

The paper is preliminary and speculative, it mainly calls the attention to the systematic study of fypercomputations.

Acknowledgements. Work supported by Proyecto de Excelencia con Investigador de Reconocida Valía, de la Junta de Andalucía, grant P08 – TIC 04200. Detailed comments by two anonymous referees are gratefully acknowledged.

References

1. Adleman, L.M.: Molecular computation of solutions to combinatorial problems. Science 226, 1021–1024 (1994)
2. Arulanandham, J.J.: Implementing Bead-Sort with P Systems. In: Calude, C.S., Dinneen, M.J., Peper, F. (eds.) UMC 2002. LNCS, vol. 2509, pp. 115–125. Springer, Heidelberg (2002)
3. Arulanandham, J.J., Calude, C.S., Dinneen, M.J.: Balance Machines: Computing = Balancing. In: Jonoska, N., Păun, G., Rozenberg, G. (eds.) Aspects of Molecular Computing. LNCS, vol. 2950, pp. 36–48. Springer, Heidelberg (2003)
4. Calude, C.S., Păun, G.: Bio-steps beyond Turing. BioSystems 77, 175–194 (2004)
5. Castellanos, J., Păun, G., Rodriguez-Patón, A.: Computing with membranes: P systems with worm-objects. In: Proc. IEEE 7th. Intern. Conf. on String Processing and Information Retrieval, SPIRE 2000, La Coruna, Spain, pp. 64–74 (2000)
6. Copeland, J.: Hypercomputation. Minds and Machines 12, 461–502 (2002)
7. Dassow, J., Păun, G.: On the power of membrane computing. J. of Universal Computer Science 5, 33–49 (1999)
8. Davis, M.: Why there is no such discipline as hypercomputation. Applied Mathematics and Computation (Special Issue on Hypercomputation) 178, 4–7 (2006)
9. Eberbach, E., Wegner, P.: Beyond Turing machines. Bulletin of the EATCS 81, 279–304 (2003)
10. Ehrenfeucht, A., Rozenberg, G.: Basic Notions of Reaction Systems. In: Calude, C.S., Calude, E., Dinneen, M.J. (eds.) DLT 2004. LNCS, vol. 3340, pp. 27–29. Springer, Heidelberg (2004)
11. Ehrenfeucht, A., Rozenberg, G.: Reaction systems. Fundamenta Informaticae 75, 263–280 (2007)
12. Ehrenfeucht, A., Rozenberg, G.: Introducing time in reaction systems. Theoretical Computer Sci. 410, 310–322 (2009)
13. Ionescu, M., Păun, G., Yokomori, T.: Spiking neural P systems. Fundamenta Informaticae 71, 279–308 (2006)
14. Ishdorj, T.-O., Leporati, A.: Uniform solutions to SAT and 3-SAT by spiking neural P systems with pre-computed resources. Natural Computing 7, 519–534 (2008)
15. Leporati, A., Gutiérrez-Naranjo, M.A.: Solving SUBSET-SUM by spiking neural P systems with pre-computed resources. Fundamenta Informaticae 87, 61–77 (2008)
16. Martin-Vide, C., Păun, G., Rodriguez-Patón, A.: On P systems with membrane creation. Computer Science J. of Moldova 9, 134–145 (2001)
17. Păun, G.: Computing with membranes. Journal of Computer and System Sciences 61, 108–143 (2000) (first circulated as Turku Center for Computer Science-TUCS Report 208 (November 1998), www.tucs.fi
18. Păun, G.: P systems with active membranes: Attacking NP-complete problems. J. Automata, Languages, and Combinatorics 6, 75–90 (2001)
19. Păun, G.: Membrane Computing. An Introduction. Springer, Berlin (2002)
20. Păun, G., Pérez-Jiménez, M.J.: Towards bridging two cell-inspired models: P systems and R systems. Theoretical Computer Sci. (to appear)
21. Păun, G., Rozenberg, G., Salomaa, A. (eds.): The Oxford Handbook of Membrane Computing. Oxford University Press (2010)
22. Pérez-Jímenez, M.J.: An Approach to Computational Complexity in Membrane Computing. In: Mauri, G., Păun, G., Jesús Pérez-Jímenez, M., Rozenberg, G., Salomaa, A. (eds.) WMC 2004. LNCS, vol. 3365, pp. 85–109. Springer, Heidelberg (2005)

23. Pérez-Jiménez, M.J., Riscos-Núñez, A., Romero-Jiménez, A., Woods, D.: Complexity – Membrane Division, Membrane Creation. In: [21], ch. 12, pp. 302–336
24. Putz, V., Svozil, K.: Can a computer be "pushed" to perform faster-than-light? In: Proc. UC 2010 Hypercomputation Workshop "HyperNet 2010", Univ. of Tokyo, June 22 (2010)
25. Siegelmann, H.: Computation beyond the Turing limit. Science 238, 632–637 (1995)
26. Sosík, P., Valík, O.: On Evolutionary Lineages of Membrane Systems. In: Freund, R., Păun, G., Rozenberg, G., Salomaa, A. (eds.) WMC 2005. LNCS, vol. 3850, pp. 67–78. Springer, Heidelberg (2006)
27. Syropoulos, A.: Hypercomputation: Computing Beyond the Church-Turing Barrier. Springer, Berlin (2008)
28. van Leeuwen, J., Wiedermann, J.: Beyond the Turing Limit: Evolving Interactive Systems. In: Pacholski, L., Ružička, P. (eds.) SOFSEM 2001. LNCS, vol. 2234, pp. 90–109. Springer, Heidelberg (2001)
29. Zandron, C.: A Model for Molecular Computing: Membrane Systems. PhD thesis, Milano Univ., Italy (2002)
30. The P Systems Website, http://ppage.psystems.eu

Undecidability of State Complexities Using Mirror Images

Arto Salomaa

Turku Centre for Computer Science,
Joukahaisenkatu 3–5 B, 20520 Turku, Finland
asalomaa@utu.fi

Abstract. We establish the undecidability of the state complexity of compositions of the operation mirror image and two other regularity-preserving operations. The undecidability of Hilbert's Tenth Problem is not needed; the weaker Davis-Putnam-Robinson Theorem suffices for the reduction. Special attention is paid to the maximal state complexity of mirror images and the maximal deterministic state complexity of nondeterministic finite automata.

Keywords: Finite automaton, State complexity, Undecidability, Mirror image, Nondeterminism, Exponential polynomials.

Dedication. This paper is dedicated to my friend and colleague *Jürgen Dassow* on the occasion of his 65th birthday. The first scientific contacts between us go back to early 70's. Jürgen was one of the first, maybe the very first, researcher in the Eastern bloc who investigated developmental languages. Our early cooperation was possible because it was easier to travel to Finland than elsewhere in the West. When life and travel became easier, many mutual visits and different forms of cooperation started to be possible between us, as well as our students. The conferences organized by Jürgen, such as the DLT-95 in Magdeburg following DLT-93 in Turku, made even further contacts possible. Jürgen was always interested in decision problems and, therefore, I believe my contribution to be appropriate for the volume. I wish Jürgen continuing success in science and life in general.

1 Introduction

It is well known that, for every regular language L, there is a unique, up to isomorphism, finite deterministic automaton accepting L which is minimal with respect to the number of states. The effect of a regularity-preserving operation on the number of states is customarily referred to as the *state complexity* of that operation. For instance, if $L_i, 1 \leq i \leq 3$, are regular languages accepted by automata with x_i states, respectively, how many states does the composition $(L_1 \cup L_2)L_3$ require in terms of the numbers x_i?

The recent study of state complexity has been motivated by many new applications of automata, e.g., in natural language and speech processing, software

H. Bordihn, M. Kutrib, and B. Truthe (Eds.): Dassow Festschrift 2012, LNCS 7300, pp. 221–235, 2012.

engineering, and parallel processing, which utilize finite automata of very large sizes. The state complexity gives a good estimate of the size of the application and a lower bound of its time and space complexities.

The effect of basic regularity-preserving operations was settled in [13]. Apart from the basic operations alone, also combined operations have been investigated, for instance, in [12,7,2,10,1]. The worst-case state complexity of the composition of two operations can be smaller than the one obtained directly from the (known) complexities of the two operations. For instance, the state complexity of the star operation on the result of the union of two regular languages, with the state complexities m and n, is $2^{m+n-1} - 2^{m-1} - 2^{n-1} + 1$. The direct composition of the two state complexities gives the result $2^{mn-1} + 2^{mn-2}$, which is much higher than the actual state complexity [7].

However, there is no general method of determining the state complexity of arbitrary compositions of operations. This undecidability result was established in [8], using compositions of two simple operations. Then a reduction of *Hilbert's Tenth Problem* could be used.

It is natural to investigate the effect and applicability of other operations within this framework. In this paper we focus the attention on the operation *mirror image*, denoted $mi(w), mi(L)$, (also called *reversal*, denoted w^r, L^r).

The state complexity of the *mirror image* of a regular language is of special interest because it is connected with the difference between nondeterminism and determinism in the following way. The mirror image of a language $L(\mathcal{A})$ is accepted by an automaton obtained from $\mathcal{A}$ by reversing all (labeled) arrows, and interchanging initial and final states. The latter automaton is nondeterministic. Thus, the (deterministic) state complexity of the mirror image is the number of states in the minimal equivalent deterministic automaton. Using the *subset construction*, [4], we see that the maximal increase in the number of states goes from n to 2^n. Thus, the state complexity of the language $mi(L)$ is between n and 2^n if the state complexity of L is n. Languages L where the mirror image $mi(L)$ actually reaches the upper bound 2^n can be used as "representations" of the exponential function. Consequently, we can, instead of Hilbert's Tenth Problem, use the weaker *Davis-Putnam-Robinson Theorem* as a basis of reduction.

A brief outline of the contents of the paper follows. In Section 2 we introduce the basics about state complexity, and discuss a special operation needed in the sequel. The next section investigates languages, with detailed proofs, whose mirror images possess the maximal state complexity. In fact, the results obtained there are interesting on their own right. They are stronger than what is actually needed for our undecidability result. Section 4 discusses exponential polynomials and modifies the Davis-Putnam-Robinson Theorem to suit for our purposes. Sections 5 and 6 present a method of associating with an exponential polynomial E a composition C of regular languages such that, for all tuples of values of the variables, the state complexity of C equals at most the value of E when the languages in C have state complexities defined by the tuple in question. For specific languages the value is actually reached. Moreover, Section 6 proves the following undecidability result. Given a sequence C_i, $i = 1, 2, \ldots$, of compositions and a sequence E_i, $i = 1, 2, \ldots$, of exponential polynomials, both effectively constructible, it is undecidable whether or not E_i is a state complexity function for C_i. Some open problems are presented in the final section.

2   State Complexity – Marked Catenations

We assume that the reader is familiar with the basics of finite automata and regular languages. Whenever necessary, the article of Sheng Yu in [4] can be consulted.

We use the customary notation

$$\mathcal{A} = (Q, \Sigma, \delta, q_0, F)$$

for *deterministic finite automata*, DFA's. The five items are, respectively, the state set, the input alphabet, the transition function, the initial state, and the set of final states. We consider only *complete* automata: $\delta(q, a)$ is defined for all $q \in Q$ and $a \in \Sigma$. Very often in this paper, n refers to the cardinality of the state set: $\sharp(Q) = n$.

A state of an automaton is called a *sink* if no sequence of transitions leads from it to a final state. (Sinks are often also referred to as *garbage states*.)

The (regular) language accepted by the DFA $\mathcal{A}$ is denoted by $L(\mathcal{A})$. The *state complexity* of a regular language L is the number of states in the minimal DFA $\mathcal{A}$ such that $L = L(\mathcal{A})$.

The DFA $\mathcal{A}$ is *functionally complete* if the transition monoid of $\mathcal{A}$, that is the monoid generated by the functions $f_a(q) = \delta(q, a)$ where a ranges over Σ, consists of all of the n^n mappings of Q into Q. The notion of *functional completeness* can be extended to sets of functions $f : Q \to Q$, where Q is an arbitrary finite set. (For more details, see [5] or [11].)

We use natural graphical representations for DFA's, where states are represented by circles and transitions by labeled arrows.

We consider also *nondeterministic* finite automata, NFA's. Our NFA's may possess several initial states. (They are actually called NNFA's in [4].)

For an NFA $\mathcal{A}$, we denote by $S(\mathcal{A})$ the DFA obtained from $\mathcal{A}$ by the *subset construction*. The initial state of $S(\mathcal{A})$ is the set of initial states of $\mathcal{A}$. As states of $S(\mathcal{A})$ we consider only subsets reachable from the initial state. If $\sharp(Q) = n$, the automaton $S(\mathcal{A})$ has at most 2^n states. It is a direct consequence of the subset construction that the automaton $S(\mathcal{A})$ is complete.

We already pointed out that, for a regular language L, there is a unique minimal automaton accepting L. The number of states in this automaton is referred to as the *state complexity* of L. The situation is more involved if we consider classes of languages and *state complexity functions*.

We are interested in *compositions* of variable regular languages. We will now give a general definition of *state complexity functions*. The definition is given for arbitrary compositions although, for the undecidability result below, we actually need it only for some special compositions. The functions we are considering will always map some power of $\mathbb{N}_0$ into $\mathbb{N}_0$. Again, only some special functions (*exponential polynomials* defined below) will be needed for our undecidability result.

In the usual state complexity considerations, each variable of the function corresponds to a unique language. We allow also the more general case, where several languages are associated with the same variable.

Definition 1. *Consider a function* $F(x_1, \ldots, x_m)$, $m \geq 1,$, *some composition* $C^n(L_1, \ldots, L_n)$, $n \geq m$, *of languages involving only regularity-preserving operations, as well as a surjective mapping* φ *of the index set* $\{1, \ldots, n\}$ *onto the index set* $\{1, \ldots, m\}$. *Then the function* $F(x_1, \ldots, x_m)$ *is a* state complexity function *of the composition* $C^n(L_1, \ldots, L_n)$ *if the following condition is satisfied. Let* $(x_1, \ldots, x_m)$ *be an arbitrary m-tuple of nonnegative integers. Whenever* $1 \leq i \leq m$ *and each* L_j, $j \in \varphi^{-1}(i)$, *is a regular language with state complexity* x_i, *then the composition* $C^n(L_1, \ldots, L_n)$ *is accepted by an automaton with at most* $F(x_1, \ldots, x_m)$ *states.*

Note that when we say that a function $F(x_1, \ldots, x_m)$ is a state complexity function of the composition $C^n(L_1, \ldots, L_n)$, this means that the value of $F(x_1, \ldots, x_m)$ gives an *upper bound* for the state complexity of the language $C^n(L_1, \ldots, L_n)$ when each variable x_i is assigned the state complexity of the languages L_j such that $\varphi(j) = i$.

Marked Catenation. $L_1 \ddagger L_2$ is a special operation needed in the sequel. It is the catenation of the languages L_1, $\ddagger$, L_2, where $\ddagger$ is a letter not appearing in the alphabets of L_1 and L_2. Similarly we consider marked catenations of arbitrarily many languages. The following result is from [8]. In view of its importance, we give the proof also here.

Theorem 1. *Assume that* L_i *are regular languages (maybe over different alphabets) with state complexities* σ_i, $1 \leq i \leq r$, $r \geq 2$. *Assume, further, that for each* i, $1 \leq i \leq r$, *the minimal automaton* $\mathcal{A}_i$ *for* L_i *has no sinks. Then the marked catenation*

$$L_1 \ddagger L_2 \ddagger \cdots \ddagger L_r$$

is accepted by an automaton $\mathcal{A}$ *with*

$$\sum_{i=1}^{r} \sigma_i + 1 = \sigma$$

states but by no automaton with fewer than σ *states. The alphabet of* $\mathcal{A}$ *consists of the union of the alphabets of* L_i *and of* $\ddagger$. *The initial state of* $\mathcal{A}_1$ *is the initial state of* $\mathcal{A}$, *and the final states of* $\mathcal{A}_r$ *constitute the set of final states of* $\mathcal{A}$.

Proof. An automaton $\mathcal{A}$ accepting the marked catenation is obtaining by joining the automata $\mathcal{A}_i$, $1 \leq i \leq r$, in the following way. From each final state of $\mathcal{A}_i$, $1 \leq i \leq r - 1$, introduce a transition labeled by $\ddagger$ to the initial state of $\mathcal{A}_{i+1}$. From all other states of $\mathcal{A}_i$, $1 \leq i \leq r - 1$, as well as from all states of $\mathcal{A}_r$, introduce a transition labeled by $\ddagger$ to an additional sink state. It is clear that $\mathcal{A}$ accepts the marked catenation and has σ states. On the other hand, no automaton with fewer states can accept the marked catenation. Each word has to have exactly $r-1$ occurrences of $\ddagger$. States in two different automata $\mathcal{A}_i$ cannot be combined because this would result into too many occurrences of the letter $\ddagger$. $\square$

If some of the automata $\mathcal{A}_i$ would possess a sink, then the various sinks can be combined, and the total number σ can be reduced accordingly.

3   Mirror Images

For a word $w = b_1 b_2 \ldots b_k$, $b_i \in \Sigma$, its *mirror image* (also called *reversal*) is defined by

$$mi(w) = b_k \ldots b_2 b_1.$$

The mirror image $mi(L)$ of a language L consists of the mirror images of its words. For a DFA $\mathcal{A} = (Q, \Sigma, \delta, q_0, F)$, we denote by $R(\mathcal{A})$ the NFA obtained from $\mathcal{A}$ by reversing all arrows and interchanging the initial and final states. It is obvious that $R(\mathcal{A})$ accepts the language $mi(L(\mathcal{A}))$. If $\sharp Q = n$, then $S(R(\mathcal{A}))$ has at most 2^n states. Consequently, the state complexity of $mi(L(\mathcal{A}))$ is at most 2^n. For a proof of the following well-known result, see [4], Vol. 1, p. 95.

Lemma 1. *If in a DFA $\mathcal{A} = (Q, \Sigma, \delta, q_0, F)$ all states of Q are reachable from q_0, then $S(R(\mathcal{A}))$ is the minimal DFA accepting $mi(L(\mathcal{A}))$.*

Thus, assuming that the state complexity of a language $L = L(\mathcal{A})$ equals n, the state complexity of $mi(L)$ equals 2^n if and only if all of the 2^n subsets of Q appear as states of $S(R(\mathcal{A}))$. We now consider a sequence of automata where this actually happens. Some of the automata were discussed, omitting many details, in [9] which was one of the very last joint works of the present author with the late *Derick Wood*. Therefore, we call them here *Wood automata*. Wood automata are investigated, from a different point of view, also in [6].

The Wood automaton $W(n)$ with n states is over the binary alphabet $\{a, b\}$. (If some other letters, say c, d, are used, this will be indicated in the notation: $W(n)(c, d)$). The state set is $Q = \{1, 2, \ldots, n\}$. The transitions $f_a(x) = \delta(x, a)$ and $f_b(x) = \delta(x, b)$ are defined as follows. The function $f_a(x)$ is the circular permutation $(123 \cdots n)$, whereas

$$f_b(1) = f_b(3) = 1, \quad f_b(4) = 3, \quad f_b(x) = x \text{ otherwise.}$$

We assume first that $n \geq 5$ and that n is not divisible by 4. Then the state 1 is both the initial and the only final state. In this case the automaton $W(n)$ is depicted in Figure 1. (Final states are marked by double circles, the incoming arrow points to the initial state.) We will return later to the remaining cases.

The essential tool in our considerations is the subset construction, and the main problem the connectedness of the resulting graph. The framework can be described in terms of *subset functions* as follows.

Consider a finite set $Q = \{1, 2, \ldots, n\}$ and mappings $f : Q \to 2^Q$. Extend such a mapping additively to a mapping from 2^Q to 2^Q. (Thus, for $X \subseteq Q$, the value $f(X)$ is the union Y of the values $f(x)$, where $x \in X$.)

Let F be a (finite) set of such *subset functions* For $X, Y \subseteq Q$, we use the notation $X \Rightarrow_F Y$ to indicate that $f(X) = Y$, for some $f \in F$. Finally, let $\Rightarrow_F^*$ be the reflexive transitive closure of the relation $\Rightarrow_F$.

Definition 2. *A set F of subset functions is* complete *if, for any $X \subseteq Q$, $X \neq \emptyset, Q$ and any $Y \subseteq Q$, we have*

$$X \Rightarrow_F^* Y.$$

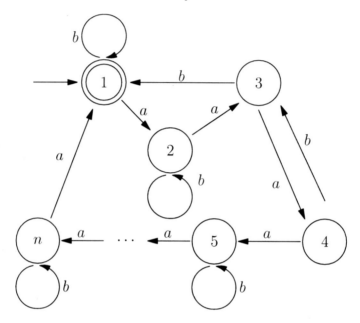

Fig. 1. Wood automaton $W(n)$, $n \geq 5$, $4 \nmid n$

If $X \Rightarrow_F^* Y$, we say that Y is *reachable* from X (via F). Although functional completeness is well understood (see [5] and the references given there), the completeness of sets of subset functions is **an open problem area**. The restrictions in Definition 2, concerning $\emptyset, Q$, become obvious below.

The following general considerations are independent of initial and final states and concern an arbitrary $n \geq 4$. For convenience, we denote the inverses of the functions f_a and f_b by A and B, respectively. We denote also $F = \{A, B\}$. Clearly, A and B are subset functions in the sense defined above. Thus, A affects the circular permutation $(n \cdots 321)$, whereas B maps 1 to $\{1, 3\}$, 3 to 4, 4 to $\emptyset$, and x to x, otherwise. In connection with the set Q, additions will be carried out *modulo n*, that is, $i + j$ stands for the smallest *positive* remainder modulo n. For $X = \{x_1, \ldots, x_k\} \subseteq Q$, we consider sets

$$X^{+i} = \{x_{1+i}, \ldots, x_{k+i}\}, \ 1 \leq i \leq n.$$

(Observe that $X^{+n} = X$.) Since A is a circular permutation, we have, for all i,

$$X \Rightarrow_A^* X^{+i}.$$

This fact will be used frequently in our following arguments.

Observe that $4 \Rightarrow_B \emptyset$. We now claim that, from any

$$X \subseteq Q, \ 2 \le \sharp X = k \le n - 1,$$

a subset Y of Q with cardinality $k - 1$ is reachable. If, for some i, $0 \le i \le n-1$, the element $i + 4$ is in X, whereas $i + 1$ is not there, then we apply B to the set $X^{+(n-i)}$, and obtain a subset Y as required. If no such pair of elements exists in X, then n is divisible by 3, and X consists of one or two residue classes modulo 3. (This follows because $X \ne Q$.) We may assume, by applying A if necessary, that the numbers $3, 6, \ldots, n$ are in X, whereas the numbers $1, 4, \ldots, n-2$ are not there. Now $B(X)$ is obtained from X by replacing 3 with 4. Hence, $Y = B(B(X))$ is of cardinality $k - 1$. This completes the proof of our claim.

We now work inductively "upwards", increasing the cardinality of the reachable sets. We will investigate which subsets Y are reachable from the singleton $\{1\}$. Reachability is immediately verified for $\emptyset$ and all singletons. We now assume inductively that every subset X' of Q with cardinality $k - 1$, $2 \le k \le n$, is reachable, and consider an arbitrary subset X of Q with cardinality k. Given X, we want to show how to choose an X' of cardinality $k - 1$ such that $X' \Rightarrow_F^* X$.

Assume first that X contains, for some i, the elements i and $i+2$. By applying A, we may assume that 1 and 3 are contained in X. If 4 is (resp. is not) in the set X thus modified, we let X' be the set obtained from X by removing the element 4 (resp. 3). Then $X' \Rightarrow_F^* X$. (Of course, we still have to use A to get the original X.)

From now on we assume that no elements i and $i + 2$ are in X. This implies that X contains no three consecutive elements $i, i + 1, i + 2$ and, whenever i is in X but $i + 1$ is not in X, then also $i + 2$ is not in X. Intuitively, X consists of singletons and pairs of two consecutive elements, all separated by at least two "non-elements". (All the time we are using the modular arithmetic: n and 1 are consecutive.)

Assume that, for some i, the element i is in X, whereas $i+1$ and $i-1$ are not. By the preceding paragraph, also $i + 2$ and $i - 2$ are not in X. By an A-shift, we may assume that 1 is in X, whereas $2, 3, n - 1, n$ are not.

Let j be the smallest element, apart from 1, in X. We know that $j \ge 4$. Construct X' by removing j from X. In the following reachability sequence we have marked down only the relevant elements in the sets. It is essential that n is not in X. (Observe that B alters elements $1, 3, 4$ only.)

$$X' \Rightarrow_B^* \{1, 3, 4\} \Rightarrow_A^* \{2, 4, 5\} \Rightarrow_B^* \{2, 5\} \Rightarrow_A^* \{1, 4\}$$
$$\Rightarrow_A^* \{n, 3, \} \Rightarrow_B^* \{n, 4\} \Rightarrow_A^* \{1, 5\}$$

This shows how X is reachable if $j = 4$ or $j = 5$. For an arbitrary j, we reach X by repeating the transformations on the second line.

Hence, we may assume that X does not contain such isolated elements. This implies, by our previous constructions, that X consists of pairs of consecutive elements, separated by at least two "non-elements". Suppose that, for some i, the elements i and $i+1$ are in X, whereas none of the elements $i-3, i-2, i-1, i+2, i+3$ is in X. By an A-shift, we may again assume that 1 and 2 are in X, whereas none of the elements $n-2, n-1, n, 3, 4$ is in X.

We now let X' be the following subset of cardinality $k-1$:

$$X' = \{1\} \cup \{j+1|\ j \in X,\ j \neq 1, 2\}.$$

Then the following reachability chain is valid:

$$\begin{aligned}
X' \Rightarrow_B^* & \{1, 3\} \cup \{j+1|\ j \in X,\ j \neq 1, 2\} = X_1 \\
\Rightarrow_A^* & \{3, 5\} \cup \{j+3|\ j \in X,\ j \neq 1, 2\} = X_2 \\
\Rightarrow_B^* & \{4, 5\} \cup \{j+3|\ j \in X,\ j \neq 1, 2\} \Rightarrow_A^* X
\end{aligned}$$

It is important to observe that neither $n-1$ nor n is in X_1 and, consequently, neither 1 nor 2 is in X_2.

Thus, we have reached the conclusion that X consists of pairs of consecutive elements, separated by exactly two "non-elements". But this means that n is divisible by 4 : $n = 4m$.

For $n = 4m$, we now define the Wood automaton $W(n)$ by choosing the set

$$WF(n) = \{4i+1, 4i+2|\ 0 \leq i \leq m-1\}$$

as the set of final states. Otherwise, the definition of $W(n)$ remains unaltered. The automaton $W(8)$ is illustrated in Figure 2.

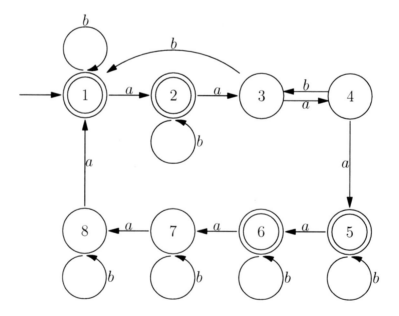

Fig. 2. Wood automaton $W(8)$

Our argument above shows, by Lemma 1 and the reachability of all subsets, that the state complexity of the language $mi(L(W(n)))$, $n \geq 5$, equals 2^n, provided n is not divisible by 4. In fact, in this case we are free, [6], to choose the initial and the set of final states. However, our choice of the final state set $WF(n)$ guarantees that the state complexity result holds true also if n is divisible by 4. (There is no change in the proof when we reduce the cardinality of reachable subsets. The argument applies also if we want to increase the cardinality or keep it $2m$ which is the cardinality of $WF(n)$.) Hence, we have established the following result.

Theorem 2. *For $n \geq 4$, the state complexity of the language $mi(L(W(n)))$ equals 2^n.*

We still have to deal with the small values of n. The automata $W(2)$ and $W(3)$ are depicted in Figure 3. It is immediately verified that the state complexities of the mirror images are 4 and 8 in these cases. Hence, we obtain the following corollary of Theorem 2.

Theorem 3. *For $n \geq 2$, the state complexity of the language $mi(L(W(n)))$ equals 2^n.*

Summarizing, we obtain the following result.

Theorem 4. *For every $n \geq 2$, the Wood automaton $W(n)$ has n states but the minimal deterministic automaton equivalent to $R(W(n))$ has 2^n states.*

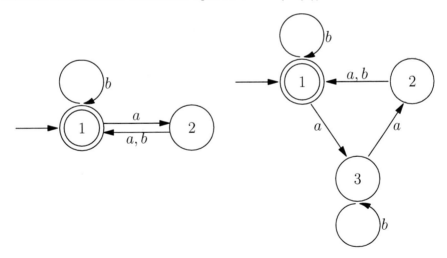

Fig. 3. Wood automata $W(2)$ and $W(3)$

4   Modifications of the Davis-Putnam-Robinson Theorem

By an *exponential polynomial*, briefly *E-polynomial*, we mean a finite sum of terms of the form $\alpha_1 \alpha_2 \cdots \alpha_n$, $n \geq 1$, where each α_i, $i \geq 1$, is a variable, or of the form 2^x, for some variable x. An exponential polynomial may contain several identical terms, which will be expressed with coefficients in $\mathbb{N}_0$. For instance,

$$2^{x_1} x_3^2 x_4 + 4x_1 x_2 + 2^{x_3} 2^{x_1} x_2 x_3 x_4$$

is an exponential polynomial.

By the *Davis-Putnam-Robinson Theorem*, for every recursively enumerable set S of nonnegative integers, there are (effectively constructible) exponential polynomials $E_i(x_0, x_1, \ldots, x_m)$, $i = 1, 2$, such that $x_0 \in S$ if and only if the equation

$$E_1(x_0, x_1, \ldots, x_m) = E_2(x_0, x_1, \ldots, x_m)$$

has a solution in nonnegative integers $(x_1, \ldots, x_m)$. (For details and a proof using register machines, see [3].) Using the universal Turing machine and the undecidability of the emptiness of recursively enumerable languages, the result can be expressed in the following form. There are (effectively constructible) exponential polynomials $E(x_0, x_1, \ldots, x_m)$ and $E'(x_0, x_1, \ldots, x_m)$ such that, given $x_0 \geq 0$, it is undecidable whether or not the equation

$$E(x_0, x_1, \ldots, x_m) = E'(x_0, x_1, \ldots, x_m)$$

has a solution in nonnegative integers $(x_1, \ldots, x_m)$. By substituting the value $x_0 = i \geq 1$ to the exponential polynomials E and E', we obtain two infinite sequences E_i and E_i', $i = 1, 2, \ldots$, such that, given $i \geq 1$, it is undecidable whether or not the equation

$$E_i(x_1, \ldots, x_m) = E_i'(x_1, \ldots, x_m)$$

has a solution in nonnegative integers $(x_1, \ldots, x_m)$. Denote

$$P_i(x_1, \ldots, x_m) = E_i(x_1, \ldots, x_m) - E_i'(x_1, \ldots, x_m), \ i \geq 1.$$

Consider the inequalities

$$0 \leq (P_i(x_1, \ldots, x_m)^2) - 1, \ i = 1, 2, \ldots$$

Clearly, for any given i, this inequality is valid for all m-tuples $(x_1, \ldots, x_m)$ of nonnegative integers exactly in case the equation

$$E_i(x_1, \ldots, x_m) = E_i'(x_1, \ldots, x_m)$$

has *no solution* in nonnegative integers. Therefore, by the Davis-Putnam-Robinson Theorem, there is no algorithm of deciding, given i, whether or not the inequality

$$0 \leq (P_i(x_1, \ldots, x_m)^2) - 1$$

holds for all m-tuples $(x_1, \ldots, x_m)$ of nonnegative integers. We now move all negative terms from the right side to the left side. This gives rise to a new inequality, equivalent to the original one,

$$E_i^{(l)}(x_1, \ldots, x_m) \leq E_i^{(r)}(x_1, \ldots, x_m),$$

where $E_i^{(l)}$ and $E_i^{(r)}$ are E-polynomials.

These considerations are summarized in the following Theorem.

Theorem 5. *There is no algorithm of deciding, given a positive integer i, whether or not the inequality*

$$E_i^{(l)}(x_1, \ldots, x_m) \le E_i^{(r)}(x_1, \ldots, x_m)$$

holds for all m-tuples $(x_1, \ldots, x_m)$ of nonnegative integers. Here $E_i^{(l)}$ and $E_i^{(r)}$ are effectively constructible E-polynomials over the set of variables $\{x_1, \ldots, x_m\}$. Moreover, there is a finite set S of terms of the form

$$y_1^{j_1} \cdots y_{2m}^{j_{2m}}, \ \ j_\mu \ge 0, \ 1 \le \mu \le 2m,$$

such that every polynomial $E_i^{(l)}$, $i = 1, 2, \ldots$, equals the sum of some terms in S, provided with positive integer coefficients.

Thus, each $E_i^{(l)}$ is a (finite) sum of terms of the form

$$y_1^{j_1} \cdots y_{2m}^{j_{2m}}, \ \ j_\nu \ge 0, \ 1 \le \nu \le 2m,$$

provided with positive integer coefficients depending on i. The choice of i affects only the multiplicity of each term, i.e., it tells how many times each term appears in the polynomial $E_i^{(l)}$. (We have $2m$ instead of m because a term may contain both x and 2^x, for some variable x.)

In the sequel we will associate the E-polynomials $E_i^{(l)}$ with specific compositions of regular operations, whereas the polynomials $E_i^{(r)}$ will constitute the proposed state complexities.

5   Special Compositions and Associated E-Polynomials

The specific compositions we will need use the three regularity-preserving operations of mirror image, intersection and marked catenation. Therefore, we will call them *three-compositions*. The operations are not nested arbitrarily. The way of nesting is specified in the following definition.

Definition 3. *A three-composition   over the set $\{L_1, \ldots, L_n\}$, $n \ge 2$, of language variables is an expression*

$$\gamma_1 \ddagger \gamma_2 \ddagger \cdots \ddagger \gamma_r, \ r \ge 2,$$

where each γ_i, $1 \le i \le r$, is of the form

$$\gamma_i = M_1 \cap \cdots \cap M_{j(i)}, \ j(i) \ge 1,$$

such that the M's are different ones among the language variables L_ν, $1 \le \nu \le n$, either appearing as plain L_ν, or in the form $mi(L_\nu)$. A language variable L_ν can appear both as such and in the form $mi(L_\nu)$ in the same γ_i.

Parentheses can be added for clarity. For instance,

$$(mi(L_2) \cap L_3)\ddagger(L_1 \cap L_2 \cap L_3)\ddagger(mi(L_1) \cap mi(L_3) \cap L_1)$$

is a three-composition over the set $\{L_1, L_2, L_3\}$ of language variables.

Above we defined the Wood automata $W(n)$, $n \geq 2$, and showed that the state complexity of the language $mi(L(W(n)))$ equals 2^n. We make the formal convention that the state complexity of $mi(L(W(1)))$ (resp. $mi(L(W(0))))$ equals 2 (resp. 1).

Let us go back to the E-polynomials $E_i^{(l)}$ defined in the preceding section. They use the fixed set of variables $\{x_1, \ldots, x_m\}$. As already pointed out, the variables may appear either by themselves or as exponents of 2. However, each variable x and each power 2^x appears only a bounded number of times in each product in each $E_i^{(l)}$. Thus, for each j, $1 \leq j \leq m$, there is a number K_j such that every exponent of x_j in every product equals at most K_j, and that 2^{x_j} appears as a factor in every product at most K_j times, no matter what $E_i^{(l)}$ we are considering. This important fact follows because, as explained above, a change of the index i in $E_i^{(l)}$ does not affect the summands in $E_i^{(l)}$, only their multiplicities.

Consider now language variables

$$L_j^\nu, \ 1 \leq j \leq m, \ 1 \leq \nu \leq K_j.$$

The variables L_j^ν, $1 \leq \nu \leq K_j$ correspond to x_j in the sense of the mapping φ in Definition 1. With each summand

$$(2^{x_1})^{\mu_1} \cdots (2^{x_m})^{\mu_m} x_1^{\nu_1} \cdots x_m^{\nu_m},$$

where $0 \leq \mu_j, \nu_j \leq K_j$, $1 \leq j \leq m$, in $E_i^{(l)}$, we associate an intersection as follows. (We consider an arbitrary index i. For readability, we do not include it in the notation.) Consider an arbitrary j, $1 \leq j \leq m$. The part $(2^{x_j})^{\mu_j}$ is associated with the intersection

$$mi(L_j^1) \cap \ldots \cap mi(L_j^{\mu_j}).$$

The part $x_j^{\nu_j}$ is associated with the intersection

$$L_j^1 \cap \ldots \cap L_j^{\nu_j}.$$

Finally, the *three-composition* $C(E_i^{(l)})$ *associated with* $E_i^{(l)}$ is the marked catenation of the summands appearing in $E_i^{(l)}$.

As an example, consider the E-polynomial

$$(2^{x_1})^2 x_1 x_3 + 2x_1 x_2^2 x_3 + 2^{x_2}.$$

Now we have $K_1 = K_2 = 2$, $K_3 = 1$. The three-composition associated with this E-polynomial is

$$(mi(L_1^1) \cap mi(L_1^2) \cap L_1^1 \cap L_3^1)\ddagger(L_1^1 \cap L_2^1 \cap L_2^2 \cap L_3^1)$$
$$\ddagger(L_1^1 \cap L_2^1 \cap L_2^2 \cap L_3^1)\ddagger(mi(L_2^1)).$$

We will need the following result from [13]. It is proved also in [8].

Theorem 6. *Assume that L_i, $1 \leq i \leq r$, are regular languages with the state complexities σ_i. Then the state complexity of the regular language $L_1 \cap \ldots \cap L_r$ is at most the product $\sigma = \sigma_1 \cdots \sigma_r$. Moreover, for any r-tuple $(\sigma_1, \ldots, \sigma_r)$ of nonnegative integers, it is possible to construct regular languages $K_i, 1 \leq i \leq r$, with the state complexity σ_i such that the intersection of the languages K_i has exactly the state complexity $\sigma = \sigma_1 \cdots \sigma_r$.*

The following theorem is an immediate corollary of Theorems 1 and 6. (The additional summand $+1$ of Theorem 1 is not needed if we assume that at least one exponential term appears in the E-polynomial.)

Theorem 7. *For each $i = 1, 2, \ldots$, the E-polynomial $E_i^{(l)}(x_1, \ldots, x_m)$ is a state complexity function of the three-composition $C(E_i^{(l)})$.*

In the next section specific languages will be substituted in three-compositions in such a way that the alphabets of the languages appearing in intersections will be pairwise disjoint. (We do not estimate here the size of the total alphabet. Such an estimation was done in an analogous situation in [8].)

6 Undecidability

We consider in the sequel an arbitrary but fixed E-polynomial $E_i^{(l)}(x_1, \ldots, x_m)$, and the numbers K_j as defined above. Let $C(E_i^{(l)})$ be the three-composition associated with $E_i^{(l)}(x_1, \ldots, x_m)$. Introduce the alphabet Σ consisting of the letters

$$a_j^\nu, b_j^\nu, c_j^\nu, \; 1 \leq j \leq m, \; 1 \leq \nu \leq K_j.$$

The specific languages defined below will be over the alphabet Σ. The languages will depend on a fixed nonnegative integer n. (It will be the value assigned for the variable x_j in our E-polynomial.) By definition,

$$A_j^\nu(n) = mi(W(n)(b_j^\nu, c_j^\nu)), \; 1 \leq j \leq m, \; 1 \leq \nu \leq K_j.$$

(Recall our way of indicating the alphabet of a Wood language.) Similarly, let $B_j^\nu(n)$ be the language over Σ consisting of all words w such that the number of occurrences of the letter a_j^ν in w is divisible by n. Finally, for each n-tuple $(x_1, \ldots, x_m)$ of nonnegative integers, let $D_i(x_1, \ldots, x_m)$ be the regular language, resulting from $C(E_i^{(l)})$ as follows. If n_j is the value assigned for x_j, $1 \leq j \leq m$, substitute every occurrence of $mi(L_j^\nu)$ (resp. L_j^ν) with $A_j^\nu(n_j)$ (resp. $B_j^\nu(n_j)$).

Consider now the example

$$(mi(L_1^1) \cap mi(L_1^2) \cap L_1^1 \cap L_3^1)\ddagger(L_1^1 \cap L_2^1 \cap L_2^2 \cap L_3^1)$$
$$\ddagger(L_1^1 \cap L_2^1 \cap L_2^2 \cap L_3^1)\ddagger(mi(L_2^1)).$$

from the preceding section, as well as the ordered triple $(8, 6, 5)$ as values of the three variables. Then the associated D_i-languages is

$$(mi(W(8)(b_1^1, c_1^1)) \cap mi(W(8)(b_1^2, c_1^2)) \cap B_2^1(6) \cap B_3^1(5))$$
$$\ddagger(B_1^1(8) \cap B_2^1(6) \cap B_2^2(6) \cap B_3^1(5))$$
$$\ddagger(B_1^1(8) \cap B_2^1(6) \cap B_2^2(6) \cap B_3^1(5))\ddagger(mi(W(6)(b_2^1, c_2^1))).$$

The proof of the following lemma is straightforward, along the lines of Theorems 1 and 6. The only additional technicality needed is to take care of the possible sinks appearing in automata for the mirror images of the languages of Wood automata. Since the alphabets are pairwise disjoint, we can introduce transitions in such a way that sinks can never be combined, and there is only one sink where the wrong transitions, including the ones involving $\ddagger$, are leading.

Lemma 2. *For all values of the variables, the state complexity of the language $D_i(x_1, \ldots, x_m)$ equals $E_i^{(l)}(x_1, \ldots, x_m)$.*

Our undecidability result now follows by Theorems 5 and 7 and Lemma 2.

Theorem 8. *For the sequence of E-polynomials $E_i^{(r)}$, $i = 1, 2, \ldots$ and three-compositions $C(E_i^{(l)})$, $i = 1, 2, \ldots$, as constructed above, it is undecidable whether or not $E_i^{(r)}$ is a state complexity function of $C(E_i^{(l)})$.*

7   Conclusion

We have investigated the operation *mirror image*, in particular, the cases where the state complexity of the language $mi(L)$ is maximal in comparison with the state complexity of L. This gives also the maximal increase in state complexity in the transition from a nondeterministic automaton to the equivalent deterministic automaton. Wood automata $W(n)$ with n states constitute good examples. If n is divisible by 4, the state complexity of $mi(W(n))$ is, for certain choices of the final state set, maximal but sometimes only $2^n - 4$. Several open problems remain in connection with mirror images, in particular, the construction of automata for languages L such that the state complexity of $mi(L)$ is close to the maximal, or close to the minimal one.

In our result concerning the undecidability of the state complexity of compositions of regular languages, we were able to use reduction to exponential polynomials, instead of polynomials. The three operations in the compositions were mirror image, intersection and marked catenation. The Davis-Putnam-Robinson Theorem provided the undecidable problem used as the basis of reduction. Other undecidable problems will in general lead to other operations. The undecidability of the state complexity will then concern composition sequences in terms of these operations. It is an interesting open problem to study the possibilities in this direction. For instance, is it possible to use the undecidability of the Post Correspondence Problem and, if this is the case, which are the regularity-preserving operations involved?

In our undecidability result above the state complexities of the languages appearing as components of the marked catenations depended on the values for the variables in a rather complicated way. On the other hand, we can study simpler cases. If the state complexity of each of the component languages L_j equals directly the value of one of the variables x_j, then the state complexity of the marked catenation is a linear function of the variables, and our problem is clearly decidable, provided an E-polynomial is the proposed state complexity function. This result can possibly be extended to the case where the state complexity of each component language is of the form x_j^t, where t is a positive integer. Each pair $((\mathcal{C}, \mathcal{F}))$, where $\mathcal{C}$ (resp. $\mathcal{F}$) is a class in compositions (resp. functions) defines in the natural way a decision problem. A general task is to find interesting pairs for which this problem is decidable.

References

1. Esik, Z., Gao, Y., Liu, G., Yu, S.: Estimation of State Complexity of Combined Operations. Theoretical Computer Science 410, 3272–3280 (2009)
2. Gao, Y., Salomaa, K., Yu, S.: State complexity of catenation and reversal combined with star. In: Descriptional Complexity of Formal Systems (DCFS 2006), pp. 153–164 (2006)
3. Rozenberg, G., Salomaa, A.: Cornerstones of Undecidability. Prentice Hall, New York (1994)
4. Rozenberg, G., Salomaa, A. (eds.): Handbook of Formal Languages, vol. 1-3. Springer, Heidelberg (1997)
5. Salomaa, A.: Composition sequences for functions over a finite domain. Theoretical Computer Science 292, 263–281 (2003)
6. Salomaa, A.: Mirror images and schemes for the maximal complexity of nondeterminism. Fundamenta Informaticae (to appear)
7. Salomaa, A., Salomaa, K., Yu, S.: State Complexity of Combined Operations. Theoretical Computer Science 383, 140–152 (2007)
8. Salomaa, A., Salomaa, K., Yu, S.: Undecidability of the State Complexity of Composed Regular Operations. In: Dediu, A.-H., Inenaga, S., Martín-Vide, C. (eds.) LATA 2011. LNCS, vol. 6638, pp. 489–498. Springer, Heidelberg (2011)
9. Salomaa, A., Wood, D., Yu, S.: On the state complexity of reversals of regular languages. Theoretical Computer Science 320, 293–313 (2004)
10. Salomaa, K., Yu, S.: On the state complexity of combined operations and their estimation. International Journal of Foundations of Computer Science 18, 683–698 (2007)
11. Stanković, R.S., Astola, J.T. (eds.): Reprints from the Early Days of Information Sciences: On the Contributions of Arto Salomaa to Multiple-Valued Logic. Tampere International Center for Signal Processing, TICSP Series 50 (2009)
12. Yu, S.: On the State Complexity of Combined Operations. In: Ibarra, O.H., Yen, H.-C. (eds.) CIAA 2006. LNCS, vol. 4094, pp. 11–22. Springer, Heidelberg (2006)
13. Yu, S., Zhuang, Q., Salomaa, K.: The state complexities of some basic operations on regular languages. Theoretical Computer Science 125, 315–328 (1994)

Asymptotic Subword Complexity

Ludwig Staiger

Institut für Informatik, Martin-Luther-Universität Halle-Wittenberg,
von-Seckendorf-Platz 1, 06099 Halle, Germany
`staiger@informatik.uni-halle.de`

Abstract. The subword complexity of an infinite word ξ is a function $f(\xi, n)$ returning the number of finite subwords (factors, infixes) of length n of ξ. In the present paper we investigate infinite words for which the set of subwords occurring infinitely often is a regular language. Among these infinite words we characterise those which are eventually recurrent.

Furthermore, we derive some results comparing the asymptotics of $f(\xi, n)$ to the information content of sets of finite or infinite words related to ξ. Finally we give a simplified proof of Theorem 6 of [18].

Following [9] the subword complexity of an infinite word ξ is a function $f(\xi, n)$ returning the number of finite subwords (infixes, factors) of ξ having length n. It was mainly investigated for infinite words of low complexity (see [2,9] or the book [1]). However [9, Question 2] asked for the general complexity of quasiperiodic infinite words. An answer on their maximally possible complexity was given in [11] showing that this complexity satisfies $f(\xi, n) \leq_{\mathrm{ae}} c \cdot t_P^n$ where t_P is the smallest Pisot number. Moreover, for quasiperiodic infinite words with maximal subword complexity the set of factors form a regular language.

The aim of our paper is to investigate in more detail those infinite words whose set of factors occurring infinitely often is a regular language. In contrast to [2] and [1] we are mainly interested in infinite words ξ whose subword complexity $f(\xi, n)$ is not bounded by a subexponential function.

In the case of exponentially growing subword complexity the results of [15] and [18] show a close connection between the growth of $f(\xi, n)$ and the Hausdorff dimension of regular ω-languages containing the infinite word ξ. Using this connection we prove that every infinite word having a regular subword language satisfies the condition $f(\xi, n) \approx c \cdot t_\xi^n$ for a suitable real number t_ξ.

As a consequence we obtain a simplified proof of Theorem 6 of [18]. This theorem states, roughly speaking, that finite automata cannot distinguish one-sided eventually recurrent infinite words having the same set of infinitely often occurring factors provided this set of factors is a regular language. A more general result for two-sided infinite words had been obtained earlier [10,12].

After introducing some necessary notation in Section 2 we derive some basic facts on infinite words having a regular language of infinitely often occurring factors. Moreover, the concept of asymptotic subword complexity of infinite words is introduced. This concept proves to be useful in the following.

The entropy of languages known from [3,7,8] is closely related to asymptotic subword complexity. In Section 3 we derive some elementary properties and

H. Bordihn, M. Kutrib, and B. Truthe (Eds.): Dassow Festschrift 2012, LNCS 7300, pp. 236–245, 2012.

also some results relating the entropy of languages to the Hausdorff dimension of ω-languages are presented (cf. also [14,15]). These facts are used to derive our results in the last section. Here we focus our considerations on eventually recurrent infinite words having a regular language of infinitely often occurring subwords. From these considerations several conditions necessary or sufficient for an infinite word to be eventually recurrent are obtained. Finally, we give a simple proof of Theorem 6 of [18].

The previous proof in [18] uses considerations involving Hausdorff measure. In the present paper we circumvent these measure-theoretic considerations confining to language-theoretic results only, although we make implicitly use of the close connection between the entropy of languages and Hausdorff dimension.

1   Notation

In this section we introduce the notation used throughout the paper. By $\mathbb{N} = \{0, 1, 2, \ldots\}$ we denote the set of natural numbers. Let X be an alphabet of cardinality $|X| = r \geq 2$. By X^* we denote the set of finite words on X, including the *empty word* e, and X^ω is the set of infinite strings (ω-words) over X. Subsets of X^* will be referred to as *languages* and subsets of X^ω as *ω-languages*.

For $w \in X^*$ and $\eta \in X^* \cup X^\omega$ let $w \cdot \eta$ be their *concatenation*. This concatenation product extends in an obvious way to subsets $W \subseteq X^*$ and $B \subseteq X^* \cup X^\omega$. For a language W let $W^* := \bigcup_{i \in \mathbb{N}} W^i$, and let $W^\omega := \{w_1 \cdots w_i \cdots : w_i \in W \setminus \{e\}\}$ the set of infinite strings formed by concatenating words in W.

We denote by $w^{[-1]}B := \{\eta : w \cdot \eta \in B\}$ the *left derivative* of the set $B \subseteq X^* \cup X^\omega$. As usual a language $W \subseteq X^*$ is *regular* provided it is accepted by a finite automaton. An equivalent condition is that its set of left derivatives $\{w^{[-1]}W : w \in X^*\}$ is finite. In the sequel we assume the reader to be familiar with basic facts of language theory.

Furthermore $|w|$ is the *length*[1] of the word $w \in X^*$ and $\mathbf{A}(B)$ is the set of all finite prefixes of strings in $B \subseteq X^* \cup X^\omega$. We shall abbreviate $w \in \mathbf{A}(\eta)$ ($\eta \in X^* \cup X^\omega$) by $w \sqsubseteq \eta$ ($w \sqsubset \eta$ if $w \neq \eta$).

$\mathbf{T}(B) := \bigcup_{w \in X^*} \mathbf{A}(w^{[-1]}B)$ is set of infixes (factors) of words in $B \subseteq X^* \cup X^\omega$, and for an infinite word $\xi \in X^\omega$ its sets of factors occurring infinitely often is $\mathbf{T}_\infty(\xi) := \bigcap_{w \sqsubseteq \xi} \mathbf{T}(w^{[-1]}\xi)$.

As usual a language $V \subseteq X^*$ is called a *code* provided $w_1 \cdots w_l = v_1 \cdots v_k$ for $w_1, \ldots, w_l, v_1, \ldots, v_k \in V$ implies $l = k$ and $w_i = v_i$. A code V is said to be a *prefix code* provided $v \sqsubseteq w$ implies $v = w$ for $v, w \in V$.

2   The Languages of Subwords

In this part, we consider, for an infinite word $\xi \in X^\omega$, the languages of subwords $\mathbf{T}(\xi)$ and of subwords occurring infinitely often $\mathbf{T}_\infty(\xi)$, respectively.

---

[1] Since there is no danger of confusion, the length $|w|$ of a word $w \in X^*$ is denoted in the same way as the cardinality $|M|$ of a set M.

For the tails (suffixes) of ξ we have the following obvious inclusion.

$$\mathbf{T}(w^{[-1]}\xi) \supseteq \mathbf{T}(v^{[-1]}\xi) \text{ whenever } w \sqsubseteq v \tag{1}$$

Thus the family $\left(\mathbf{T}(w^{[-1]}\xi)\right)_{w \sqsubseteq \xi}$ is an infinite decreasing chain of languages, and the infinite intersection $\mathbf{T}_\infty(\xi) := \bigcap_{w \sqsubseteq \xi} \mathbf{T}(w^{[-1]}\xi)$ consists of all subwords occurring infinitely often in ξ.

It depends on the ω-word ξ whether the chain in Eq. (1) is stationary or not. If the family $\left(\mathbf{T}(v^{[-1]}\xi)\right)_{v \sqsubseteq \xi}$ is stationary, that is, there is a prefix $v \sqsubseteq \xi$ such that $\mathbf{T}(v^{[-1]}\xi) = \mathbf{T}_\infty(\xi)$, we will refer to the ω-word $\xi \in X^\omega$ as *eventually recurrent*[2] (see [21]).

Next we consider the case when one of the languages $\mathbf{T}(w^{[-1]}\xi)$ is a regular language. To this end we derive the following relation between $v^{[-1]}\mathbf{T}(\xi)$ and $\mathbf{T}(v^{[-1]}\xi)$.

Lemma 1. *Let $v \sqsubset \xi$. Then $v^{[-1]}\mathbf{T}(\xi) \subseteq \mathbf{T}(v^{[-1]}\xi) = \mathbf{T}(v^{[-1]}\mathbf{T}(\xi))$.*

Proof. If $u \in v^{[-1]}\mathbf{T}(\xi)$ then $vu \in \mathbf{T}(\xi)$ and thus there is a w such that $wvu \sqsubset \xi$. Since $v \sqsubset \xi$, we have also $v \sqsubseteq wv$. Consequently, $wv = v\bar{w}$ for some $\bar{w}$, and we obtain $v\bar{w}u \sqsubset \xi$, that is, $u \in \mathbf{T}(v^{[-1]}\xi)$.

$v^{[-1]}\mathbf{T}(\xi) \subseteq \mathbf{T}(v^{[-1]}\xi)$ implies $\mathbf{T}(v^{[-1]}\mathbf{T}(\xi)) \subseteq \mathbf{T}(v^{[-1]}\xi)$, so it suffices to show $\mathbf{T}(v^{[-1]}\xi) \subseteq \mathbf{T}(v^{[-1]}\mathbf{T}(\xi))$. Let $u \in \mathbf{T}(v^{[-1]}\xi)$. Then there is a $\bar{w} \in X^*$ such that $v\bar{w}u \sqsubset \xi$. Consequently, $\bar{w}u \in v^{[-1]}\mathbf{T}(\xi)$, whence $u \in \mathbf{T}(v^{[-1]}\mathbf{T}(\xi))$. $\square$

As in [18] we refer to an ω-word $\xi \in X^\omega$ as *infix-regular* provided there is a prefix $w \sqsubset \xi$ such that $\mathbf{T}(w^{[-1]}\xi)$ is a regular language. The following lemma yields a connection between infix-regular ω-words and eventually recurrent ω-words.

Lemma 2. *An ω-word $\xi \in X^\omega$ is an infix-regular ω-word if and only if ξ is eventually recurrent and $\mathbf{T}_\infty(\xi)$ is a regular language.*

Proof. Let $\xi \in X^\omega$ be infix-regular. Then in Lemma 5 of [18] it is shown that there is a $w' \sqsubset \xi$ such that $\mathbf{T}(w'^{[-1]}\xi)$ is a regular language and $\mathbf{T}(w'^{[-1]}\xi) = \mathbf{T}_\infty(\xi)$.

The other direction is follows from the definition and the fact that $\mathbf{T}_\infty(\xi)$ is a regular language. $\square$

Corollary 1. *If $\mathbf{T}(\xi)$ is regular then $\mathbf{T}_\infty(\xi)$ is also regular.*

It should be noted that not every ω-word ξ for which $\mathbf{T}_\infty(\xi)$ is a regular language is eventually recurrent. The following example shows that $\mathbf{T}_\infty(\xi)$ might be regular, although none of the sets $\mathbf{T}(w^{[-1]}\xi)$, $w \sqsubset \xi$, is regular.

Example 1. Consider $\xi_0 := \prod_{i=1}^\infty a^i \cdot b$. Then $\mathbf{T}_\infty(\xi_0) = a^* \cup a^* \cdot b \cdot a^*$, but, for every $w \sqsubset \xi_0$, the intersection $\mathbf{T}(w^{[-1]}\xi_0) \cap b \cdot a^* \cdot b \cdot a^* \cdot b$ is a non-regular language of the form $\{b \cdot a^i \cdot b \cdot a^{i+1} \cdot b : i \in \mathbb{N} \wedge i \geq c_w\}$, hence $\mathbf{T}(w^{[-1]}\xi_0)$ is also non-regular. $\square$

---

[2] An ω-word ξ is referred to as *recurrent* iff $\mathbf{T}_\infty(\xi) = \mathbf{T}(\xi)$. This resembles the notion of recurrence for $\mathbb{Z}$-words as considered in [10,12].

2.1 Subword Complexity and Asymptotic Subword Complexity of ω-Words

The *subword complexity* of an infinite word ξ is the function $f(\xi, n) := |\mathbf{T}(\xi) \cap X^n|$. In this section we focus on the growth of the function $f(\xi, n)$, in particular, on the real number λ_ξ for which $\lim_{n \to \infty} f(\xi, n)/(\lambda_\xi + \varepsilon)^n = 0$ and $\lim_{n \to \infty} f(\xi, n)/(\lambda_\xi - \varepsilon)^n = \infty$. [3].

First observe the following simple property for eventually recurrent ω-words.

Lemma 3. *If $\xi \in X^\omega$ and $\mathbf{T}(w_0^{[-1]}\xi) = \mathbf{T}_\infty(\xi)$ then $f(\xi, n) \leq |w_0| + |\mathbf{T}_\infty(\xi) \cap X^n|$.*

Proof. This follows from the fact that every factor of length n of ξ is a factor of $w_0^{[-1]}\xi$ or a factor of the length $|w_0| + n - 1$ prefix of ξ. □

Along with the subword complexity we consider the *asymptotic subword complexity* $\tau(\xi)$ of an ω-word ξ. This quantity is defined as the logarithm of the real number λ_ξ.

Definition 1 (Asymptotic subword complexity)

$$\tau(\xi) := \lim_{n \to \infty} \frac{\log_{|X|} f(\xi, n)}{n}$$

Since $f(\xi, n + m) \leq f(\xi, n) \cdot f(\xi, m)$, the limit in Definition 1 exists and equals $\tau(\xi) = \inf\left\{\frac{\log_{|X|} f(\xi, n)}{n} : n \in \mathbb{N}\right\}$. Moreover, we have the following relation between $f(\xi, n)$ and $|\mathbf{T}_\infty(\xi) \cap X^n|$ (see [15, Eq. (5.2)]).

$$\tau(\xi) = \lim_{n \to \infty} \frac{\log_{|X|} |\mathbf{T}_\infty(\xi) \cap X^n|}{n} \tag{2}$$

3 The Entropy of Languages

Closely related with the asymptotic subword complexity is the concept of the entropy of languages introduced in [3]. Let $W \subseteq X^*$. Then the quantity

$$\mathsf{H}_W := \limsup_{n \to \infty} \frac{\log_{|X|} \max\{1, |W \cap X^n|\}}{n} \tag{3}$$

is referred to as the *entropy* of the language W. Eq. (3) strongly resembles Eq. (2). Since the limit need not exist, we use the limit superior instead, and the additional 1 in the numerator is added to ensure that $\mathsf{H}_W = 0$ for finite languages W. For more details on the entropy of languages see also [7,8,19].

---

[3] We have to express this fact in the complicated manner because the growth of $f(\xi, n)$ need not behave like $c \cdot \lambda_\xi^n$.

3.1   The Entropy of Regular Languages

Next we derive some properties of the entropy of regular languages (cf. also [5,15]).

We start with some easily derived relations between the number of words in a regular language and the number of its subwords.

Lemma 4. *If $W \subseteq X^*$ is a regular language then there is a $k \in \mathbb{N}$ such that*

$$|W \cap X^n| \leq |\mathbf{T}(W) \cap X^n| \leq \tfrac{k}{2} \cdot \sum_{i=0}^{k} |W \cap X^{n+i}| .$$

As a suitable k one may choose twice the number of states of an automaton accepting the language $W \subseteq X^*$.

A first consequence of Lemma 4 is the following.

Corollary 2. *Let $W \subseteq X^*$ be a non-empty regular language. Then $\mathsf{H}_{\mathbf{T}(W)} = \mathsf{H}_{\mathbf{A}(W)} = \mathsf{H}_W$.*

Corollary 4 of [13] shows a more precise bound for the number of words in regular star languages $W^* \subseteq X^*$.

Lemma 5. *For every regular language $W \subseteq X^*$ there are constants $c_1, c_2 > 0$ and a λ, $0 \leq \lambda \leq |X|$, such that*

$$c_1 \cdot \lambda^n \leq |\mathbf{A}(W^*) \cap X^n| \leq c_2 \cdot \lambda^n .$$

A consequence of Lemma 4 is that $|\mathbf{T}(W) \cap X^n| \leq k \cdot |\mathbf{A}(W) \cap X^{n+k}|$. Thus Lemma 5 holds also (with constant $k \cdot c_2 \cdot |X|^k$ instead of c_2) for $\mathbf{T}(W^*)$.

In order to obtain a relation between H_W and H_{W^*} we consider, for a language $W \subseteq X^*$, the generating function $\mathsf{S}_W(t) := \sum_{i \in \mathbb{N}} |W \cap X^i| \cdot t^i$. It is well-known (cf. [8]) that $\mathsf{H}_W = -\log_{|X|} \sup\{t : 0 \leq t \leq 1 \wedge \mathsf{S}_W(t) < \infty\}$. Moreover, for regular languages W, the function $\mathsf{S}_W(t)$ is a rational function [3,5], that is, in particular, if $W \neq \emptyset$ there is always a value $\mathbf{t}_1 < |X|^{-\mathsf{H}_W}$ such that $\mathsf{S}_W(\mathbf{t}_1) = 1$.

For codes $V \subseteq X^*$ we have $\mathsf{S}_{V^*}(t) = (1 - \mathsf{S}_V(t))^{-1}$, and consequently, $\mathsf{H}_{V^*} = -\log_{|X|} \mathbf{t}_1$ whenever $\mathbf{t}_1 < |X|^{-\mathsf{H}_V}$. Thus we have the following.

Lemma 6. *Let $\emptyset \neq V \subseteq X^*$ be a regular language and simultaneously a code. Then $\mathsf{H}_{V^*} > \mathsf{H}_V$.*

Proposition 1. *If V is a regular code, $v \in V$ and $W = V \setminus \{v\}$ then $\mathsf{H}_{W^*} < \mathsf{H}_{V^*}$.*

Proof. Since V is regular, there is a value $\mathbf{t}_1$ such that $\mathsf{S}_V(\mathbf{t}_1) = 1$, that is, $\mathsf{H}_{V^*} = -\log_{|X|} \mathbf{t}_1$.

We use the inequality $\mathsf{S}_W(t) < \mathsf{S}_V(t)$ which holds for $0 \leq t < |X|^{-\mathsf{H}_V}$ and the fact that W is also a regular code. Then the value $\mathbf{t}_1'$ for which $\mathsf{S}_W(\mathbf{t}_1') = 1$ satisfies $\mathbf{t}_1 < \mathbf{t}_1'$, and the assertion follows.   □

We conclude this part with the following connection between the asymptotic subword complexity $\tau(\xi)$ and the entropy of regular languages containing $\mathbf{A}(\xi)$.

Theorem 1. $\tau(\xi) = \inf\{\mathsf{H}_W : W \text{ is regular } \wedge \mathbf{A}(\xi) \subseteq \mathbf{A}(W)\}$

Proof. The inequality $\tau(\xi) \leq \mathsf{H}_W$ follows from $\tau(\xi) = \mathsf{H}_{\mathbf{T}(\xi)}$, $\mathbf{T}(\xi) \subseteq \mathbf{T}(W)$ and Corollary 2.

Since $\tau(\xi) = \inf\left\{\frac{\log_{|X|} f(\xi,n)}{n} : n \in \mathbb{N}\right\}$, the relations $\mathbf{A}(\xi) \subseteq \mathbf{A}((\mathbf{T}(\xi) \cap X^n)^*)$, for $n > 0$, and $\mathsf{H}_{(\mathbf{T}(\xi) \cap X^n)^*} = \frac{\log_{|X|} f(\xi,n)}{n}$ show the other inequality. $\qquad\square$

3.2   Entropy of Languages and Hausdorff Dimension

In the next sections we will see that the asymptotic subword complexity of an ω-word ξ is closely related to the Hausdorff dimension of certain ω-languages containing ξ. To this end we derive here some properties of the entropy of languages and the Hausdorff dimension of related ω-languages.

The usual definition of Hausdorff dimension (see e.g. [6,15]) is based on measure theoretical notions. Here we avoid this and refer instead to a characterisation via the entropy of languages given in Eq. (3.11) of [15].

Definition 2. Let $F \subseteq X^\omega$. Then

$$\dim_\mathrm{H} F := \inf\{\mathsf{H}_W : W \subseteq X^* \wedge F \subseteq \{\xi : |\mathbf{A}(\xi) \cap W| = \infty\}\}$$

is referred to as the *Hausdorff dimension* of the set F.

We mention the following well-known stability property of the Hausdorff dimension.

$$\dim_\mathrm{H} \bigcup_{i \in \mathbb{N}} F_i = \sup\{\dim_\mathrm{H} F_i : i \in \mathbb{N}\} \tag{4}$$

In what follows we shall use Eq. (4) mainly to show that $F' \subseteq F$ implies $\dim_\mathrm{H} F' \leq \dim_\mathrm{H} F$ or that $\dim_\mathrm{H} W \cdot F = \dim_\mathrm{H} F$ when $W \neq \emptyset$.

Next we consider the *limit* (or *adherence*) $\mathbf{ls}\, W := \{\xi : \mathbf{A}(\xi) \subseteq \mathbf{A}(W)\} \subseteq X^\omega$ of a language $W \subseteq X^*$.

For languages of the form $\mathbf{T}(V)$ the language itself and its limit $\mathbf{ls}\, \mathbf{T}(V)$ satisfy $\mathbf{A}(\mathbf{ls}\, \mathbf{T}(V)) = \mathbf{T}(V)$, $\mathbf{T}(V) \supseteq v^{[-1]}\mathbf{T}(V)$ and $\mathbf{ls}\, \mathbf{T}(V) \supseteq v^{[-1]}(\mathbf{ls}\, \mathbf{T}(V))$, for $v \in X^*$. Then one can apply Theorem 6 of [14] and obtains

$$\dim_\mathrm{H} \mathbf{ls}\, \mathbf{T}(V) = \mathsf{H}_{\mathbf{T}(V)}. \tag{5}$$

In view of Corollary 2 our Eq. (5) implies $\dim_\mathrm{H} \mathbf{ls}\, W \leq \mathsf{H}_W$ for regular languages $W \subseteq X^*$. Furthermore, the Hausdorff dimension of the ω-power V^ω equals the entropy of V^* (see Eq. (6.2) of [15]).

$$\dim_\mathrm{H} V^\omega = \mathsf{H}_{V^*} \tag{6}$$

Now Corollary 2, Eqs. (5), (6) and Lemma 6 yield the following.

Corollary 3. *Let* $V \subseteq X^*$ *be a regular language. Then* $\dim_\mathrm{H} \mathbf{ls}\, V \leq \dim_\mathrm{H} V^\omega$, *and if, moreover,* V *is a code then* $\dim_\mathrm{H} \mathbf{ls}\, V < \dim_\mathrm{H} V^\omega$.

4 Maximum Subword Complexity in Regular ω-Languages

In this section we derive the announced above results on eventually recurrent ω-words having a regular language of infinitely often occurring subwords. To this end we investigate the relations between the asymptotic subword complexity $\tau(\xi)$ of an ω-word ξ and its containment in ω-languages of a special shape. Here we consider the class of regular ω-languages (see [16,20]), that is, the class of ω-languages accepted by finite automata. This class of regular ω-languages is closely related to regular languages.

As usual an ω-language $F \subseteq X^\omega$ is referred to as *regular* provided there are an $n \in \mathbb{N}$ and regular languages $W_i, V_i \subseteq X^*$ such that

$$F = \bigcup_{i=1}^{n} W_i \cdot V_i^\omega .$$

Here the languages V_i can be chosen to be prefix codes (see [4]). We mention still that the class of regular ω-languages is closed under Boolean operations (see [16,20]).

In the sequel we need the identity

$$\boldsymbol{ls}\, V^* = V^\omega \cup V^* \cdot \boldsymbol{ls}\, V \text{ for } V \subseteq X^* \tag{7}$$

which can be found in [17] and the fact that $\boldsymbol{ls}\, V$ is a regular ω-language whenever V is a regular language (see [15,16]).

Then the following relation between the asymptotic subword complexity and the Hausdorff dimension of regular ω-languages can be proved.

$$\tau(\xi) = \inf\{\dim_{\mathrm{H}} F : F \subseteq X^\omega \wedge F \text{ is regular } \wedge \xi \in F\} \tag{8}$$

Proof. Since $\xi \in \boldsymbol{ls}\, W$ if and only if $\mathbf{A}(\xi) \subseteq \mathbf{A}(W)$ and $\boldsymbol{ls}\, W$ is regular provided W is regular, the inequality "$\geq$" follows from Theorem 1 and Eq. (5), and the reverse inequality is Proposition 5.4 of [15]. □

We proceed with a relation between $\mathbf{T}_\infty(\xi)$ and an ω-power V^ω containing a tail of ξ.

Lemma 7. *1. If $\xi \in w \cdot V^\omega$ for some $w \in X^*$ then $\mathbf{T}_\infty(\xi) \subseteq$*
     $\mathbf{T}(V^*) \subseteq \mathbf{T}(V) \cdot V^* \cdot \mathbf{T}(V)$.
  2. If η is eventually recurrent then there is a $w \in X^*$ such that $\eta \in w \cdot \boldsymbol{ls}\, \mathbf{T}_\infty(\eta)$.

Proof. The first assertion is immediate.

Since η is eventually recurrent, $\mathbf{T}_\infty(\eta) = \mathbf{T}(w^{[-1]}\eta)$ for some $w \sqsubseteq \eta$. Thus $\{\eta\} = \boldsymbol{ls}\, w \cdot \mathbf{A}(w^{[-1]}\eta) \subseteq w \cdot \boldsymbol{ls}\, \mathbf{T}_\infty(\eta)$. □

This yields an obvious upper bound on $\tau(\xi)$ when $\xi \in w \cdot V^\omega$.

Corollary 4. *If $\xi \in w \cdot V^\omega$ then $\tau(\xi) \leq \mathsf{H}_{\mathbf{T}(V^*)}$.*

For regular codes $V \subseteq X^*$ we have a stronger property.

Theorem 2. *Let $V \subseteq X^*$ be a regular code, $\xi \in w \cdot V^\omega$ for some $w \in X^*$ and $\tau(\xi) = \mathsf{H}_{V^*}$. Then $V^* \subseteq \mathbf{T}_\infty(\xi) = \mathbf{T}(V^*)$.*

Proof. The inclusion $\mathbf{T}_\infty(\xi) \subseteq \mathbf{T}(V^*)$ is Lemma 7.1, and together with $V^* \subseteq \mathbf{T}_\infty(\xi)$ it implies $\mathbf{T}_\infty(\xi) = \mathbf{T}(V^*)$. Thus, it remains to show $V^* \subseteq \mathbf{T}_\infty(\xi)$.

Assume the contrary, that is, there is a $v_0 \in V^*$ such that $v_0 \notin \mathbf{T}_\infty(\xi)$. Since, for $n > 0$, $V^\omega = (V^n)^\omega$ and V^n is also a regular code whenever V is a regular code, we may assume $v_0 \in V$. Set $W := V \setminus \{v_0\}$.

Then $\xi \in w \cdot W^\omega$, and according to Corollary 4 and Proposition 1 we have $\tau(\xi) \leq \mathsf{H}_{W^*} < \mathsf{H}_{V^*}$. This contradicts our assumption. □

4.1   Eventually Recurrent ω-Words with Regular $\mathbf{T}_\infty(\xi)$

Theorem 2 allows us to derive conditions necessary or sufficient for an ω-word ξ with a regular language $\mathbf{T}_\infty(\xi)$ to be eventually recurrent.

The first condition is a sufficient one.

Theorem 3. *Let $F \subseteq X^\omega$ be a regular ω-language. If $\xi \in F$ and $\tau(\xi) = \dim_{\mathrm{H}} F$ then ξ is eventually recurrent and $\mathbf{T}_\infty(\xi)$ is a regular language.*

Proof. Since F is regular and $\xi \in F$ there are a word $w \in X^*$ and a regular prefix code $V \subseteq X^*$ such that $\xi \in w \cdot V^\omega \subseteq F$. Corollaries 4 and 2 and Eq. (6) show that $\tau(\xi) \leq \mathsf{H}_{V^*} = \dim_{\mathrm{H}} V^\omega \leq \dim_{\mathrm{H}} F$.

Now the assertion follows with Theorem 2. □

The next two conditions are necessary ones.

Lemma 8. *If ξ is eventually recurrent and $\mathbf{T}_\infty(\xi)$ is a regular language then there is a regular prefix code $V \subseteq X^*$ such that $\mathbf{T}_\infty(\xi) = \mathbf{T}(V^*)$.*

Proof. Lemma 7.2 shows $\xi \in w \cdot \mathbf{ls}\, \mathbf{T}_\infty(\xi)$ for a suitable $w \sqsubseteq \xi$. By assumption, the ω-language $w \cdot \mathbf{ls}\, \mathbf{T}_\infty(\xi) = \mathbf{ls}\,(w \cdot \mathbf{T}_\infty(\xi))$ is regular. Thus there is a regular prefix code such that $\xi \in w' \cdot V^\omega \subseteq \mathbf{ls}\,(w \cdot \mathbf{T}_\infty(\xi))$ and according to Theorem 2 we have $\mathbf{T}_\infty(\xi) = \mathbf{T}(V^*)$. □

Together with Lemmata 5 and 3 we obtain the following.

Corollary 5. *If ξ is eventually recurrent and $\mathbf{T}_\infty(\xi)$ is a regular language then there are constants $c_1, c_2 > 0$ such that*
$$c_1 \cdot |X|^{\tau(\xi) \cdot n} \leq |\mathbf{T}_\infty(\xi) \cap X^n| \leq |\mathbf{T}(\xi) \cap X^n| \leq c_2 \cdot |X|^{\tau(\xi) \cdot n}.$$

The conditions in Lemma 8 and Corollary 5 are, however, not sufficient as will be seen in the subsequent example. To this end we derive a relation between $\mathbf{T}(\xi)$ and $\mathbf{T}_\infty(\xi)$.

Lemma 9. *Let $M_\xi := \mathrm{Min}_{\mathrm{infix}}(\mathbf{T}(\xi) \setminus \mathbf{T}_\infty(\xi))$ the set of minima w.r.t. to the infix relation of $\mathbf{T}(\xi) \setminus \mathbf{T}_\infty(\xi)$. If every $w \in M_\xi$ occurs only once as a factor in ξ then $|\mathbf{T}(\xi) \cap X^n| \leq |\mathbf{T}_\infty(\xi) \cap X^n| + \sum_{w \in M_\xi} \max\{0, n - |w| + 1\}$.*

Proof. If $v \in \mathbf{T}(\xi) \setminus \mathbf{T}_\infty(\xi)$ then some $w \in M_\xi$ is a subword of v. Since w occurs only once as a factor in ξ, v is one of the $|v| - |w| + 1$ factors of length $|v|$ of ξ containing w. $\qquad\square$

Example 2. Let $V := (aa)^* \cdot ab$. Then $\mathsf{H}_{V^*} = \frac{1}{2}$. We use an enumeration $\{v_i : i \in \mathbb{N}\}$ of $V^* \setminus \{e\}$ and set $\xi_1 := \prod_{i \in \mathbb{N}} v_i a^{2i} b$. Then $\mathbf{T}_\infty(\xi_1) = \mathbf{T}(V^*)$, $M_{\xi_1} = b(aa)^* b$ and every word of M_{ξ_1} occurs only once as a factor in ξ_1.

Using Lemma 9 we calculate $|\mathbf{T}(\xi_1) \cap X^n| \leq |\mathbf{T}_\infty(\xi_1) \cap X^n| + n^2$, and thus the inequality of Corollary 5 is satisfied although every $\mathbf{T}(w^{[-1]}\xi_1) \setminus \mathbf{T}_\infty(\xi_1)$, $w \sqsubset \xi_1$, contains infinitely many words from $b(aa)^* b$. $\qquad\square$

It should be mentioned that the ω-word ξ_0 from Example 1 satisfies $\mathbf{T}_\infty(\xi_0) = a^* b a^* \cup a^*$, whence $|\mathbf{T}_\infty(\xi_0) \cap X^n| = n + 1$ and $\tau(\xi_0) = 0$. Thus Corollary 5 yields another proof that ξ_0 is not eventually recurrent.

4.2   A New Proof of Theorem 6 of [18]

Theorem 2 and Lemma 7 allow us to simplify the proof of Theorem 6 in [18]. We start with an auxiliary lemma.

Lemma 10. *Let $F \subseteq X^\omega$ be regular, $\xi \in F$ and $\tau(\xi) = \dim_H F$. If η is eventually recurrent and $\mathbf{T}_\infty(\xi) = \mathbf{T}_\infty(\eta)$ then there are $u, u' \in X^*$ such that $u' \cdot (u^{[-1]}\eta) \in F$.*

Proof. First Theorem 3 shows that ξ is eventually recurrent and $\mathbf{T}_\infty(\xi)$ is a regular language. Thus, for a suitable $w \sqsubset \xi$, $F \cap w \cdot \boldsymbol{ls}\,\mathbf{T}_\infty(\xi)$ is a regular language containing ξ. Consequently, there are a $u' \sqsubset \xi$ and a regular prefix code $V \subseteq X^*$ such that $\xi \in u' \cdot V^\omega \subseteq F \cap w \cdot \boldsymbol{ls}\,\mathbf{T}_\infty(\xi)$. Now, it suffices to prove $\eta \in X^* \cdot V^\omega$. Then $\eta \in u \cdot V^\omega$ and, consequently, $u' \cdot (u^{[-1]}\eta) \in u' \cdot V^\omega \subseteq F$.

To this end observe that in view of $\mathsf{H}_{V^*} = \dim_H V^\omega \geq \tau(\xi) = \dim_H F$ Theorem 2 and Lemma 7.2 imply $\mathbf{T}(V^*) = \mathbf{T}_\infty(\xi) = \mathbf{T}_\infty(\eta)$ and $\eta \in v \cdot \boldsymbol{ls}\,\mathbf{T}(V^*)$ for a suitable $v \sqsubset \eta$. From $\mathbf{T}(V^*) \subseteq \mathbf{T}(V) \cdot V^* \cdot \mathbf{T}(V)$ and Eq. (7) we obtain $\boldsymbol{ls}\,\mathbf{T}(V^*) \subseteq \mathbf{T}(V) \cdot V^* \cdot \boldsymbol{ls}\,\mathbf{T}(V) \cup \mathbf{T}(V) \cdot V^\omega$. Since V is a regular prefix code, in view of Corollary 3 we have $\dim_H \boldsymbol{ls}\,\mathbf{T}(V) < \dim_H V^\omega = \tau(\eta)$. This shows $\eta \in v \cdot \mathbf{T}(V) \cdot V^\omega$. $\qquad\square$

Now we can drop the assumption that $\xi \in F$ but have to ensure that ξ is eventually recurrent and $\mathbf{T}_\infty(\xi)$ is regular.

Theorem 4. *Let $F \subseteq X^\omega$ be regular, ξ, η be eventually recurrent and $\mathbf{T}_\infty(\xi) = \mathbf{T}_\infty(\eta)$ be a regular language.*

If $\xi \in F$ then there are $u, u' \in X^*$ such that $u' \sqsubset \xi$ and $u' \cdot (u^{[-1]}\eta) \in F$.

Proof. Since ξ is eventually recurrent and $\mathbf{T}_\infty(\xi)$ is regular there is a $u' \sqsubset \xi$ such that $\xi \in u' \cdot \boldsymbol{ls}\,\mathbf{T}_\infty(\xi)$ and $\boldsymbol{ls}\,\mathbf{T}_\infty(\xi)$ is a regular ω-language. Moreover, $\tau(\xi) = \dim_H \boldsymbol{ls}\,\mathbf{T}_\infty(\xi)$. Now apply Lemma 10 to the ω-language $F \cap u' \cdot \boldsymbol{ls}\,\mathbf{T}_\infty(\xi)$. $\qquad\square$

Our Example 2 shows that the assumption in Theorem 4 and Lemma 10 that η be eventually recurrent cannot be dropped. Take e.g. $F := ((aa)^* \cdot ab)^\omega$, $\xi := \prod_{i \in \mathbb{N}} v_i$ and $\eta := \xi_1$.

References

1. Allouche, J.P., Shallit, J.: Automatic sequences. Cambridge University Press, Cambridge (2003)
2. Berstel, J., Karhumäki, J.: Combinatorics on words: a tutorial. Bulletin of the EATCS 79, 178–228 (2003)
3. Chomsky, N., Miller, G.A.: Finite state languages. Information and Control 1, 91–112 (1958)
4. Choueka, Y.: Theories of automata on ω-tapes: a simplified approach. J. Comput. System Sci. 8, 117–141 (1974)
5. Eilenberg, S.: Automata, languages, and machines, vol. A. Academic Press, New York (1974)
6. Falconer, K.: Fractal geometry. John Wiley & Sons Ltd., Chichester (1990)
7. Hansel, G., Perrin, D., Simon, I.: Compression and Entropy. In: Finkel, A., Jantzen, M. (eds.) STACS 1992. LNCS, vol. 577, pp. 515–528. Springer, Heidelberg (1992)
8. Kuich, W.: On the entropy of context-free languages. Information and Control 16, 173–200 (1970)
9. Marcus, S.: Quasiperiodic infinite words. Bulletin of the EATCS 82, 170–174 (2004)
10. Perrin, D., Schupp, P.E.: Automata on the integers, recurrence distinguishability, and the equivalence and decidability of monadic theories. In: Proceedings of Symposium on Logic in Computer Science, June 16-18, pp. 301–304. IEEE Computer Society, Cambridge (1986)
11. Polley, R., Staiger, L.: The maximal subword complexity of quasiperiodic infinite words. Electronic Proceedings in Theoretical Computer Science 31, 169–176 (2010), http://arxiv.org/abs/1008.1659
12. Semenov, A.L.: Decidability of Monadic Theories. In: Chytil, M.P., Koubek, V. (eds.) MFCS 1984. LNCS, vol. 176, pp. 162–175. Springer, Heidelberg (1984)
13. Staiger, L.: The entropy of finite-state ω-languages. Problems Control Inform. Theory/Problemy Upravlen. Teor. Inform. 14(5), 383–392 (1985)
14. Staiger, L.: Combinatorial properties of the Hausdorff dimension. J. Statist. Plann. Inference 23(1), 95–100 (1989)
15. Staiger, L.: Kolmogorov complexity and Hausdorff dimension. Inform. and Comput. 103(2), 159–194 (1993)
16. Staiger, L.: ω-Languages. In: Rozenberg, G., Salomaa, A. (eds.) Handbook of Formal Languages, vol. 3, pp. 339–387. Springer, Berlin (1997)
17. Staiger, L.: On ω-Power Languages. In: Păun, G., Salomaa, A. (eds.) New Trends in Formal Languages. LNCS, vol. 1218, pp. 377–394. Springer, Heidelberg (1997)
18. Staiger, L.: Rich ω-Words and Monadic Second-Order Arithmetic. In: Nielsen, M., Thomas, W. (eds.) CSL 1997. LNCS, vol. 1414, pp. 478–490. Springer, Heidelberg (1998)
19. Staiger, L.: The entropy of Łukasiewicz-languages. Theor. Inform. Appl. 39(4), 621–639 (2005)
20. Thomas, W.: Automata on infinite objects. In: van Leeuwen, J. (ed.) Handbook of Theoretical Computer Science, vol. B, pp. 133–191. Elsevier Science Publishers B.V., Amsterdam (1990)
21. Thomsen, K.: Languages of finite words occurring infinitely many times in an infinite word. Theor. Inform. Appl. 39(4), 641–650 (2005)

On Grammars Controlled by Parikh Vectors

Ralf Stiebe

Fakultät für Informatik, Otto-von-Guericke-Universität Magdeburg,
Postfach 4120, 39106 Magdeburg, Germany
stiebe@iws.cs.uni-magdeburg.de

Abstract. We suggest a concept of grammars with controlled derivations where the Parikh vectors of all intermediate sentential forms have to be from a given restricting set. For several classes of restricting sets, we investigate set-theoretic and closure properties of the corresponding language families.

1 Introduction

Grammars with restricted numbers of nonterminal symbols in the sentential forms in the course of the derivation process have been investigated for a long time. Most prominent are the grammars of finite index, introduced by Brainerd [2], where every word of the generated language can be generated using only sentential forms with a bounded number of nonterminal symbols. These grammars (and regulated grammars of finite index, like matrix grammars, as well) have been studied in numerous publications. Ginsburg and Spanier [4] discussed the slightly different concept of a derivation-bounded grammar where only those derivations are permitted that use sentential forms with a bounded number of nonterminals. While being of finite index is a combinatorial property of the grammar and context-free grammars of finite index can by definition only generate context-free languages, the latter concept provides a kind of control for the derivation process and could potentially lead to the generation of languages not in the original language class. However, it has been shown in [4] that derivation-bounded context-free grammars can only generate context-free languages of finite index.

More recently, Stiebe and Turaev [9] introduced capacity-bounded grammars where a capacity function associates to each nonterminal symbol a bound. A derivation is valid if in every sentential form the number of appearances of each symbol is at most its capacity. It could be shown that context-free capacity-bounded grammars generate non-context-free grammars and are strictly weaker than matrix grammars of finite index.

To overcome the limitations of capacity-bounded grammars, in particular the restriction to sentential forms with a bounded number of nonterminal symbols, we will discuss in this paper some more general conditions for the nonterminals in the sentential forms. Probably the most straightforward extension is to allow infinite capacities for some nonterminal symbols. More generally, we will

H. Bordihn, M. Kutrib, and B. Truthe (Eds.): Dassow Festschrift 2012, LNCS 7300, pp. 246–264, 2012.

demand that, for every sentential form in a derivation process, the Parikh vector (restricted to the nonterminal symbols) has to be in a given restricting set. Grammars with such conditions will be called *Parikh vector controlled grammars* in what follows. Depending on the properties of the restricting sets, several language classes can be defined. We will study the relations of these language classes among each other and to known families of languages as well. When encountering previously unknown language classes, we will also investigate closure properties.

Beside this introduction, the paper contains two sections. The necessary definitions and notations are given in Section 2, in its end introducing the notion of Parikh vector controlled grammars. Section 3 contains the results.

2 Definitions

Throughout the paper, we assume that the reader is familiar with basic concepts of formal language theory; for details we refer to [7]. An introduction to regulated rewriting can be found in [3].

The sets of integers and non-negative integers are denoted by $\mathbb{Z}$ and $\mathbb{N}$, respectively. The cardinality of a set S is denoted by $|S|$, and the power set of a set S by $\mathcal{P}(S)$. We use the symbols $\subseteq$ for inclusion and $\subset$ for proper inclusion. In the vector space $\mathbb{Z}^k$, the zero vector is denoted by $\mathbf{0}$ and the i-th unit vector, $1 \leq i \leq k$, by $\boldsymbol{e}_i$ (a reference to the dimension k will usually not be necessary, as it is clear from the context). A subset $M \subseteq \mathbb{N}^k$ is called *linear* if it can be written as $M = \{\boldsymbol{c} + \sum_{i=1}^{n} a_i \boldsymbol{p}_i : a_i \in \mathbb{N}, 1 \leq i \leq n\}$, for appropriate $\boldsymbol{c}, \boldsymbol{p}_1, \ldots, \boldsymbol{p}_n \in \mathbb{N}^k$. A set is *semilinear* if it is the union of a finite number of linear sets.

A *system of linear inequalities* in n variables is a finite set of inequalities

$$\sum_{j=1}^{n} a_{i,j} x_j \leq b_i, \ (1 \leq i \leq m) \text{ with } a_{i,j}, b_i \in \mathbb{Z}, \text{ for } 1 \leq i \leq m, 1 \leq j \leq n.$$

A non-negative and integral *solution* of above the system of linear inequalities is a vector $(x_1, x_2, \ldots, x_n) \in \mathbb{N}^n$ that satisfies all inequalities. The set of all non-negative and integral solutions of a system of linear inequalities will, for the sake of brevity, simply be referred to as the *solution set* of the given system. In this paper, a system of linear inequalities as above will be called

- *positive* if $a_{i,j} \geq 0$, for all $1 \leq i \leq m, 1 \leq j \leq n$;
- *strictly positive* if furthermore $\sum_{i=1}^{m} a_{i,j} > 0$, for all $1 \leq j \leq n$.

The solution sets of systems of linear inequalities have some useful properties utilized in this paper. The simple proofs are left to the reader.

1. The solution set of a system of linear equations is semilinear.
2. If S_1 and S_2 are solution sets of systems of linear inequalities with disjoint sets of variables then $S_1 \times S_2$ is the solution set of the system containing all inequalities of both systems. Moreover, if both S_1 and S_2 are (strictly) positive, the resulting system is (strictly) positive, too.

3. The solution set of a strictly positive system of linear inequalities is finite.
4. After an appropriate renaming of the variables, the solution set of a positive system of linear inequalities can be written as $S_1 \times \mathbb{N}^k$ where S_1 is the solution set of a strictly positive system of linear inequalities.
5. If $\boldsymbol{x} \in \mathbb{N}^n$ is a solution of a positive system of linear inequalities then every $\boldsymbol{y} \in \mathbb{N}^n$ with $\boldsymbol{y} \leq \boldsymbol{x}$ is a solution of the same system.

The set of finite strings over an alphabet X is denoted by X^*, the *length* of a string $w \in X^*$ by $|w|$, the number of occurrences of a symbol a in w by $|w|_a$ and the number of occurrences of symbols from $Y \subseteq X$ in w by $|w|_Y$. The *empty* string is denoted by λ. Given an ordered alphabet $X = \{a_1, a_2, \ldots, a_n\}$, the *Parikh mapping* is the homomorphism $\Psi : X^* \to \mathbb{N}^n$ sending a_i, $1 \leq i \leq n$, to the i-th unit vector. For a string $w \in X^*$, $\Psi(w)$ is referred to as the *Parikh vector* of w; for a language $L \subseteq X^*$, the *Parikh set* of L is $\Psi(L) = \{\Psi(w) : w \in L\}$. For a subset Y of X with $Y = \{a_{i_1}, a_{i_2}, \ldots, a_{i_m}\}$, $i_1 < i_2 < \ldots < i_m$, let $\Psi_Y : X^* \to \mathbb{N}^m$ be the homomorphism sending a_{i_j} to the i-th unit vector (of $\mathbb{N}^m$) and $x \in X \setminus Y$ to the zero vector (of $\mathbb{N}^m$). In what follows, for any alphabet, an order will be tacitly assumed so that Parikh mappings are used without explicitly mentioning the order.

Besides the AFL operations (union, concatenation, homomorphisms, inverse homomorphisms, intersection with regular sets, Kleene closure) we will consider *nested iterated substitutions* which were extensively investigated by Greibach [5,6]. A substitution is a homomorphism $\tau : X^* \to \mathcal{P}(Y^*)$ where X and Y are alphabets. We extend τ to $(X \cup Y)^*$, where $X \cap Y = \emptyset$, by defining $\tau(a) = \{a\}$ for all $a \in Y$. For $n \geq 0$, τ^n is the substitution defined by $\tau^0(a) = \{a\}$ and $\tau^{n+1}(a) = \tau(\tau^n(a))$, for $a \in X \cup Y$. The *iterated substitution* defined by τ is the substitution τ^∞ defined by $\tau^\infty(a) = \bigcup_{n=0}^\infty \tau^n(a)$, for $a \in X \cup Y$. Moreover, τ^∞ is called a *nested iterated substitution* if $a \in \tau(a)$, for all $a \in X \cup Y$. A family of languages $\mathcal{L}$ is closed under nested iterated substitutions if $L \in \mathcal{L}$ and $\tau(a) \in \mathcal{L}$ for every $a \in X$ imply $\tau^\infty(L) \in \mathcal{L}$. It has been shown in [6] that the family of semilinear languages is closed under nested iterated substitutions.

A *finite automaton* is a tuple $\mathcal{A} = (Z, X, z_0, F, \delta)$ where Z is a finite set of states, X is a finite input alphabet, $z_0 \in Z$ is the initial state, $F \subseteq Z$ is the set of accepting states, and $\delta \subseteq Z \times X \times Z$ is the transition relation. The successor relation $\vdash$ over $Z \times X^*$ is defined as $(z, v) \vdash (z', v')$ iff $v = av'$ and $(z, a, z') \in \delta$. The reflexive and transitive closure of $\vdash$ is denoted by $\vdash^*$. The language accepted by $\mathcal{A}$ is $L(\mathcal{A}) = \{w \in X^* : (z_0, w) \vdash^* (z_f, \lambda), \text{ for some } z_f \in F\}$.

A *grammar* is a quadruple $G = (V, \Sigma, S, R)$ where V and Σ are two finite disjoint alphabets of *nonterminal* and *terminal* symbols, respectively, $S \in V$ is the *start symbol* and $R \subseteq (V \cup \Sigma)^* V (V \cup \Sigma)^* \times (V \cup \Sigma)^*$ is a finite set of *rules*. G is called a *GS grammar*[1] if $R \subseteq V^+ \times (V \cup \Sigma)^*$ and a *context-free grammar* if $R \subseteq V \times (V \cup \Sigma)^*$. A string $x \in (V \cup \Sigma)^*$ *directly derives* a string $y \in (V \cup \Sigma)^*$ in G, written as $x \Rightarrow_G y$, if and only if there is a rule $\alpha \to \beta \in R$ such that $x = x_1 \alpha x_2$ and $y = x_1 \beta x_2$ for some $x_1, x_2 \in (V \cup \Sigma)^*$. The reflexive

---

[1] This kind of grammar was introduced by Ginsburg and Spanier and for this reason named GS grammar here.

and transitive closure of the relation $\Rightarrow_G$ is denoted by $\Rightarrow_G^*$. A derivation using the sequence of rules $\pi = r_1 r_2 \cdots r_k$, $r_i \in R$, $1 \leq i \leq k$, is denoted by $\overset{\pi}{\Rightarrow}_G$ or $\xrightarrow{r_1 r_2 \cdots r_k}_G$. The *language* generated by G, denoted by $L(G)$, is defined by $L(G) = \{w \in \Sigma^* : S \Rightarrow_G^* w\}$. If G is clear from the context, the subscript G will be omitted in the notation. The family of languages generated by context-free grammars is denoted **CF**.

We next give some prerequisites concerning grammars with controlled derivations. Unless stated otherwise, extensive explanations, proofs and reference to the original literature can be found in [3]. A *matrix grammar* is a quadruple $G = (V, \Sigma, S, M)$ where V, Σ, S are defined as for a context-free grammar, M is a finite set of *matrices* which are finite strings (or finite sequences) over a set R of context-free rules. The *language* generated by a matrix grammar G consists of all strings $w \in \Sigma^*$ such that there is a derivation $S \xrightarrow{r_1 r_2 \cdots r_n} w$ where $r_1 r_2 \cdots r_n$ is a concatenation of some matrices $m_{i_1}, m_{i_2}, \ldots, m_{i_k} \in M$, $k \geq 1$. The family of languages generated by matrix grammars is denoted by **MAT**.

A *grammar with regular control* is a quintuple $G = (V, \Sigma, S, R, L)$ where $G' = (V, \Sigma, S, R)$ is a context-free grammar and $L \subseteq R^*$ is a regular language. The language of G is defined by $L(G) = \{w \in \Sigma^* : S \overset{\pi}{\Rightarrow} w, \text{ for some } \pi \in L\}$. It is known that the family of languages generated by grammars with regular control is **MAT**.

A *valence grammar over* $\mathbb{Z}^k$ is a quintuple $G = (V, \Sigma, S, R, \mathbb{Z}^k)$ where V, Σ, S are defined as in a context-free grammar, and R is a finite set of valence rules $(A \to \beta, \boldsymbol{r})$, where $A \to \beta$ is a rule and $\boldsymbol{r} \in \mathbb{Z}^k$. The direct derivation relation $\Rightarrow$ over $(V \cup \Sigma)^* \times \mathbb{Z}^k$ is defined by:

$(\gamma, \boldsymbol{z}) \Rightarrow (\gamma', \boldsymbol{z}')$ iff
$$\gamma = \gamma_1 A \gamma_2, \gamma' = \gamma_1 \beta \gamma_2 \text{ and } \boldsymbol{z}' = \boldsymbol{z} + \boldsymbol{r} \text{ for some } (A \to \beta, \boldsymbol{r}) \in P.$$

The language generated by G is $L(G) = \{w \in T^* : (S, \boldsymbol{0}) \Rightarrow^* (w, \boldsymbol{0})\}$.

A *positive valence grammar over* $\mathbb{Z}^k$ [8] is defined like a valence grammar with the additional condition $\boldsymbol{z} \geq \boldsymbol{0}$ in the definition of the derivation relation $(\gamma, \boldsymbol{z}) \Rightarrow (\gamma', \boldsymbol{z}')$.[2] It has be shown in [8] that the family of languages generated by positive valence grammars is **MAT**.

A *programmed grammar with appearance checking* is defined as a sextuple $G = (V, \Sigma, S, R, \sigma, \phi)$ where (V, Σ, S, R) is a context-free grammar, and σ and ϕ are mappings from R into $\mathcal{P}(R)$. For a rule $r \in R$, $\sigma(r)$ and $\phi(r)$ are called the *success field* and the *failure field* of r, respectively. The derivation relation over $(V \times \Sigma)^* \times R$ is defined as follows. If $r : A \to \alpha$ is a rule in R then $(\beta, r) \Rightarrow (\beta', r')$ iff either $\beta = \beta_1 A \beta_2$, $\beta' = \beta_1 \alpha \beta_2$ and $r' \in \sigma(r)$ or $|\beta|_A = 0$ and $r' \in \phi(r)$. The language generated by G is $L(G) = \{w \in T^* : (S, r) \Rightarrow^* (w, r'), r, r' \in R\}$. It is known that programmed grammars with appearance checking generate the family of recursively enumerable languages [3].

---

[2] Actually, $\boldsymbol{z} \geq \boldsymbol{0}$ and $\boldsymbol{z}' \geq \boldsymbol{0}$ were required in [8]. The definition given here is equivalent to the previous one, since the zero vector has to be reached in the final step. It will be technically useful later.

A context-free grammar G is of *index* $k \in \mathbb{N}$ if every word $w \in L(G)$ has a derivation with at most k nonterminal symbols in every sentential form. G is of *finite index* if such a k exists. The family of languages generated by context-free grammars of finite index is denoted by $\mathbf{CF}_{fi}$. For grammars with regulated rewriting, the concept of finite index is defined analogously. It is known that matrix grammars of finite index and programmed grammars of finite index generate the same family of languages $\mathbf{MAT}_{fi}$.

A *capacity-bounded grammar* [9] is a quintuple $G = (V, \Sigma, S, R, \kappa)$ where $G' = (V, \Sigma, S, R)$ is a grammar and $\kappa : V \to \mathbb{N}$ is a capacity function assigning to each nonterminal a bound. The direct derivation relation $\Rightarrow$ over $(V \cup \Sigma)^* \times \mathbb{Z}^k$ is defined by $\alpha \Rightarrow_G \beta$ iff $\alpha \Rightarrow_{G'} \beta$ and $|\alpha|_A \leq \kappa(A)$ and $|\beta|_A \leq \kappa(A)$, for all $A \in V$. It has been shown that capacity-bounded GS grammars are equivalent to matrix grammars of finite index while capacity-bounded context-free grammars generate a proper subset of $\mathbf{MAT}_{fi}$.

Finally, we give the definition of the generative device to be investigated.

Definition 1. *A* Parikh vector controlled grammar *is defined as a quintuple* $G = (V, \Sigma, S, R, C)$ *where* $G' = (V, \Sigma, S, R)$ *is a grammar and* $C \subseteq \mathbb{N}^n$, $n = |V|$ *is a set of admitted nonterminal Parikh vectors, referred to as the* restricting set *of* G. *The derivation relation* $\Rightarrow_G$ *is defined as* $\alpha \Rightarrow_G \beta$ *iff* $\alpha \Rightarrow_{G'} \beta$ *and* $\Psi_V(\alpha) \in C$. *The language of* G *is defined as* $L(G) = \{w \in \Sigma^* : S \Rightarrow_G^* w\}$.

Note that by this definition only the Parikh vectors of the *nonterminal* sentential forms have to be within the restricting set.

The main objective of this paper is to study the generative power of Parikh vector controlled grammars with respect to properties of the restricting sets. To avoid complicated notations, we just enumerate the types of restricting sets and the respective language families. Let $G = (V, \Sigma, S, R, C)$ be a Parikh vector controlled grammar with $|V| = n$. G is of

- *type 1* if $C = [0, k_1] \times [0, k_2] \times \cdots \times [0, k_n]$, $k_1, k_2, \ldots, k_n \in \mathbb{N}$;
- *type 2* if C is the solution set of a strictly positive system of linear inequalities;
- *type 3* if C is finite;
- *type 4* if $C = C_1 \times \mathbb{N}^{n-j}$ where $j \in \{0, 1, \ldots n\}$ and $C_1 = [0, k_1] \times [0, k_2] \times \cdots \times [0, k_j]$, $k_1, k_2, \ldots, k_j \in \mathbb{N}$;
- *type 5* if C is the solution set of a positive system of linear inequalities; (equivalently, if $C = C_1 \times \mathbb{N}^{n-j}$ where $j \in \{0, 1, \ldots n\}$ and $C_1 \subseteq \mathbb{N}^j$ is the solution set of a strictly positive system of linear inequalities;)
- *type 6* if $C = C_1 \times \mathbb{N}^{n-j}$ where $j \in \{0, 1, \ldots n\}$ and C_1 is a finite subset of $\mathbb{N}^j$;
- *type 7* if C is the solution set of a system of linear inequalities;
- *type 8* if C is semilinear.

The restricting sets are usually given by defining conditions. Instead of explicitly giving a nonterminal Parikh vector $\Psi_V(\beta)$ we will often refer to its components $|\beta|_A$. In particular, a system of inequalities for a nonterminal alphabet V will sometimes be written as $\sum_{A \in V} a_{i,A} |\beta|_A \leq b_i$ $(1 \leq i \leq m)$.

Note that Parikh vector controlled grammars of type 1 are simply the capacity-bounded context-free grammars and that the types 4,5,6 are extensions of the types 1,2,3, respectively, by adjoining nonterminals that are no subject to any restrictions. In what follows, let $\mathcal{L}_i$, $1 \leq i \leq 8$, denote the family of languages generated by Parikh vector controlled grammars of type i.

3   Results

We will mainly investigate the generative power of Parikh vector controlled grammars of the different types. In cases where the language families do not coincide with previously known families we will also study closure properties with respect to the AFL operations.

Lemma 1. *The following inclusions hold for the language families $\mathbf{CF}_{fi}$, $\mathbf{CF}$ and $\mathcal{L}_1, \ldots, \mathcal{L}_8$ (a dotted arrow represents a not necessarily proper inclusion; disconnected families need not to be incomparable).*

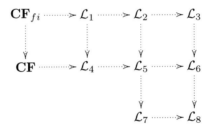

Proof. All inclusions follow easily from the definitions and some elementary properties. More specifically,

- $\mathbf{CF}_{fi} \subseteq \mathbf{CF}, \mathcal{L}_1 \subseteq \mathcal{L}_4, \mathcal{L}_2 \subseteq \mathcal{L}_5, \mathcal{L}_3 \subseteq \mathcal{L}_6$, $\mathbf{CF} \subseteq \mathcal{L}_4$ and $\mathcal{L}_5 \subseteq \mathcal{L}_7$ hold directly by definition;
- $\mathbf{CF}_{fi} \subseteq \mathcal{L}_1$ holds as a grammar (V, Σ, R, S) of finite index k generates the same language as the Parikh vector controlled grammar (V, Σ, R, S, C) where $C = [0, k]^{|V|}$, see [9];
- $\mathcal{L}_1 \subseteq \mathcal{L}_2$ and $\mathcal{L}_4 \subseteq \mathcal{L}_5$ are valid because a set $[0, k_1] \times [0, k_2] \times \cdots \times [0, k_n]$ is the solution set of the system of linear inequalities $x_i \leq k_i$ $(i = 1, 2, \ldots, k)$;
- $\mathcal{L}_2 \subseteq \mathcal{L}_3$ and $\mathcal{L}_5 \subseteq \mathcal{L}_6$ are valid because the solution set of a strictly positive system of linear inequalities is finite;
- $\mathcal{L}_6 \subseteq \mathcal{L}_8$ and $\mathcal{L}_7 \subseteq \mathcal{L}_8$ are true as the restricting sets of grammars of type 6 and 7 are semilinear.

$\square$

It is of course known that the proper inclusion $\mathbf{CF}_{fi} \subset \mathbf{CF}$ holds. Moreover, in [9] the proper inclusions $\mathbf{CF}_{fi} \subset \mathcal{L}_1 \subset \mathbf{MAT}_{fi}$ have been shown. We will now investigate the properness of the remaining inclusions and try to relate the $\mathcal{L}_i$ to known families of languages.

3.1   The Families $\mathcal{L}_1$, $\mathcal{L}_2$ and $\mathcal{L}_3$

We will first study the grammars with a finite restricting set. While $\mathcal{L}_1$ is known to be a proper subfamily of $\mathbf{MAT}_{fi}$, it turns out that $\mathcal{L}_2$ and $\mathcal{L}_3$ coincide with $\mathbf{MAT}_{fi}$. To this end, we will prove the inclusions $\mathcal{L}_3 \subseteq \mathbf{MAT}_{fi}$ and $\mathbf{MAT}_{fi} \subseteq \mathcal{L}_2$.

Lemma 2. $\mathcal{L}_3 \subseteq \mathbf{MAT}_{fi}$.

Proof. Let $G = (V, \Sigma, S, R, C)$ be a Parikh vector controlled grammar with $V = \{A_1, A_2, \ldots, A_n\}$, $S = A_1$ and C a finite subset of $\mathbb{N}^n$. The idea is to construct a grammar with regular control, where the control language is given by a finite automaton whose state keeps track of the nonterminal Parikh vector of the derived sentential form. More specifically, let $\mathcal{A} = (C \cup \{\mathbf{0}\}, R, \mathbf{e}_1, \{\mathbf{0}\}, \delta)$ be the deterministic finite automaton with the transition function δ defined as follows. If $r : A_i \to \alpha$ is a rule in R, $\boldsymbol{x} = (x_1, \ldots, x_n)$ is in C and $\boldsymbol{y}$ is defined by $\boldsymbol{y} = \boldsymbol{x} - \boldsymbol{e}_i + \Psi_V(\alpha)$ then

$$\delta(\boldsymbol{x}, r) = \begin{cases} \boldsymbol{y}, & \text{if } x_i > 0 \text{ and } \boldsymbol{y} \in C \cup \{\mathbf{0}\} \\ \text{undefined}, & \text{otherwise.} \end{cases}$$

It is easy to prove by induction that a sequence $\rho = r_1 r_2 \cdots r_m$ reaches a state $\boldsymbol{z} \in C \cup \{\mathbf{0}\}$ iff ρ is a possible derivation sequence in G and leading to a sentential form with nonterminal Parikh vector $\boldsymbol{z}$. Hence, $L(\mathcal{A})$ is the set of all correct terminal derivation sequences in G and the grammar with regular control $G' = (V, \Sigma, S, R, L(\mathcal{A}))$ generates the same language as G.   □

Lemma 3. $\mathbf{MAT}_{fi} \subseteq \mathcal{L}_2$.

Proof. In [9] it has been shown that matrix grammars of finite index are equivalent to capacity-bounded GS grammars. We will therefore show how to simulate a capacity-bounded GS grammar by a Parikh vector controlled grammar with a restricting set defined by a strictly positive system of linear inequalities. Let $G = (V, \Sigma, S, R, \kappa)$ be a capacity-bounded GS grammar. As proved in [9, Lemma 3], we can assume that any word from $L(G)$ can be derived replacing in each derivation step a maximal nonterminal block. (A maximal nonterminal block is a substring over V which cannot be extended to a longer substring over V. As G is capacity-bounded there is only a finite number of maximal nonterminal blocks.) Then we construct the equivalent Parikh vector controlled grammar $G' = (V', \Sigma, [S], R', C)$ where we devise the nonterminal alphabet V' and the set of rules R' like in [9] as

$V' = \{[\alpha] : \alpha \in V^+$ is a maximal nonterminal block in $G\}$,
$R' = \{[\alpha] \to w_0[\beta_1]w_1[\beta_2] \cdots w_{k-1}[\beta_k]w_k : \alpha \to w_0\beta_1 w_1\beta_2 \cdots w_{k-1}[\beta_k]w_k \in R,$
$\qquad \text{where } \beta_1, \ldots, \beta_k \in V^+, w_0, w_k \in \Sigma^*, w_1, \ldots, w_{k-1} \in \Sigma^+\}$.

With $V' = \{[\alpha_1], [\alpha_2] \ldots, [\alpha_n]\}$, the restricting set C is defined as the solution set of the system of linear inequalities

$$\sum_{i=1}^{n} |\alpha_i|_A \cdot x_i \leq \kappa(A), \qquad A \in V.$$

For a word $\beta \in (V \cup \Sigma)^*$ with all maximal blocks in V', let $[\beta] \in (V' \cup \Sigma)^*$ be the word obtained by replacing every maximal block α in β by $[\alpha]$. It is now easily checked by induction on the derivation steps that a sentential form $\beta \in (V \cup \Sigma)^*$ can be derived in G iff all its maximal blocks are in V' and $[\beta]$ can be derived in G'.   □

Corollary 1. $\mathcal{L}_1 \subset \mathcal{L}_2 = \mathcal{L}_3 = \mathbf{MAT}_{fi}$.

3.2   The Families $\mathcal{L}_4$, $\mathcal{L}_5$ and $\mathcal{L}_6$

The Parikh vector controlled grammars of types 4,5,6 can be seen as Parikh vector controlled grammars of types 1,2,3, respectively, extended by sets of non-restricted nonterminal symbols. In other words, the nonterminal set V of a Parikh vector controlled grammar of type 4,5 or 6 can be decomposed as $V = V_1 \cup V_2$ where V_1 is subject to the restrictions as in Parikh vector controlled grammars of types 1,2,3, respectively, and V_2 is not at all restricted.

   Essentially, we will show that $\mathcal{L}_4$ and $\mathcal{L}_5$ are obtained from $\mathcal{L}_1$ and $\mathcal{L}_2$ by nested iterated substitutions, while $\mathcal{L}_6$ is equal to the family of matrix languages $\mathbf{MAT}$. In particular, $\mathcal{L}_4$ and $\mathcal{L}_5$ are proper subfamilies of $\mathbf{MAT}$.

   We start with the following "replacement lemma" for languages from $\mathcal{L}_4$, which is virtually the same result as for capacity-bounded grammars given in [9].

Lemma 4. *For any infinite language $L \in \mathcal{L}_4$, there are a constant n and a finite set M of infinite languages from $\mathcal{L}_4$ such that, for every word $z \in L$ with $|z| \geq n$, there are a decomposition $z = uvw$, $|v| \leq n$, and a language $L' \in M$ such that $uv'w \in L$, for all $v' \in L'$.*

Proof. As the claim of the lemma, the proof is virtually the same as that for capacity-bounded grammars in [9]. Consider some Parikh vector controlled grammar $G = (V, \Sigma, S, R, C)$ of type 4 such that $L = L(G)$. For $A \in V$, let $G_A = (V, \Sigma, R, A, C)$ and $L_A = L(G_A)$; clearly, $L_A = \{w \in \Sigma^* : A \Rightarrow_G^* w\}$. The following assertions hold for any derivation in G involving A:

- If $\alpha A \beta \Rightarrow_G^* uvw$, where $\alpha, \beta \in (V \cup \Sigma)^*$, $u, v, w \in \Sigma^*$ and v is the yield of A, then $v \in L_A$. (Given a derivation $\alpha A \beta \Rightarrow_G^* uvw$, construct a derivation of v from A by keeping the derivation steps arising from A. The Parikh vectors of the sentential forms in the second derivation are less or equal to those of the corresponding sentential forms in the first derivation, hence the derivation $A \Rightarrow_G^* v$ is valid.)

– On the other hand, for all $u, v, w \in \Sigma^*$ such that $v \in L_A$, the relation $uAw \Rightarrow^*_G uvw$ holds. (Given a derivation $A \Rightarrow^*_G v$, do the same derivation steps starting from uAw. The nonterminal Parikh vectors of the sentential forms in the second derivation are equal to those of the corresponding sentential forms in the first derivation, hence the derivation $uAw \Rightarrow^*_G uvw$ is valid.)

The nonterminal set V can be decomposed as $V = V_{inf} \cup V_{fin}$, where

$$V_{inf} = \{A \in V : L_A \text{ is infinite}\},$$
$$V_{fin} = \{A \in V : L_A \text{ is finite}\}.$$

We choose $\mathcal{M} = \{L_A : A \in V_{inf}\}$ and $n = r \cdot \max\{|w| : w \in \bigcup_{A \in V_{fin}} L_A\}$, where r is the longest length of a right side in a rule of R. For a derivation of $z \in L$ with $|z| > n$, consider the last sentential form α with a symbol from V_{inf}. Let this symbol be A. All other nonterminals in α are from V_{fin}, and none of them generates a subword containing A in the further derivation process. We get thus another derivation of z in G by postponing the rewriting of A until all other nonterminals have vanished by applying on them the derivation sequence of the original derivation. This new derivation has the form

$$S \Rightarrow^* \alpha \Rightarrow^* uAw \Rightarrow^* uvw = z.$$

The length of v can be estimated by $|v| \leq n$, as A is in the first step replaced by a word over $(\Sigma \cup V_{fin})$ of length at most r. By the remarks in the beginning of the proof, any word $uv'w$ with $v' \in L_A$ can be derived in G. $\square$

The replacement lemma can be used to show that certain languages are not in $\mathcal{L}_4$. This implies some limitations for $\mathcal{L}_4$, similar to those of $\mathcal{L}_1$ shown in [9].

Corollary 2. $\mathcal{L}_4$ and $\mathbf{MAT}_{fi}$ are incomparable, while $\mathcal{L}_4$ is a proper subset of $\mathcal{L}_5$.

Proof. Using the same arguments as in [9], it can be shown that the language $L = \{a^n b^n c^n : n \geq 1\}$ does not satisfy the consequence of the replacement lemma, hence it is not in $\mathcal{L}_4$. On the other hand, L is a language from $\mathbf{MAT}_{fi} = \mathcal{L}_2$ and thus in $\mathcal{L}_5$. Together with the inclusions $\mathbf{CF} \subseteq \mathcal{L}_4 \subseteq \mathcal{L}_5$, this proves the claims. $\square$

Next we prove a useful result concerning derivations in Parikh vector controlled grammars of type 4 or 5.

Lemma 5. Let $G = (V, \Sigma, S, R, C)$ be a Parikh vector controlled grammar of type 4 or 5, and let $V = V_1 \cup V_2$ be a partition of V such that the appearance of symbols from V_2 is unrestricted. Then every word in $L(G)$ can be derived such that whenever the current sentential form contains a symbol from V_1, a symbol from V_1 will be replaced.

Proof. Consider a derivation $S \Rightarrow^* \gamma \Rightarrow^* w$ of a word $w \in L(G)$ where the first derivation step replacing a symbol from V_2 although a symbol from V_1 is present, is after generating γ. We will construct a derivation of w with the same number of derivation steps such that a symbol from V_1 is replaced in γ. The claim of the lemma then follows by induction.

We can decompose γ and w as

$$\gamma = \alpha_0 A_1 \alpha_1 A_2 \cdots \alpha_{m-1} A_m \alpha_m,$$
$$w = u_0 v_1 u_1 v_2 \cdots u_{m-1} v_m u_m,$$

where $A_1, A_2, \ldots, A_m$ are the symbols from V_2 in γ, $u_0, u_1, \ldots, u_m$ are the subwords of w derived from $\alpha_0, \alpha_1, \ldots, \alpha_m$, $v_1, v_2, \ldots, v_m$ are the subwords of w derived from $A_1, A_2, \ldots, A_m$.

Consider the derivation $S \Rightarrow^* \gamma \Rightarrow^* \gamma' \Rightarrow^* w$ with

$$\gamma' = u_0 A_1 u_1 A_2 \cdots u_{m-1} A_m u_m,$$

where the derivation steps replacing the A_i and their derivatives are first omitted, thus yielding γ', and then executed in the same sequence, yielding w. This derivation is valid as $u_0, u_1, \ldots, u_m$ and $A_1, A_2, \ldots, A_m$ do not contain symbols from V_1.   □

Lemma 6. *Every language $L \in \mathcal{L}_5$ over the alphabet Σ can be represented as $L = L' \cap \Sigma^*$ where L' is the nested iterated substitution of languages from $\mathcal{L}_2$.*

Proof. Let $G = (V, \Sigma, S, R, C_1 \times \mathbb{N}^{n-l})$ be a Parikh vector controlled grammar where $V = \{A_1, A_2, \ldots, A_n\}$ and $C_1 \subseteq \mathbb{N}^l$ is the solution set of a strictly positive system of linear inequalities. The nonterminal set V is partitioned as $V = V_1 \cup V_2$, where $V_1 = \{A_1, \ldots, A_l\}$ is the set of restricted symbols and $V_2 = \{A_{l+1}, \ldots, A_n\}$ is the set of non-restricted symbols. Without loss of generality, we assume that S does not appear on the right-hand side of any rule and belongs to V_2. For every $A \in V_2$, let L_A be the language generated by the Parikh vector controlled grammar $G_A = (V_1 \cup \{A'\}, V_2 \cup \Sigma, A', R_A, C_1 \times [0,1])$ where

$$R_A = \{A' \to \alpha : A \to \alpha \in R\} \cup \{B \to \alpha : B \to \alpha \in R, B \in V_1\}.$$

G_A is of type 2 as $C_1 \times [0,1] \subseteq \mathbb{N}^{l+1}$ is the solution set of the system of inequalities for C_1 (in the variables $x_1, \ldots, x_n$) with the additional inequality $x_{l+1} \leq 1$. Obviously, L_A is the set of all sentential forms that can be derived in G from A by replacing, except for the first step, only symbols from V_1. We claim that $L(G) = \Sigma^* \cap \tau^\infty(S)$ where $\tau(A) = L_A \cup \{A\}$ for $A \in V_2$ and $\tau(a) = \{a\}$ for $a \in \Sigma$.

To prove the inclusion $L(G) \supseteq \Sigma^* \cap \tau^\infty(S)$ we show by induction that every word from $\tau^\infty(S)$ is derivable in G. The induction basis is correct, as $\tau^0(S) = \{S\}$ and S is derivable. Now assume that every word in $\tau^n(S)$ is derivable. By definition, a word $w \in \tau^{n+1}(S)$ can be written as $w = w_1 w_2 \cdots w_m$ where $w_i \in \tau(X_i)$, $X_i \in \Sigma \cup V_2$, and $w' = X_1 X_2 \cdots X_m \in \tau^n(S)$. Now w can be derived in G by first generating $w' \in \tau^n(S)$ and then deriving sequentially from each X_i the subword w_i. The subderivations $w_1 \cdots w_{i-1} X_i X_{i+1} \cdots X_m \Rightarrow^* w_1 \cdots w_{i-1} w_i X_{i+1} \cdots X_m$ are valid as $w_1, \ldots, w_{i-1}, X_{i+1}, \ldots, X_m$ contain no symbols from V_2.

To prove $L(G) \subseteq \Sigma^* \cap \tau^\infty(S)$ we can restrict to derivations where a symbol from V_1 is replaced when present. We will show by induction that every sentential form over $(V_2 \cup \Sigma)$ obtained in such a derivation is from $\tau^\infty(S)$. The claim is true for S. Now consider some sentential form $\alpha \in (V_2 \cup \Sigma)^*$ with $\alpha \in \tau^\infty(S)$. It is decomposed as $\alpha = \alpha_1 A \alpha_2$ where $A \in V_2$ is the next symbol to be replaced. The next sentential form α' over $(V_2 \cup \Sigma)$ is reached when all symbols from V_1 that originate from the A replaced in the first step are rewritten. Hence, it has the shape $\alpha' = \alpha_1 \beta \alpha_2$ where $\beta \in L_A \subseteq \tau(A)$. By $\alpha_1 \in \tau(\alpha_1)$, $\alpha_2 \in \tau(\alpha_2)$ and the induction hypothesis $\alpha \in \tau^\infty(S)$ we conclude $\alpha' \in \tau^\infty(S)$. $\square$

If the grammar G in the proof is of type 4 then all grammars constructed in the further course are of type 1. This implies:

Corollary 3. *Every language $L \in \mathcal{L}_4$ over the alphabet Σ can be written as $L = L' \cap \Sigma^*$ where L' is the nested iterated substitution of languages from $\mathcal{L}_1$.*

Since all languages in $\mathcal{L}_2 = \mathbf{MAT}_{fi}$ are semilinear and by the closure of the semilinear languages under nested iterated substitution, we can furthermore conclude:

Corollary 4. *Any language in $\mathcal{L}_5$ is semilinear.*

Let us now study the closure properties of $\mathcal{L}_4$ and $\mathcal{L}_5$. As regards $\mathcal{L}_5$, the well-known constructions to show the closure of the context-free languages under the AFL operations can be adapted.

Theorem 1. *The family $\mathcal{L}_5$ is a full AFL.*

Proof. We need to show closure under union, concatenation, Kleene closure, homomorphisms, inverse homomorphisms and intersection with regular languages. Let $G_1 = (V_1, \Sigma, S_1, R_1, C_1)$ and $G_2 = (V_2, \Sigma, S_2, R_2, C_2)$ be Parikh vector controlled grammars of type 5. Without loss of generality, suppose that $V_1 \cap V_2 = \emptyset$. For the mentioned operations, we give now the respective constructions.

Union. Let $G' = (V', \Sigma, S', R', C')$ where $V' = V_1 \cup V_2 \cup \{S'\}$, $R' = \{S' \to S_1, S' \to S_2\} \cup R_1 \cup R_2$ and $C' = C_1 \times C_2 \times \mathbb{N}$.
The first derivation step produces either S_1 or S_2. In the first case, only rules from R_1 are used in the rest of the derivation. Since $\mathbf{0} \in C_2$, the derivation is valid iff every encountered Parikh vector is from $C_1 \times \{\mathbf{0}\} \times \{\mathbf{0}\}$, i.e., iff from the second step on it is valid in G_1. Analogously, if S_2 is produced, the derivation is valid iff from the second step on it is valid in G_2. Hence, $L(G') = L(G_1) \cup L(G_2)$.

Concatenation. Set $G' = (V', \Sigma, S', R', C')$ where $V' = V_1 \cup V_2 \cup \{S'\}$, $R' = \{S' \to S_1 S_2\} \cup R_1 \cup R_2$ and $C' = C_1 \times C_2 \times \mathbb{N}$. The first derivation step produces $S_1 S_2$. Since S_1 and S_2 derive only sentential forms over $\Sigma \cup V_1$ and $\Sigma \cup V_2$, respectively, and since $\mathbf{0} \in C_1$, we can restrict to derivations where symbols from V_1 are replaced as long as they are present. Such a derivation is of the form $S' \Rightarrow S_1 S_2 \Rightarrow^*_{G_1} w_1 S_2 \Rightarrow^*_{R_2} w_1 w_2$ with $w_1 w_2 \in \Sigma^*$. The subderivation $S_1 S_2 \Rightarrow^*_{G_1} w_1 S_2$ is valid iff every Parikh vector is in $C_1 \times \{e_1\} \times \{0\}$, i.e., iff $w_1 \in L(G_1)$ and $e_1 \in C_2$. The subderivation $w S_2 \Rightarrow^*_{R_2} w_1 w_2$ is valid iff every Parikh vector is in $\mathbf{0} \times C_2 \times \{0\}$, i.e., iff $w_2 \in L(G_2)$. Since $w_2 \in L(G_2)$ implies $e_1 \in C_2$, the complete derivation is valid iff $w_1 \in L(G_1)$ and $w_2 \in L(G_2)$.

Kleene Closure. Let $G' = (V', \Sigma, S', R', C')$ where $V' = V_1 \cup \{S'\}$, $R' = \{S' \to S_1 S', S' \to \lambda\} \cup R_1$ and $C' = C_1 \times \mathbb{N}$. We can restrict to derivations where a symbol from V_1 is replaced if present. Such a derivation has the form

$$S' \Rightarrow S_1 S' \Rightarrow^*_{G_1} w_1 S' \Rightarrow w_1 S_1 S' \Rightarrow^*_{G_1} w_1 w_2 S' \Rightarrow^* w_1 w_2 \cdots w_n S' \Rightarrow w_1 w_2 \cdots w_n$$

with $w_1, w_2, \ldots, w_n \in \Sigma^*$. A subderivation

$$w_1 \cdots w_{i-1} S' \Rightarrow w_1 \cdots w_{i-1} S_1 S' \Rightarrow^*_{G_1} w_1 \cdots w_{i-1} w_i S'$$

is valid iff every encountered sentential form is in $C_1 \times \{1\}$, i.e., iff $w_i \in L(G_1)$.

Homomorphisms. Let $h : \Sigma^* \to \Delta^*$ be a homomorphism. We extend h to a mapping from $(\Sigma \cup V_1)^*$ to $(\Delta \cup V_1)^*$ by setting $h(A) = A$, for all $A \in V_1$. Now set $G' = (V_1, \Delta, S_1, R', C)$ where $R' = \{A \to h(\alpha) : A \to \alpha \in R_1\}$. A sentential form β can be derived in the context-free grammar associated with G_1 iff $h(\beta)$ can be derived in the context-free grammar associated with G'. Moreover, the nonterminal Parikh vectors of β and $h(\beta)$ are equal. Hence, β is derivable in G_1 iff $h(\beta)$ is so in G' and thus $L(G') = h(L(G_1))$.

Inverse Homomorphisms. It suffices to show closure under inverse alphabetic homomorphisms (see, e.g., [1]). Let $h : \Delta^* \to \Sigma^*$ be an alphabetic homomorphism, i.e., a homomorphism sending each $a \in \Delta$ to a word from $\Sigma \cup \{\lambda\}$. Without loss of generality, assume that $\Sigma \cap \Delta = \emptyset$. Then G' is constructed as $G' = (V_1 \cup \Sigma \cup \{\Lambda\}, \Delta, S_1, R', C \times \mathbb{N}^{|\Sigma|+1})$ with $\Lambda \notin V_1 \cup \Sigma \cup \Delta$ and the set of rules

$$R' = R_1 \cup \{a' \to \Lambda a, a' \to a\Lambda : a' \in \Sigma, a \in \Delta, h(a) = a'\} \cup$$
$$\{\Lambda \to a\Lambda : a \in \Delta, h(a) = \lambda\} \cup \{\Lambda \to \lambda\}.$$

By Lemma 5 we can restrict to derivations where a symbol from V_1 is replaced if present. This way we apply in the first phase the rules from R_1 generating a word $w \in \Sigma^*$. Then the remaining rules can be used to generate an arbitrary word in $h^{-1}(w)$. Since the same restrictions as in G_1 apply to V_1, exactly the words from $L(G_1)$ can be generated in the first phase. In the second phase, a word $w \in L(G_1)$ can be transformed to any of its preimages under h by replacing every symbol in w by one of its preimages under h and inserting symbols from Δ whose images under h is λ. Hence, $L(G') = h^{-1}(L(G_1))$.

Intersection with Regular Sets. Let $\mathcal{A} = (Z, \Sigma, z_0, Q, \delta)$ be a finite automaton. Without loss of generality, assume that Q contains only the single state q.

Construct $G' = (V', \Sigma, S', R', C')$ such that $V' = Z \times (V_1 \cup \Sigma) \times Z$, $S' = (z_0, S, q)$,

$$R' = \{(z, A, z') \to (z, x_1, z_1)(z_1, x_2, z_2) \cdots (z_{r-1} x_r z') : A \to x_1 x_2 \cdots x_r \in R_1\} \cup$$
$$\{(z, A, z) \to \lambda : A \to \lambda \in R\} \cup \{(z, a, z') \to a : (z, a, z') \in \delta\},$$

and C' is defined such that in the defining system of inequalities for C every term of the form $k \cdot |\alpha|_A$, $k \in \mathbb{N}$, $A \in V_1$, is replaced by $k \cdot \sum\limits_{z,z' \in Z} |\alpha'|_{(z,A,z')}$.

Again, we can restrict to derivations where a symbol from $Z \times V_1 \times Z$ is replaced, if possible. So, in a first derivation phase we generate a word $\beta = (z_0, a_1, z_1)(z_1, a_2, z_2) \cdots (z_{n-1}, a_n, q)$, $a_1, a_2, \ldots, a_n \in \Sigma$. It can be shown by induction that a sentential form can be generated in the first phase iff it has the shape $\alpha' = (z_0, x_1, z_1')(z_1', x_2, z_2') \cdots (z_{m-1}', x_m, q)$, $x_1, x_2, \ldots, x_m \in V_1 \cup \Sigma$ and $\alpha = x_1 x_2 \cdots x_m$ can be derived in G_1. In particular, α' satisfies the constraints of G' iff α satisfies the constraints of G_1 because $|\alpha|_A = \sum\limits_{z,z' \in Z} |\alpha'|_{(z,A,z')}$, for all $A \in V_1$. In a second phase, the intermediate word β can be transformed to $w = a_1 a_2 \cdots a_n$ iff it describes a successful run of $\mathcal{A}$ on w. Hence, w can be generated by G' iff it is in $L(G_1)$ and $L(\mathcal{A})$. □

All constructions but the last result in a Parikh vector controlled grammar of type 4 if both G_1 and G_2 are of type 4. We can conclude:

Corollary 5. *The family $\mathcal{L}_4$ is closed under union, concatenation, Kleene closure, homomorphisms and inverse alphabetic homomorphisms.*

Regarding the remaining two AFL operations, we can prove nonclosure of $\mathcal{L}_4$ by help of the "replacement lemma".

Corollary 6. *$\mathcal{L}_4$ is not closed under intersection with regular sets and inverse homomorphisms.*

Proof. Using Lemma 4, it can be shown that the languages

$$L_1 = \{a^{3n} x^3 y b^{3n} a^{3n} x^2 y b^{3n} : n \geq 1\},$$
$$L_2 = \{a^{3n} c b^{3n} a^{3n} d b^{3n} : n \geq 1\}$$

are not in $\mathcal{L}_4$. However, as discussed in [9], there are a language $L \in \mathcal{L}_1 \subseteq \mathcal{L}_4$, a regular set M and a homomorphism g such that $L_1 = L \cap M$ and $L_2 = g^{-1}(L)$. □

By a slight modification of the proof of Lemma 6, we can also show that $\mathcal{L}_4$ and $\mathcal{L}_5$ are closed under nested iterated substitutions.

Theorem 2. *$\mathcal{L}_4$ and $\mathcal{L}_5$ are closed under nested iterated substitutions.*

A full AFL which is closed under nested iterated substitutions has been termed a *superAFL* by Greibach [6]. We can therefore give the following characterization of $\mathcal{L}_5$.

Corollary 7. $\mathcal{L}_5$ *is the least superAFL containing* $\mathbf{MAT}_{fi}$.

Finally, we are going to prove that grammars of type 6 generate exactly the family of matrix languages. This is achieved by giving simulations showing equivalence to grammars with regular control.

Lemma 7. $\mathcal{L}_6 \subseteq \mathbf{MAT}$.

Proof. The construction is similar to that in the proof of Lemma 2.
Let $G = (V, \Sigma, S, R, C_1 \times \mathbb{N}^{n-l})$ be a Parikh vector controlled grammar where $V = \{A_1, A_2, \ldots, A_n\}$ and $C_1 \subseteq \mathbb{N}^l$ is finite. Without loss of generality we can assume that S does not appear on the right-hand of any rule, $S = A_1$ and $C_1 \subseteq [0, 1] \times \mathbb{N}^{l-1}$ (the last assumption implies $l > 0$). We set $V_1 = \{A_1, \ldots, A_l\}$, $V_2 = \{A_{l+1}, \ldots, A_n\}$. The automaton for the regular control language keeps track of the nonterminals from V_1, hence its state set is basically C_1, a subset of $\mathbb{N}^l$. Let $\mathcal{A} = (C_1 \cup \{\mathbf{0}\}, R, \mathbf{e}_1, \{\mathbf{0}\}, \delta)$ be the deterministic finite automaton with the transition function δ defined as follows. If $r : A_i \rightarrow \alpha$ is a rule in R with $1 \leq i \leq l$, $\mathbf{x} = (x_1, \ldots, x_l)$ is in C_1 and $\mathbf{y}$ is defined by $\mathbf{y} = \mathbf{x} - \mathbf{e}_i + \Psi_{V_1}(\alpha)$ then

$$\delta(\mathbf{x}, r) = \begin{cases} \mathbf{y}, & \text{if } x_i > 0 \text{ and } \mathbf{y} \in C_1 \cup \{\mathbf{0}\} \\ \text{undefined}, & \text{otherwise.} \end{cases}$$

If $r : A_i \rightarrow \alpha$ is a rule in R with $l < i \leq n$, $\mathbf{x} = (x_1, \ldots, x_l)$ is in C_1 and $\mathbf{y}$ is defined by $\mathbf{y} = \mathbf{x} + \Psi_{V_1}(\alpha)$ then

$$\delta(\mathbf{x}, r) = \begin{cases} \mathbf{y}, & \text{if } \mathbf{y} \in C_1 \cup \{\mathbf{0}\} \\ \text{undefined}, & \text{otherwise.} \end{cases}$$

It is easy to prove by induction that a sequence $\rho = r_1 r_2 \cdots r_m$ reaches a state $\mathbf{z} \in C$ iff ρ is a possible derivation sequence in G and leading to a sentential form α with $\Psi_{V_1}(\alpha) = \mathbf{z}$. Hence, $L(\mathcal{A})$ is the set of all correct terminal derivation sequences in G and the grammar with regular control $G' = (V, \Sigma, S, R, L(\mathcal{A}))$ generates the same language as G. $\square$

Lemma 8. $\mathbf{MAT} \subseteq \mathcal{L}_6$.

Proof. Let $G = (V, \Sigma, S, R, L)$ be a grammar with regular control and let $\mathcal{A} = (Z, R, z_0, F, \delta)$ be a finite automaton accepting L. The proof strategy is to construct a Parikh vector controlled grammar of type 6 that simulates the steps of G while simultaneously keeping track of the state of the automaton. Formally, we construct $G' = (V', \Sigma, S', R', C)$ where the set of nonterminals is

$$V' = V \cup \{S'\} \cup Z \cup V_R \cup V_\delta \text{ with } V_R = \{A_r, B_r : r \in R\}, V_\delta = \{X_t, Y_t : t \in \delta\},$$

R' contains the following rules:

- $S' \rightarrow z_0 S$;
- for each rule $r : A \rightarrow \alpha$ in R, the rules $A \rightarrow A_r$, $A_r \rightarrow B_r$, $B_r \rightarrow \alpha$;

- for each transition $t = (z, r, z')$ in δ, the rules $z \to X_t$, $X_t \to Y_t$, $Y_t \to z'$;
- for each $z_f \in F$, the rule $z_f \to \lambda$;

and C is defined by the following constraints on each nonterminal sentential form in a derivation process:

1. Exactly one symbol from $\{S'\} \cup Z \cup V_\delta$ is present.
2. At most one symbol from V_R is present.
3. If a symbol from Z is present, then no symbol A_r, $r \in R$, is allowed.
4. If a symbol of the form X_t with $t \in \delta$, $t = (z, r, z')$ is present, then the only admissible symbol from V_R is A_r.
5. If a symbol of the form Y_t with $t \in \delta$ is present, then at least one symbol from V_R is present.

First note that G' is indeed of type 6, since the total number of symbols from $\{S'\} \cup Z \cup V_\delta \cup V_R$ is bounded by 2, while the symbols from V are unrestricted.

In the first step of a derivation in G', the rule $S' \to z_0 S$ is applied; the last derivation step of each successful derivation is of the form $z_f w \Rightarrow w$ with $z_f \in F, w \in \Sigma^*$. Now consider a sentential form $z\beta$ with $z \in Z$ and $\beta \in (V \cup \Sigma)^*$ where β contains at least one nonterminal symbol. Because of restriction 3, the symbol z has to be rewritten in the first step using some rule $z \to X_t$; let $t = (z, r, z')$ and $r : A \to \alpha$. In the next step, the rule $X_t \to Y_t$ cannot be applied by restriction 5, so restriction 4 requires the rule $A \to A_r$ to be used. This implies that β can be decomposed as $\beta_1 A \beta_2$ and the sentential form reached after the second step is $X_t \beta_1 A_r \beta_2$. By restriction 2 no other symbol from V can be rewritten in the next derivation steps. Restriction 4 forbids the application of $A_r \to B_r$, so the next applied rule has to be $X_t \to Y_t$ yielding $Y_t \beta_1 A_r \beta_2$. In the next step Y_t cannot be replaced by z' because of restriction 3; hence A_r must be rewritten to reach the sentential form $Y_t \beta_1 B_r \beta_2$. Now the only admissible rule is $Y_t \to z'$ due to restriction 5, giving $z' \beta_1 B_r \beta_2$. Finally, restriction 4 requires the application of $B_r \to \alpha$ which derives $z' \beta_1 \alpha \beta_2$. Hence, every sentential form reachable from $z\beta$ in six steps has the form $z'\beta'$ where β' can be directly derived in G from β using rule r and z can be transferred by r to z' in $\mathcal{A}$. On the other hand, every such sentential form $z'\beta'$ can be derived from $z\beta$ using the above derivation sequence thus completing the proof.   □

Corollary 8. $\mathcal{L}_4 \subset \mathcal{L}_5 \subset \mathcal{L}_6 = \mathbf{MAT}$.

3.3   The Families $\mathcal{L}_7$ and $\mathcal{L}_8$

Finally, we discuss grammars whose restricting sets are solution sets of arbitrary systems of linear inequalities (type 7) or semilinear sets (type 8). While the first variant turns out to be equivalent to matrix grammars, the second can generate all recursively enumerable languages.

Lemma 9. $\mathbf{MAT} \subseteq \mathcal{L}_7$.

Proof. The same construction as in the proof of Lemma 8 can be used. We need just to verify that the restrictions on the Parikh sets can be established by a system of linear inequalities. Indeed, the five restrictions can be reformulated as follows:

1. $|\alpha|_{S'} + \sum\limits_{z \in Z} |\alpha|_z + \sum\limits_{X \in V_\delta} |\alpha|_X = 1.$

2. $\sum\limits_{X \in V_R} |\alpha|_X \leq 1.$

3. $\sum\limits_{z \in Z} |\alpha|_z + \sum\limits_{r \in R} |\alpha|_{A_r} \leq 1.$

4. $|\alpha|_{X_t} + \sum\limits_{Y \in V_R \setminus \{A_r\}} |\alpha|_Y \leq 1$, for every $t \in \delta$ where $t = (z, r, z').$

5. $|\alpha|_{Y_t} - \sum\limits_{Y \in V_R} |\alpha|_Y \leq 0$, for every $t \in \delta$ where $t = (z, r, z').$

Only the last kind of inequalities is not obvious. It follows since the count of Y_t is limited by one (in condition 2). □

The reverse inclusion $\mathcal{L}_7 \subseteq \mathbf{MAT}$ can be quite easily shown by the construction of a positive valence grammar where the compliance with each of the inequalities is accomplished by a dimension of the valence vector. However, the construction below (Lemma 11) will lead to a positive valence grammar with a slightly different acceptance condition than the usual one. We will therefore first show a technical result regarding positive valence grammars. Let $G = (V, \Sigma, S, R, \mathbb{Z}^k)$ be a positive valence grammar and $t \in \mathbb{Z}^k$ a vector. Then we define $L(G, t)$ as the set of all words w for which a derivation

$$(S, \mathbf{0}) \Rightarrow (\alpha_1, z_1) \Rightarrow \cdots \Rightarrow (\alpha_r, z_r) \Rightarrow (w, t) \text{ with } z_i \geq \mathbf{0}, 1 \leq i \leq r,$$

exists. Note that t needs not to be in $\mathbb{N}^k$.

Lemma 10. *For every positive valence grammar G over $\mathbb{Z}^k$ and every $t \in \mathbb{Z}^k$, there is a positive valence grammar G' (over $\mathbb{Z}^{k+1}$) such that $L(G') = L(G, t)$.*

Proof. Let $G = (V, \Sigma, S, R, \mathbb{Z}^k)$ be a positive valence grammar and $t \in \mathbb{Z}^k$. The idea of the construction is to add an extra vector $-t$ in the final derivation step. To this end the simulating grammar G' needs for its nonterminal alphabet a copy of V and one additional dimension in the valence vectors. Hence, the nonterminal alphabet of G' is $V \cup V' \cup \{S_0\}$ where V' is a disjoint copy of V and $S_0 \notin V \cup V'$ is the new start symbol. In what follows, the copy of a nonterminal symbol $A \in V$ in V' will be denoted by A'; moreover, the vectors in $\mathbb{Z}^{k+1}$ will be written in the form (y, z) where $y \in \mathbb{Z}^k$ and $z \in \mathbb{Z}$. The set R' of valence rules in G' is defined as

$$R' = \{(S_0 \to S, (\mathbf{0}, 1))\} \cup \{(A \to A', (\mathbf{0}, -1)) : A \in V\} \cup$$
$$\{(A' \to \alpha, (z, 1) : (A \to \alpha, z) \in R\} \cup$$
$$\{(A' \to \alpha, (z - t, 0)) : (A \to \alpha, z) \in R\}.$$

We will prove by induction over n that a pair $(\beta, (y, z))$ with $z \geq 0$ is derivable in G' in $2n + 1$ steps, iff either

- $(\beta, \boldsymbol{y})$ is derivable in G in n steps and $z = 1$ or
- $(\beta, \boldsymbol{y} + \boldsymbol{t})$ is derivable in G in n steps and $z = 0$.

The assertion is true for $n = 0$ since the only pair derivable in one step in G' is $(S, (\boldsymbol{0}, 1))$. The first step in a derivation in G' is $(S_0, (\boldsymbol{0}, 0)) \Rightarrow (S, (\boldsymbol{0}, 1))$. Now suppose that the assertion has been shown for $n = k$. Consider some pair $(\beta, (\boldsymbol{y}, z))$ derived in G' in $2k+1$ steps. By induction hypothesis, $\beta \in (V \cup \Sigma)^*$ and $z \in \{0, 1\}$ hold. The next derivation step has to apply a rule $(A \to A', (\boldsymbol{0}, -1))$ yielding $(\beta_1 A' \beta_2, (\boldsymbol{y}, z - 1))$ where $\beta = \beta_1 A \beta_2$ is a decomposition of β. If $z = 0$, no further derivation step is possible. If $z = 1$, the next step must use a rule of either of the forms $(A' \to \alpha, (\boldsymbol{z}, 1))$ or $(A' \to \alpha, (\boldsymbol{z} - \boldsymbol{t}, 0))$ in order to keep the last component non-negative. In the first case, the resulting pair is $(\beta_1 \alpha \beta_2, (\boldsymbol{y} + \boldsymbol{z}, 1))$. The step is valid iff $\boldsymbol{y} + \boldsymbol{z} \geq \boldsymbol{0}$, i.e., iff $(\beta_1 \alpha \beta_2, \boldsymbol{y} + \boldsymbol{z})$ is directly derivable from $(\beta, \boldsymbol{y})$ in G. In the second case, the resulting pair is $(\beta_1 \alpha \beta_2, (\boldsymbol{y} + \boldsymbol{z} - \boldsymbol{t}, 0))$. Hence, the induction hypothesis is true for $n = k + 1$.

Since all sentential forms generated in $2n$ steps by G' are nonterminal, the language of G' is found as

$$L(G') = \{w \in \Sigma^* : (w, (\boldsymbol{0}, 0)) \text{ derivable in } G' \text{ in } 2n + 1 \text{ steps }, n \geq 0\}$$
$$= \{w \in \Sigma^* : (w, \boldsymbol{t}) \text{ derivable in } G \text{ in } n \text{ steps }, n \geq 0\} = L(G, \boldsymbol{t}),$$

as claimed. □

Lemma 11. $\mathcal{L}_7 \subseteq \textbf{MAT}$.

Proof. Let $G = (V, \Sigma, S, R, C)$ be a Parikh vector controlled grammar where $V = \{A_1, A_2, \ldots, A_n\}$, $S = A_1$ and $C \subseteq \mathbb{N}^n$ is the solution set of a system of m linear inequalities

$$\sum_{j=1}^{n} a_{i,j} x_j + b_i \geq 0, \ (1 \leq i \leq m).$$

We construct the positive valence grammar $G' = (V \cup \{S'\}, \Sigma, S', R', \mathbb{Z}^m)$ with the start symbol $S' \notin V$ and the set of valence rules R' constructed as follows.

- The starting rule is $(S' \to A_1, (z_1, \ldots, z_m))$ with $z_i = a_{i,1} + b_i$, $1 \leq i \leq m$.
- For any rule $A_r \to \alpha$ with $\Psi_V(\alpha) = (y_1, \ldots, y_n)$, R' contains the valence rule $(A_r \to \alpha, (z_1, \ldots, z_m))$ with $z_i = \sum_{j=1}^{n} a_{i,j} y_j - a_{i,r}$.

It is easy to verify by induction on the number of derivation steps that G' can generate a pair $(\alpha, (z_1, \ldots, z_m))$ iff G can generate α and $\Psi_V(\alpha) = (x_1, \ldots, x_n)$ satisfies $z_i = \sum_{j=1}^{n} a_{i,j} x_j + b_i$, for $1 \leq i \leq m$. Hence, G produces the same language as G' with the target vector $(b_1, \ldots, b_m)$. □

Lemma 12. $\mathcal{L}_8$ *is the family of recursively enumerable languages.*

Proof. We will simulate a programmed grammar with appearance checking by a Parikh vector controlled grammar with a semilinear restricting set. Let $G = (V, \Sigma, S, R, \sigma, \phi)$ be a programmed grammar with appearance checking. Then we construct the Parikh vector controlled grammar $G' = (V', \Sigma, S', R', C)$ where

$$V' = V \cup \{S'\} \cup \{X_r, Y_r, Z_r, F_r, A_r, B_r : r \in R\},$$
$$\begin{aligned} R' = \{S' &\Rightarrow SX_r : r \in R\} \cup \\ &\{X_r \to Y_r, Y_r \to Z_r : r \in R\} \cup \{Z_r \to X_s : s \in \sigma(r)\} \cup \\ &\{X_r \to F_r : r \in R\} \cup \{F_r \to X_f : f \in \phi(r)\} \cup \\ &\{A \to A_r, A_r \to B_r, B_r \to \alpha : (r : A \to \alpha) \in R\}, \end{aligned}$$

and C is defined by the following constraints on the Parikh vector for a nonterminal sentential form:

1. One symbol from $\{S'\} \cup \{X_r, Y_r, Z_r, F_r : r \in R\}$ is present.
2. At most one symbol from $\{A_r, B_r : r \in R\}$ is present.
3. If X_s is present then no symbol A_r is present, $r, s \in R$.
4. If Y_s is present then no symbol B_r is present, $r, s \in R$.
5. If Z_r is present then one of the symbols A_r, B_r is present.
6. If F_s is present, $s \in R$, then no symbol from $\{A_r, B_r : r \in R\}$ is present.
7. If F_r is present, for $r : A \to \alpha$, then A is not present.

It is easy to see that each of the constraints describes a semilinear set. The set C is the intersection of all these sets and thus semilinear, too. The correctness proof is similar to that in Lemma 7. The last constraint models the appearance checking case. It is the only one that cannot be described by a system of linear inequalities. □

4   Conclusions

We have introduced Parikh vector controlled grammars and investigated several restrictions on the Parikh sets of sentential forms. The results concerning the generative power with respect to the different restrictions can be summarized as follows (arrows indicating strict inclusions, disconnected families being incomparable).

$$\begin{array}{ccccc}
\mathbf{CF}_{fi} & \longrightarrow & \mathcal{L}_1 & \longrightarrow & \mathbf{MAT}_{fi} = \mathcal{L}_2 = \mathcal{L}_3 \\
\downarrow & & \downarrow & & \downarrow \\
\mathbf{CF} \longrightarrow \mathcal{L}_4 & & \longrightarrow \mathcal{L}_5 & \longrightarrow & \mathbf{MAT} = \mathcal{L}_6 = \mathcal{L}_7 \longrightarrow \mathbf{RE} = \mathcal{L}_8
\end{array}$$

A particularly interesting family is $\mathcal{L}_5$ defined by grammars whose restricting sets are solutions of positive systems of linear inequalities. This language family does not coincide with any of the formerly known classes and is a superAFL of semilinear languages.

It remains to study the power of non-erasing grammars of the respective types. It might be also worthwhile to investigate connections to other variants of regulated rewriting. For instance, a characterization of random context grammars by an appropriate Parikh vector control could be helpful to settle the longstanding question if random context grammars are equivalent to matrix grammars.

References

1. Berstel, J.: Transductions and Context-Free Languages. Teubner-Verlag, Stuttgart (1979)
2. Brainerd, B.: An analog of a theorem about context-free languages. Information and Control 11, 561–567 (1968)
3. Dassow, J., Păun, G.: Regulated Rewriting in Formal Language Theory. Springer (1989)
4. Ginsburg, S., Spanier, E.: Derivation bounded languages. Journal of Computer and System Sciences 2, 228–250 (1968)
5. Greibach, S.: Full AFLs and nested iterated substitution. Information and Control 16(1), 7–35 (1970)
6. Greibach, S.: A generalization of Parikh's semilinear theorem. Discrete Mathematics 2, 347–355 (1972)
7. Hopcroft, J.E., Ullman, J.D.: Introduction to Automata Theory, Languages, and Computation. Addison-Wesley, Reading (1979)
8. Stiebe, R.: Positive valence grammars. In: Csuhaj-Varjú, E., Kintala, C., Wotschke, D., Vaszil, G. (eds.) Fifth International Workshop Descriptional Complexity of Formal Systems, pp. 186–197. MTA SZTAKI, Budapest (2003)
9. Stiebe, R., Turaev, S.: Capacity bounded grammars. Journal of Automata, Languages and Combinatorics 15, 175–194 (2010)

On the Nonterminal Complexity
of Tree Controlled Grammars

György Vaszil

Department of Computer Science, Faculty of Informatics,
University of Debrecen,
P.O. Box 12, 4010 Debrecen, Hungary
vaszil.gyorgy@inf.unideb.hu

Abstract. A tree controlled grammar is a regulated rewriting device which can be given as a pair (G, R) where G is a context-free grammar and R is a regular set over the terminal and nonterminal alphabets of G. The language generated by the tree controlled grammar contains those words of $L(G)$ which have a derivation tree where all the words obtained by reading the symbols labeling the nodes belonging to the different levels of the tree, from left to right, belong to the language R. The nonterminal complexity of tree controlled grammars can be given as the number of nonterminals of the context-free grammar G, and the number of nonterminals that a regular grammar needs to generate the control language R. Here we improve the currently known best upper bound on the nonterminal complexity of tree controlled grammars from seven to six, that is, we show that a context-free grammar with five nonterminals and a control language which can be generated by a grammar with one nonterminal is sufficient to generate any recursively enumerable language.

1 Introduction

The aim of the area called "regulated rewriting" is to add some kind of a control mechanism to ordinary (usually context-free) grammars which restricts the application of the rules in such a way that some of the derivations which are possible in the usual derivation process are eliminated from the controlled variant. This means that the set of words generated by the controlled device is a subset of the original (context-free) language generated without the control mechanism. As these generated subsets can be non-context-free languages, these mechanisms are usually more powerful than ordinary (context-free) grammars.

The need for rewriting devices which use rules of a simple form but still have a considerable generative power is justified by the study of phenomena occurring in different areas of mathematics, linguistics, or even developmental biology. To study problems in these areas which cannot be described by the capabilities of context-free languages, it is often desirable to construct generative mechanisms which have as many context-free-like properties as possible, but are also able to describe the non-context-free features of the specific problem in question. See [2]

H. Bordihn, M. Kutrib, and B. Truthe (Eds.): Dassow Festschrift 2012, LNCS 7300, pp. 265–272, 2012.

for a discussion about non-context-free phenomena in different areas, and [3] for regulated rewriting in general.

Tree controlled grammars were introduced in [1] as a pair (G, R) where G is a context-free grammar and R is a regular set, called the control language. The control language contains words composed of the terminal and nonterminal alphabets of G, and it is used to control the work of G by restricting the set of derivations which G is allowed to make. Only those words belong to the generated language of the tree controlled grammar which can be generated by the context-free grammar G, and moreover, which have a derivation tree where all the strings obtained by reading from left to right the symbols labeling nodes which belong to the same level of the tree (with the exception of the last level) are elements of the regular set R.

As it was already shown in [1], tree controlled grammars are able to generate any recursively enumerable language. Variants of the notion with control sets which are not regular but belong to different classes from the Chomsky hierarchy were studied in [7], the power of subregular control sets were examined in more detail in [4].

Investigations concerning the nonterminal complexity of tree controlled grammars began in [11] where this measure was defined as the sum of the number of nonterminals of the context-free grammar and the number of nonterminals which are necessary to generate the regular control language. They showed in [11] that nine nonterminals altogether are sufficient to generate any recursively enumerable language with a tree controlled grammar. Then they improved this bound to seven in [10] by simulating a phrase structure grammar being in the so called Geffert normal form (see [6]), that is, having four nonterminals and only linear context-free productions, together with one non-context-free production in addition which is able to erase three neighboring nonterminals as it is of the form $ABC \to \lambda$.

Here we improve the bound from seven to six using a similar technique as in [10], but simulating a grammar being in a different version of the Geffert normal form: Instead of four nonterminals and the non-context-free production $ABC \to \lambda$, we use a variant which has five nonterminals and two non-context-free productions $AB \to \lambda$, $CD \to \lambda$. Since the nonterminals, A, B, C, D, will be encoded in the simulating tree controlled grammar by strings over just two symbols (as it was done also in [10]), it does not matter that instead of the symbols A, B, C we need to simulate a grammar which uses more symbols, A, B, C, D. On the other hand, the fact that instead of three, only two neighboring symbols have to be deleted simultaneously, helps to construct a simulating tree controlled grammar which uses one nonterminal less than the one in [10], thus helps to reduce the currently known best bound from seven to six.

2   Preliminaries and Definitions

The reader is assumed to be familiar with the basics of formal language theory. In the following we list some notions and notations we will use in the subsequent parts of the paper. For more information, see for example [9,2,8].

A finite set of symbols V is called an alphabet. The cardinality, that is, the number of elements of V is denoted by $|V|$. The set of non-empty words over the alphabet V is denoted by V^+; the empty word is λ, and $V^* = V^+ \cup \{\lambda\}$. A set $L \subseteq V^*$ is called a language over V. For a word $w \in V^*$ and a set of symbols $A \subseteq V$, we denote the length of w by $|w|$, and the number of occurrences of symbols from A in w by $|w|_A$. If A is a singleton set, i.e., $A = \{a\}$, then we write $|w|_a$ instead of $|w|_{\{a\}}$. The concatenation of two sets $L_1, L_2 \subseteq V^*$, denoted as $L_1 L_2$, is defined as $L_1 L_2 = \{w_1 w_2 \mid w_1 \in L_1, w_2 \in L_2\}$.

A generative grammar G is a quadruple $G = (N, T, S, P)$ where N and T are the disjoint sets of nonterminal and terminal symbols, $S \in N$ is the initial nonterminal, and P is a set of rewriting rules (or productions) of the form $\alpha \to \beta$ where $\alpha, \beta \in (N \cup T)^*$ with $|\alpha|_N \geq 1$. A string v can be derived from a string u, denoted as $u \Rightarrow v$ for some $u, v \in (N \cup T)^*$, if they can be written as $u = u_1 \alpha u_2$, $v = v_1 \beta v_2$ for a rewriting rule $\alpha \to \beta \in P$. The reflexive and transitive closure of the relation $\Rightarrow$ is denoted by $\Rightarrow^*$. The language generated by the grammar G is the set of terminal strings which can be derived from the initial nonterminal, that is, $L(G) = \{w \in T^* \mid S \Rightarrow^* w\}$. It is known that any recursively enumerable language can be generated by a generative grammar defined as above.

A generative grammar is context-free, if the rewriting rules $\alpha \to \beta$ are such, that $\alpha \in N$. A context-free grammar is regular, if in addition to the property that $\alpha \in N$, it also holds, that $\beta \in T^* \cup T^* N$. The classes of context-free and regular grammars and languages are denoted by CF, REG, $\mathcal{L}(\text{CF})$, and $\mathcal{L}(\text{REG})$, respectively.

The number of nonterminals of a generative grammar $G = (N, T, S, P)$ is denoted by $\text{Var}(G)$, that is, $\text{Var}(G) = |N|$. For a language L and a class of grammars $X \in \{\text{REG}, \text{CF}\}$, we denote by $\text{Var}_X(L)$ the minimal number of nonterminals necessary to generate L with a grammar of type X, that is, $\text{Var}_X(L) = \min\{\text{Var}(G) \mid L = L(G) \text{ and } G \text{ is of type } X \in \{\text{REG}, \text{CF}\}\}$.

An ordered tree is a derivation tree of a context-free grammar $G = (N, T, S, P)$ if its nodes are labeled with symbols from $N \cup T \cup \{\lambda\}$ in a way which satisfies the following properties: (a) The root is labeled with S, (b) the leaves are labeled with symbols from $T \cup \{\lambda\}$, and (c) every interior vertex is labeled from N in such a way that if a vertex has a label $A \in N$ and its children are labeled from left to right with $x_1, x_2, \ldots, x_m$, $x_i \in N \cup T \cup \{\lambda\}$, $1 \leq i \leq m$, then there is a production $A \to x_1 x_2 \ldots x_m$ in P. A derivation tree corresponds to a terminal word w from $L(G)$ if the concatenation of the symbols labeling the leaves of the tree from left to right coincide with w.

The distance of a vertex t from the root is the length of the shortest path leading to t from the root node. The string corresponding to the ith level of a derivation tree for some $i \geq 0$ is the word obtained by concatenating from left to right the symbols labeling those nodes of the tree which are at distance i from the root.

A tree controlled grammar G is a pair $G = (G', R)$ where $G' = (N, T, S, P)$ is a context-free grammar and R is a regular language over the alphabet $N \cup T$. The language $L(G)$ generated by the tree controlled grammar G contains all words

$w \in T^*$ from $L(G')$ which have a derivation tree where the strings corresponding to each different level, except the last one, belong to the regular set R.

The nonterminal complexity of a tree controlled grammar $G = (G', R)$ is the number of nonterminals of G' plus the minimal number of nonterminals that a regular grammar needs for generating the language R, that is, $\text{Var}(G) = \text{Var}(G') + \text{Var}_{\text{REG}}(R)$.

To illustrate the notion of tree controlled grammars, let us recall an example from [2].

Example 1. Let $G = (G', R)$ where $G' = (\{S\}, \{a\}, S, \{S \rightarrow SS, S \rightarrow a\})$ and $R = \{S\}^*$. As the control language R contains words which are sequences of the nonterminal symbol S, all the nodes of every level (except the last one) of the derivation tree of a word $w \in L(G)$ are labeled by the symbol S. This means that all the nonterminals, with the exception of the ones labeling the nodes directly above the last level of the tree, are rewritten by the rule $S \rightarrow SS$, and the nonterminals of the level directly above the last one are rewritten by $S \rightarrow a$. Thus, the language generated by G is $L(G) = \{a^{2^n} \mid n \geq 0\}$.

Now we present a normal form result for generative grammars from [6] where it is shown that any recursively enumerable language can be generated by a grammar $G = (N, T, S, P)$ where $N = \{S, A, B, C, D\}$ and $P = P_1 \cup P_2$ where P_1 contains linear productions of the form $S \rightarrow zSa, S \rightarrow uSv$ and $S \rightarrow \lambda$ where $z, u \in \{A, C\}^*, v \in \{B, D\}^*, a \in T$, and P_2 contains the two non-context-free productions $P_2 = \{AB \rightarrow \lambda, CD \rightarrow \lambda\}$. A grammar of the above form is said to be in the Geffert normal form.

The derivations of a Geffert normal form grammar consists of three phases. First, rules of the form $S \rightarrow zSa, z \in \{A, C\}^*, a \in T$ are used to generate a sentential form from $\{A, C\}^* S T^*$, this is the first phase. Then the rules $S \rightarrow uSv, u \in \{A, C\}^*, v \in \{B, D\}^*$ and $S \rightarrow \lambda$ are used to produce a string from $\{A, C\}^* \{B, D\}^* T^*$, this is the second phase. Finally, in the third phase, the erasing rules $AB \rightarrow \lambda, CD \rightarrow \lambda$ are used to produce a terminal word from T^*.

We will also need the notion of a unique-sum set which was introduced in [5] as follows. A set of natural numbers $U = \{u_1, \ldots, u_p\}$ having the sum $\sigma_U = \Sigma_{i=1}^p u_i$ is said to be a unique-sum set, if the equation $\Sigma_{i=1}^p c_i u_i = \sigma_U$ for $c_i \in \mathbb{N}$ has the only solution $c_i = 1, 1 \leq i \leq p$. Examples of unique-sum sets are $\{2, 3\}, \{4, 6, 7\}$, or $\{8, 12, 14, 15\}$, while the set $\{4, 5, 6\}$ is not unique-sum, as $4 + 5 + 6 = 15 = 5 + 5 + 5$. It is clear that any subset of a unique-sum set is unique-sum, and that the sum of any two numbers from the set, $\sigma_{i,j} = u_i + u_j$, cannot be produced as the linear combination of elements of the set in any other way.

3   The Number of Nonterminals

Now we show that every recursively enumerable language can be generated by a tree controlled grammar with six nonterminals.

Theorem 1. *For any recursively enumerable language L, there exists a tree controlled grammar G with $L = L(G)$, such that $\text{Var}(G) = 6$.*

Proof. Let $L \subseteq T^*$ be a recursively enumerable language generated by the Geffert normal form grammar $G_1 = (\{S, A, B, C, D\}, T, S, P)$ where $T = \{a_1, a_2, \ldots, a_t\}$ and $P = \{AB \to \lambda, CD \to \lambda, S \to \lambda\} \cup \{S \to z_i S a_i, S \to u_j S v_j \mid z_i, u_j \in \{A, C\}^*, v_j \in \{B, D\}^*, 1 \le i \le t, 1 \le j \le s\}$.

Let us define the morphism $h : \{A, B, C, D\}^* \to \{0, \$\}^*$ by $h(A) = \$0^6\$$, $h(B) = \$0^{10}\$$, $h(C) = \$0^{12}\$$, $h(D) = \$0^{13}\$$ which encodes four of the nonterminals of the grammar G_1 as strings over two symbols. Notice that the length of the coding sequences forms the unique-sum set $\{8, 12, 14, 15\}$.

Let us now construct the tree controlled grammar $G = (G', R)$ where $G' = (N, T, S, P')$ with $N = \{S, S', \$, 0, \#\}$,

$$P' = \{S \to h(z)Sa \mid S \to zSa \in P, a \in T, z \in \{A, C\}^*\} \cup$$
$$\{S \to S'\} \cup$$
$$\{S' \to h(u)S'h(v) \mid S \to uSv \in P, u \in \{A, C\}^*, v \in \{B, D\}^*\} \cup$$
$$\{S' \to \lambda, \$ \to \$, \$ \to \#, 0 \to 0, 0 \to \#, \# \to \lambda\},$$

and

$$R = (\{S, S'\} \cup T \cup X_1 \cup X_2)^*\{\#^{20}, \#^{29}, \lambda\},$$

where

$$X_1 = \{\$0^6\$, \$0^{10}\$, \$0^{12}\$, \$0^{13}\$\}, \tag{1}$$
$$X_2 = \{\$0^6\$, \$0^{10}\$\}\{\#^{20}, \#^{29}\}\{\$0^{12}\$, \$0^{13}\$\}. \tag{2}$$

First we show that any terminal derivation of G_1 can be simulated by the tree controlled grammar G, that is, $L(G_1) \subseteq L(G)$. Let $w \in L(G_1)$ and let

$$S \Rightarrow^* zSw \Rightarrow^* zuSvw \Rightarrow zuvw \tag{3}$$

be the first and the second phases of a derivation of w in G_1 where $z, u \in \{A, C\}^*, v \in \{B, D\}^*$. We can generate $h(zuv)w$, the encoded version of $zuvw$ with the rules of G as follows

$$S \Rightarrow^* h(z)Sw \Rightarrow h(z)S'w \Rightarrow^* h(zu)S'h(v)w \Rightarrow h(zuv)w, \tag{4}$$

$h(zu) \in \{\$0^6\$, \$0^{12}\$\}^*, h(v) \in \{\$0^{10}\$, \$0^{13}\$\}^*$. If we use the chain rules, $\$ \to \$$ and $0 \to 0$, we can make sure that the word corresponding to each level of the derivation tree belongs to the regular set R, and moreover, that $h(zuv)$ is the string corresponding to the last level of the derivation tree which belongs to the derivation (4) of G above simulating the first two phases of the derivation of the word w in G_1 depicted at (3).

Now w can be derived in G_1 if zuv can be erased by using the rules $AB \to \lambda$ and $CD \to \lambda$. If AB or CD is a substring of zuv, then $h(AB) = \$0^6\$\$0^{10}\$$ or $h(CD) = \$0^{12}\$\$0^{13}\$$ is a substring of $h(zuv)$, thus, one of the derivations

$$h(zuv) \Rightarrow h(zu')\#^{20}h(v')w \Rightarrow h(zu'v')w,$$

or

$$h(zuv) \Rightarrow h(zu')\#^{29}h(v')w \Rightarrow h(zu'v')w$$

can be executed in G using the chain rules as above, and the rules $0 \to \#$, $\$ \to \#$, $\# \to \lambda$ in such a way that $h(zu'v')$ is again the string which corresponds to the last level of the derivation tree of $h(zu'v')w$.

It is clear, that whenever zuv can be erased in G_1, then $h(zuv)$ can also be erased in G, thus, w can also be generated by G which means that $w \in L(G)$.

Now we show that $L(G) \subseteq L(G_1)$. To see this, we have to show that any $w \in L(G)$ can also be generated by G_1. Consider the derivation tree corresponding to a derivation of $w \in L(G)$ in G and look at the words corresponding to the different levels of the tree.

Notice the following: (A) There is no symbol $\#$ appearing in the levels as long as S or S' is present. This statement holds because the words in R have a special form: They are the concatenations of "complete" coding sequences of $A, B, C,$ or D, that is, each subword over $\{\$, 0\}$ is a concatenation of coding strings of the form $\$0^i\$$ (for some $i \in \{6, 10, 12, 13\}$). Thus, if $\#$ symbols appear in a word corresponding to a level of the derivation tree, then either all symbols of such a coding subword are rewritten to $\#$, or no symbol of such a coding subword is rewritten to $\#$. Recall that the lengths of these coding sequences form a unique-sum set, $\{8, 12, 14, 15\}$, thus, 20 and 29 can only arise through some linear combination of the elements as $20 = 8 + 12$, and $29 = 14 + 15$. This, together with the above considerations, means that $\#^{20}$ or $\#^{29}$ can only be obtained by rewriting all symbols of $\$0^6\$\$0^{10}\$$ or $\$0^{12}\$\$0^{13}\$$ to $\#$. Notice that when S or S' is present, then no sequence over $\{\$0^6\$, \$0^{12}\$\}$ can be followed directly by a sequence over $\{\$0^{10}\$, \$0^{13}\$\}$, thus, when S or S' is present no neighboring code sequences of length 20 or 29 can occur which means that the words cannot contain $\#^{20}$ or $\#^{29}$ as a subsequence.

Statement (A) above implies that as long as S or S' is present in the words corresponding to the levels of the derivation tree, the chain rules $\$ \to \$$ and $0 \to 0$ have to be used on the symbols $\$, 0$ when passing to the next level of the derivation tree. This is also true for the word corresponding to the first level in which S' disappears after using a rule of the form $S' \to h(u)h(v)$, since $uv \neq \lambda$. Note that the part of the derivations of G with the presence of S and the presence of S' corresponds to the first and the second phases of the derivations of the Geffert normal form grammar G_1, respectively.

Consider now the first such level of the derivation tree corresponding to a derivation of w in G in which none of the symbols S or S' are present. From the above considerations it follows that the string corresponding to this level has the form $h(zu)h(v)$ where $h(zu) \in \{\$0^6\$, \$0^{12}\$\}^*$, $h(v) \in \{\$0^{10}\$, \$0^{13}\$\}^*$, and $zuvw$ can also be derived in the grammar G_1.

Note also: (B) There cannot be two distinct subsequences of the symbols $\#$ in any of the words corresponding to any level of the derivation tree of the word $w \in L(G)$. To see this, consider the first level of the tree which is without S and S', and denote the string corresponding to this level as $h(zuv)$. Recall that

$h(zuv) = \alpha_1\alpha_2$ where $\alpha_1 \in \{\$0^6\$, \$0^{12}\$\}^*$, $\alpha_2 \in \{\$0^{10}\$, \$0^{13}\$\}^*$, so subwords of the form $\{\$0^6\$, \$0^{12}\$\}^*\{\#^{20}, \#^{29}\}\{\$0^{10}\$, \$0^{13}\$\}^*$ can only be present in the words corresponding to subsequent levels of the tree in such a way that the sequence of $\#$ symbols is the result of rewriting a suffix of α_1 and a prefix of α_2 to $\#$.

Property (B) above implies that in order to be in the control set R, a word which corresponds to some level of the derivation tree and also contains $\#$, must be of the form $\{\$0^6\$, \$0^{12}\$\}^*\{\#^{20}, \#^{29}\}\{\$0^{10}\$, \$0^{13}\$\}^*$ where $\#^{20}$ or $\#^{29}$ is obtained from the word corresponding to the previous level of the tree by rewriting each symbol in a substring $\$0^6\$\$0^{10}\$$ or $\$0^{12}\$\$0^{13}\$$ to $\#$, respectively. Therefore, the word corresponding to the previous level of the tree is either $\alpha_1'\$0^6\$\$0^{10}\$\alpha_2'$ or $\alpha_1'\$0^{12}\$\$0^{13}\$\alpha_2'$ where α_1' and α_2' satisfy either $h^{-1}(\alpha_1')AB\ h^{-1}(\alpha_2') = zuv$ or $h^{-1}(\alpha_1')CD\ h^{-1}(\alpha_2') = zuv$ provided that $\alpha_1\alpha_2 = h(zuv)$.

This means that the uncoded version of the word corresponding to the next level of the derivation tree, where the $\#$ symbols are erased, can also be derived in G_1 by the rules $AB \to \lambda$ or $CD \to \lambda$. More precisely, the word corresponding to the next level of the derivation tree is either of the form $\alpha_1'\alpha_2'$ or $\alpha_1''\{\#^{20}, \#^{29}\}\alpha_2''$, all of them corresponding to the sentential form $h^{-1}(\alpha_1')h^{-1}(\alpha_2')w$ which can also be derived in G_1.

Continuing the above reasoning, we obtain that the word corresponding to the level which is above the last one in the derivation tree of $w \in L(G)$ is of the form $\#^{20}$ or $\#^{29}$, corresponding to the sentential form ABw or CDw in G_1, thus, if w can be generated by the tree controlled grammar G with the control set R, then w can also be generated by the Geffert normal form grammar G_1.

This means that $L(G) \subseteq L(G_1)$, and since we have already shown the that the opposite inclusion holds, we have $L(G) = L(G_1)$. As the control set R can be generated by the regular grammar $G_2 = (\{A\}, T \cup \{0, \$, \#, S, S'\}, A, P_2)$ with $P_2 = \{A \to xA, A \to \#^{20}, A \to \#^{29}, A \to \lambda \mid x \in \{S, S'\} \cup T \cup X_1 \cup X_2$ where X_1 and X_2 is defined as above at (1) and (2), respectively, and this grammar has just one nonterminal, we have proved the statement of the theorem.

4   Conclusion

We have shown how to improve a descriptional complexity result from [10] by reducing the nonterminal complexity of tree controlled grammars from seven to six. We have used a similar technique as was used in [10], namely, we have shown how to simulate a phrase structure grammar in the so called Geffert normal form by tree controlled grammars. Instead of the normal form with the single erasing rule $ABC \to \lambda$, we have used the variant with two erasing rules $AB \to \lambda$, $CD \to \lambda$, thus we needed to simulate the simultaneous erasing of only two nonterminals, as opposed to the simultaneous erasing of three symbols simulated in [10]. This simplification made it possible to realize the simulation with six nonterminals which number is one less than needed in the proof of the previously known best result.

References

1. Čulik II, K., Maurer, H.: Tree controlled grammars. Computing 19, 129–139 (1977)
2. Dassow, J., Păun, G.: Regulated Rewriting in Formal Language Theory. Springer, Berlin (1989)
3. Dassow, J., Păun, G., Salomaa, A.: Grammars with controlled derivations. In: Salomaa, A., Rozenberg, G. (eds.) Handbook of Formal Languages, vol. II, pp. 101–154. Springer, Heidelberg (1997)
4. Dassow, J., Stiebe, R., Truthe, B.: Generative capacity of subregularly tree controlled grammars. International Journal of Foundations of Computer Science 21 (2010)
5. Frisco, P.: Computing with Cells. Advances in Membrane Computing. Oxford University Press, New York (2009)
6. Geffert, V.: Context-Free-Like Forms for the Phrase Structure Grammars. In: Chytil, M.P., Janiga, L., Koubek, V. (eds.) MFCS 1988. LNCS, vol. 324, pp. 309–317. Springer, Heidelberg (1988)
7. Păun, G.: On the generative capacity of tree controlled grammars. Computing 21, 213–220 (1979)
8. Rozenberg, G., Salomaa, A. (eds.): Handbook of Formal Languages. Springer, Berlin (1997)
9. Salomaa, A.: Formal Languages. Academic Press, New York (1973)
10. Turaev, S., Dassow, J., Selamat, M.: Language Classes Generated by Tree Controlled Grammars with Bounded Nonterminal Complexity. In: Holzer, M., Kutrib, M., Pighizzini, G. (eds.) DCFS 2011. LNCS, vol. 6808, pp. 289–300. Springer, Heidelberg (2011)
11. Turaev, S., Dassow, J., Selamat, M.: Nonterminal complexity of tree controlled grammars. Theoretical Computer Science 412(41), 5789–5795 (2011)

One-Way Finite Automata
with Quantum and Classical States⋆

Shenggen Zheng[1], Daowen Qiu[1,3,4,⋆⋆], Lvzhou Li[1], and Jozef Gruska[2]

[1] Department of Computer Science, Sun Yat-sen University,
Guangzhou 510006, China
{zhengshenggen,lilvzhou}@gmail.com, issqdw@mail.sysu.edu.cn
[2] Faculty of Informatics, Masaryk University,
Brno, 602 00, Czech Republic
gruska@fi.muni.cz
[3] SQIG–Instituto de Telecomunicações, Departamento de Matemática,
Instituto Superior Técnico, TULisbon, Av. Rovisco Pais 1049-001, Lisbon, Portugal
[4] The State Key Laboratory of Computer Science, Institute of Software,
Chinese Academy of Sciences, Beijing 100080, China

Abstract. In this paper, we introduce and explore a new model of *quantum finite automata* (QFA). Namely, *one-way finite automata with quantum and classical states* (1QCFA), a one way version of *two-way finite automata with quantum and classical states* (2QCFA) introduced by Ambainis and Watrous in 2002 [3]. First, we prove that *coin-tossing one-way probabilistic finite automata* (coin-tossing 1PFA) [23] and *one-way quantum finite automata with control language* (1QFACL) [6] as well as several other models of QFA, can be simulated by 1QCFA. Afterwards, we explore several closure properties for the family of languages accepted by 1QCFA. Finally, the state complexity of 1QCFA is explored and the main succinctness result is derived. Namely, for any prime m and any $\varepsilon_1 > 0$, there exists a language L_m that cannot be recognized by any *measure-many one-way quantum finite automata* (MM-1QFA) [12] with bounded error $\frac{7}{9} + \epsilon_1$, and any 1PFA recognizing it has at last m states, but L_m can be recognized by a 1QCFA for any error bound $\epsilon > 0$ with $\mathbf{O}(\log m)$ quantum states and 12 classical states.

⋆ This work is supported in part by the National Natural Science Foundation of China (Nos. 60873055, 61073054,61100001), the Natural Science Foundation of Guangdong Province of China (No. 10251027501000004), the Fundamental Research Funds for the Central Universities (Nos. 10lgzd12,11lgpy36), the Research Foundation for the Doctoral Program of Higher School of Ministry of Education (Nos. 20100171110042, 20100171120051) of China, the Czech Ministry of Education (No. MSM0021622419), the China Postdoctoral Science Foundation project (Nos. 20090460808, 201003375), and the project of SQIG at IT, funded by FCT and EU FEDER projects projects QSec PTDC/EIA/67661/2006, AMDSC UTAustin/MAT/0057/2008, NoE Euro-NF, and IT Project QuantTel, FCT project PTDC/EEA-TEL/103402/2008 Quant-PrivTel.
⋆⋆ Corresponding author.

H. Bordihn, M. Kutrib, and B. Truthe (Eds.): Dassow Festschrift 2012, LNCS 7300, pp. 273–290, 2012.
© Springer-Verlag Berlin Heidelberg 2012

1   Introduction

An important way to get a deeper insight into the power of various quantum resources and features for information processing is to explore power of various quantum variations of the basic models of classical automata. Of a special interest and importance is to do that for various quantum variations of classical finite automata because quantum resources are not cheap and quantum operations are not easy to implement. Attempts to find out how much one can do with very little of quantum resources and consequently with the most simple quantum variations of classical finite automata are therefore of particular interest. This paper is an attempt to contribute to such line of research.

There are two basic approaches how to introduce quantum features to classical models of finite automata. The first one is to consider quantum variants of the classical *one-way (deterministic) finite automata* (1FA or 1DFA) and the second one is to consider quantum variants of the classical *two-way finite automata* (2FA or 2DFA). Already the very first attempts to introduce such models, by Moore and Crutchfields [20] and Kondacs and Watrous [12] demonstrated that in spite of the fact that in the classical case, 1FA and 2FA have the same recognition power, this is not so for their quantum variations. Moreover, already the first important model of *two-way quantum finite automata* (2QFA), namely that introduced by Kondacs and Watrous, demonstrated that very natural quantum variants of 2FA are much too powerful - they can recognize even some non-context free languages and are actually not really finite in a strong sense. It started to be therefore of interest to introduce and explore some "less quantum" variations of 2FA and their power [1–6, 8, 14–19, 22, 26–30].

A very natural "hybrid" quantum variations of 2FA, namely, *two-way quantum automata with quantum and classical states* (2QCFA) were introduced by Ambainis and Watrous [3]. Using this model they were able to show in an elegant way that an addition of a single qubit to a classical model can enormously increase power of automata. A 2QCFA is essentially a classical 2FA augmented with a quantum memory of constant size (for states in a fixed Hilbert space) that does not depend on the size of the (classical) input. In spite of such a restriction, 2QCFA have been shown to be more powerful than *two-way probabilistic finite automata* (2PFA) [3].

Because of the simplicity, elegance and interesting properties of the 2QCFA model, as well as its natural character, it seems to be both useful and interesting to explore what such a new "hybrid" approach will provide in case of one-way finite automata and this we will do in this paper by introducing and exploring 1QCFA.

In the first part of the paper, 1QCFA are introduced formally and it is shown that they can be used to simulate a variety of other models of finite automata. Namely, 1DFA, coin-tossing 1PFA, measure-once 1QFA (MO-1QFA) [12], measure-many 1QFA (MM-1QFA) [12] and *one-way quantum finite automata with control language* (1QFACL) [6]. Of a special interest is the way how 1QCFA can simulate 1QFACL - an interesting model the behavior of which is, however, quite special. Our simulation of 1QFACL by 1QCFA allows to see behavior of 1QFACL in a quite transparent way. We also explore several closure properties of the family of languages accepted by 1QCFA. Finally, we derive a result concerning the state complexity of 1QCFA that also demonstrates a merit of this new model. Namely we show that for any prime m and any $\varepsilon_1 > 0$, there exists a language L_m than cannot be recognized by any MM-1QFA with bounded error $\frac{7}{9} + \epsilon_1$, and any 1PFA recognizing it has at last m states, but L_m can be recognized by a 1QCFA for any error bound $\epsilon > 0$ with $\mathbf{O}(\log m)$ quantum states and 12 classical states.

The rest of the paper is organized as follows. Definitions of all automata models explored in the paper are presented in Section 2. In Section 3 we show how several other models of finite automata can be simulated by 1QCFA. We also explore several closure properties of the family of languages accepted by 1QCFA in Section 4. In Section 5 the above mentioned succinctness result is proved and the last section contains just few concluding remarks.

2   Basic Models of Classical and Quantum Finite Automata

In the first part of this section we formally introduce those basic models of finite automata we will refer to in the rest of the paper and in the second part of this section, we formally introduce as a new model 1QCFA. Concerning the basics of quantum computation we refer the reader to [9, 21] and concerning the basic properties of the automata models introduced in the following we refer the reader to [9–11, 23, 25].

2.1   Basic Models of Classical and Quantum Finite Automata

In this subsection, we recall the definitions of DFA, 1PFA, MO-1QFA, MM-1QFA and 1QFACL.

Definition 1. *A deterministic finite automaton (DFA) $\mathcal{A}$ is specified by a 5-tuple*

$$\mathcal{A} = (S, \Sigma, \delta, s_0, S_{acc}), \tag{1}$$

where:

1. *S is a finite set of classical states;*
2. *Σ is a finite set of input symbols;*
3. *$s_0 \in S$ is the initial state of the machine;*
4. *$S_{acc} \subset S$ is the set of accepting states;*
5. *δ is the transition function:*

$$\delta : S \times \Sigma \to S. \tag{2}$$

Let $w = \sigma_1 \sigma_2 \cdots \sigma_n$ be a string over the alphabet Σ. The automaton $\mathcal{A}$ accepts the string w if a sequence of states, $r_0, r_1, \cdots, r_n$, exists in S with the following conditions:

1. $r_0 = s_0$;
2. $r_{i+1} = \delta(r_i, \sigma_{i+1})$, for $i = 0, \cdots, n-1$;
3. $r_n \in S_{acc}$.

DFA recognize exactly the set of *regular languages* (RL).

Definition 2. *A one-way probabilistic finite automata (1PFA) $\mathcal{A}$ is specified by a 5-tuple*

$$\mathcal{A} = (S, \Sigma, \delta, s_1, S_{acc}), \tag{3}$$

where:

1. $S = \{s_1, s_2, \cdots, s_n\}$ *is a finite set of classical states;*
2. Σ *is a finite set of input symbols; Σ is then extended to the tape symbol set $\Gamma = \Sigma \cup \{ \, \xi, \$ \}$, where $\xi \notin \Sigma$ is called the left end-marker and $\$ \notin \Sigma$ is called the right end-marker;*
3. $s_1 \in S$ *is the initial state;*
4. $S_{acc} \subset S$ *is the set of accepting states;*
5. δ *is the transition function:*

$$\delta : S \times \Gamma \times S \to [0,1]. \tag{4}$$

For example, $\delta(s, \sigma, t)$ means that if $\mathcal{A}$ is in the state s with the tape head scanning the symbol σ, then the automaton enters the state t with probability $\delta(s, \sigma, t)$.

Note: A 1 PFA is a coin-tossing 1PFA if the range of its transition function δ is $\{0, 1/2, 1\}$. For any $s \in S$ and any $\sigma \in \Gamma$, $\delta(s, \sigma, t)$ is a so-called coin-tossing distribution[1] on S such that $\sum_{t \in S} \delta(s, \sigma, t) = 1$. It is not hard to see that rational transition probabilities can be obtained by repeating coin-flip.

For an input string $\omega = \sigma_1 \ldots \sigma_l$, the probability distribution on the states of $\mathcal{A}$ during its acceptance process can be traced using n-dimensional vectors. It is assumed that $\mathcal{A}$ starts to process the input word written on the input tape as $w = \xi \omega \$$ and let $v_0 = (1, 0, \ldots, 0)^T_{n \times 1}$ denote the initial probability distribution on states. If, during the acceptance process, the current probability distribution vector is v and a tape symbol σ is read, then the new state probability distribution vector will be, after the automaton step, $u = A_\sigma v$, where A_σ is such a matrix that $A_\sigma(i, j) = \delta(s_j, \sigma, s_i)$. We then use $v_{|w|} = A_\$ A_{\sigma_l} \cdots A_{\sigma_1} A_{\xi} v_0$ to denote the final probability distribution on states in case of the input ω. The accepting probability of $\mathcal{A}$ with input ω is then

$$Pr[\mathcal{A} \text{ accepts } \omega] = \sum_{s_i \in S_{acc}} v_{|w|}(i), \tag{5}$$

where $v_{|w|}(i)$ denotes the ith entry of $v_{|w|}$.

---

[1] A coin-tossing distribution on a finite set Q is a mapping ϕ from Q to $\{0, 1/2, 1\}$ such that $\sum_{q \in Q} \phi(q) = 1$.

Definition 3. *A measurement-once one-way quantum automaton (MO-1QFA) $\mathcal{A}$ is specified by a 5-tuple*

$$\mathcal{A} = (Q, \Sigma, \Theta, |q_0\rangle, Q_{acc}), \tag{6}$$

where:

1. *Q is a finite set of quantum orthogonal states;*
2. *Σ is a finite set of input symbols; Σ is then extended to the tape symbol set $\Gamma = \Sigma \cup \{\phi, \$\}$, where $\phi \notin \Sigma$ is called the left end-marker and $\$ \notin \Sigma$ is called the right end-marker;*
3. *$|q_0\rangle \in Q$ is the initial quantum state;*
4. *$Q_{acc} \subset Q$ is the set of accepting quantum states;*
5. *For each $\sigma \in \Gamma$, a unitary transformation Θ_σ is defined on the Hilbert space spanned by the states from Q.*

We describe the acceptance process of $\mathcal{A}$ for any given input string $\omega = \sigma_1 \cdots \sigma_l$ as follows. The automaton $\mathcal{A}$ states with the initial state $|q_0\rangle$, reading the left-marker ϕ. Afterwards, the unitary transformation Θ_ϕ is applied on $|q_0\rangle$. After that, $\Theta_\phi|q_0\rangle$ becomes the current state and the automaton reads σ_1. The process continues until $\mathcal{A}$ reads $\$$ and ends in the state $|\psi_\omega\rangle = \Theta_\$\Theta_{\sigma_l} \cdots \Theta_{\sigma_1}\Theta_\phi|q_0\rangle$. Finally, a measurement is performed on $|\psi_\omega\rangle$ and the accepting probability of $\mathcal{A}$ on the input ω is equal to

$$Pr[\mathcal{A} \ accepts \ \omega] = \langle\psi_\omega|P_a|\psi_\omega\rangle = ||P_a|\psi_\omega\rangle||^2, \tag{7}$$

where $P_a = \sum_{q \in Q_{acc}} |q\rangle\langle q|$ is the projection onto the subspace spanned by $\{|q\rangle : |q\rangle \in Q_{acc}\}$.

Definition 4. *A measurement-many one-way quantum automaton (MM-1QFA) $\mathcal{A}$ is specified by a 6-tuple*

$$\mathcal{A} = (Q, \Sigma, \Theta, |q_0\rangle, Q_{acc}, Q_{rej}), \tag{8}$$

where Q, Σ, Θ, $|q_0\rangle$, Q_{acc}, and the tape symbol set Γ are the same as those defined above in an MO-1QFA. $Q_{rej} \subset Q$ is the set of rejecting states.

For any given input string $\omega = \sigma_1 \cdots \sigma_l$, the acceptance process is similar to that of MO-1QFA except that after every transition, MM-1QFA $\mathcal{A}$ measures its state with respect to the three subspaces that are spanned by the three subsets Q_{acc}, Q_{rej} and Q_{non}, respectively, where $Q_{non} = Q \setminus (Q_{acc} \cup Q_{rej})$. In other words, the projective measurement consists of $\{P_a, P_r, P_n\}$, where $P_a = \sum_{q \in Q_{acc}} |q\rangle\langle q|$, $P_r = \sum_{q \in Q_{rej}} |q\rangle\langle q|$ and $P_n = \sum_{q \in Q_{non}} |q\rangle\langle q|$. The accepting and rejecting probability are given as follows (for convenience, we denote $\sigma_0 = \phi$ and $\sigma_{l+1} = \$$):

$$Pr[\mathcal{A} \ accepts \ \omega] = \sum_{k=0}^{l+1} ||P_a\Theta_{\sigma_k} \prod_{i=0}^{k-1}(P_n\Theta_{\sigma_i})|q_0\rangle||^2, \tag{9}$$

$$Pr[\mathcal{A} \; reject \; \omega] = \sum_{k=0}^{l+1} ||P_r \Theta_{\sigma_k} \prod_{i=0}^{k-1} (P_n \Theta_{\sigma_i}) |q_0\rangle||^2. \tag{10}$$

An important convention: In this paper we define $\prod_{i=1}^{n} A_i = A_n A_{n-1} \cdots A_1$, instead of the usual one $A_1 A_2 \cdots A_n$.

Definition 5. *A one-way quantum finite automata with control language (1QFACL) $\mathcal{A}$ is specified by as a 6-tuple*

$$\mathcal{A} = (Q, \Sigma, \Theta, |q_0\rangle, \mathcal{O}, \mathcal{L}), \tag{11}$$

where:

1. *$Q, \Sigma, \Theta, |q_0\rangle$ and the tape symbol set $\Gamma = \Sigma \cup \{ \cent, \$ \}$ are the same as those defined above in an MO-1QFA;*
2. *$\mathcal{O}$ is an observable with the set of possible eigenvalues $\mathcal{C} = \{c_1, \cdots, c_s\}$ and the projector set $\{P(c_i) : i = 1, \cdots, s\}$ where $P(c_i)$ denotes the projector onto the eigenspace corresponding to c_i;*
3. *$\mathcal{L} \subset \mathcal{C}^*$ is a regular language (called here a control language).*

The input word $\omega = \sigma_1 \cdots \sigma_l$ to 1QFACL $\mathcal{A}$ is in the form: $w = \cent w \$$ (for convenience, we denote $\sigma_0 = \cent$ and $\sigma_{l+1} = \$$). Now, we define the behavior of $\mathcal{A}$ on the word w. The computation starts in the state $|q_0\rangle$, and then the transformations associated with symbols in the word w are applied in succession. The transformation associated with any symbol $\sigma \in \Gamma$ consists of two steps:

1. Firstly, Θ_σ is applied to the current state $|\phi\rangle$ of $\mathcal{A}$, yielding the new state $|\phi'\rangle = \Theta_\sigma |\phi\rangle$.
2. Secondly, the observable $\mathcal{O}$ is measured on $|\phi'\rangle$. According to quantum mechanics principle, this measurement yields result c_k with probability $p_k = ||P(c_k)|\phi'\rangle||^2$, and the state of $\mathcal{A}$ collapses to $P(c_k)|\phi'\rangle/\sqrt{p_k}$.

Thus, the computation on the word w leads to a string $y_0 y_1 \ldots y_{l+1} \in \mathcal{C}^*$ with probability $p(y_0 y_1 \ldots y_{l+1} | \sigma_0 \sigma_1 \ldots \sigma_{l+1})$ given by

$$p(y_0 y_1 \ldots y_{l+1} | \sigma_0 \sigma_1 \ldots \sigma_{l+1}) = || \prod_{i=0}^{l+1} (P(y_i) \Theta_{\sigma_i}) |q_0\rangle||^2. \tag{12}$$

A computation leading to a word $y \in \mathcal{C}^*$ is said to be accepted if $y \in \mathcal{L}$. Otherwise, it is rejected. Hence, the accepting probability of 1QFACL $\mathcal{A}$ is defined as:

$$Pr[\mathcal{A} \; accepts \; \omega] = \sum_{y_0 y_1 \ldots y_{l+1} \in \mathcal{L}} p(y_0 y_1 \ldots y_{l+1} | \sigma_0 \sigma_1 \ldots \sigma_{l+1}) \tag{13}$$

2.2   Definition of 1QCFA

In this subsection we introduce 1QCFA and its acceptance process formally and in details.

2QCFA were first introduced by Ambainis and Watrous [3], and then studied by Qiu, Yakaryilmaz and etc. [24, 28, 32–34]. 1QCFA are the one-way version of 2QCFA. Informally, we describe a 1QCFA as a DFA which has access to a quantum memory of a constant size (dimension), upon which it performs quantum transformations and measurements. Given a finite set of quantum states Q, we denote by $\mathcal{H}(Q)$ the Hilbert space spanned by Q. Let $\mathcal{U}(\mathcal{H}(Q))$ and $\mathcal{O}(\mathcal{H}(Q))$ denote the sets of unitary operators and projective measurements over $\mathcal{H}(Q)$, respectively.

Definition 6. *A one-way finite automata with quantum and classical states (1QCFA) $\mathcal{A}$ is specified by a 10-tuple*

$$\mathcal{A} = (Q, S, \Sigma, \Theta, \Delta, \delta, |q_0\rangle, s_0, S_{acc}, S_{rej}) \tag{14}$$

where:

1. *Q is a finite set of quantum states;*
2. *S, Σ and the tape symbol set $\Gamma = \Sigma \cup \{ \phi, \$ \}$ are the same as those defined above in a 1PFA;*
3. *$|q_0\rangle \in Q$ is the initial quantum state;*
4. *$s_0 \in S$ is the initial classical state;*
5. *$S_{acc} \subset S$ and $S_{rej} \subset S$ are the sets of classical accepting and rejecting states, respectively;*
6. *Θ is the mapping:*
   $$\Theta : S \times \Gamma \to \mathcal{U}(\mathcal{H}(Q)), \tag{15}$$
   assigning to each pair (s, γ) a unitary transformation;
7. *Δ is the mapping:*
   $$\Delta : S \times \Gamma \to \mathcal{O}(\mathcal{H}(Q)), \tag{16}$$
   where each $\Delta(s, \gamma)$ corresponds to a projective measurement (a projective measurement will be taken each time a unitary transformation is applied; if we do not need a measurement, we denote that $\Delta(s, \gamma) = I$, and we assume the result of the measurement to be ε with certainty);
8. *δ is a special transition function of classical states. Let the results set of the measurement be $\mathcal{C} = \{c_1, c_2, \ldots, c_s\}$, then*
   $$\delta : S \times \Gamma \times \mathcal{C} \to S, \tag{17}$$
   where $\delta(s, \gamma)(c_i) = s'$ means that if a tape symbol $\gamma \in \Gamma$ is being scanned and the projective measurement result is c_i, then the state s is changed to s'.

Given an input $\omega = \sigma_1 \cdots \sigma_l$, the word on the tape will be $w = \phi \, \omega \$ $ (for convenience, we denote $\sigma_0 = \phi$ and $\sigma_{l+1} = \$$). Now, we define the behavior of 1QCFA $\mathcal{A}$ on the word w. The computation starts in the classical state s_0 and the quantum state $|q_0\rangle$, then the transformations associated with symbols in the word $\sigma_0 \sigma_1 \cdots, \sigma_{l+1}$ are applied in succession. The transformation associated with a state $s \in S$ and a symbol $\sigma \in \Gamma$ consists of three steps:

1. Firstly, $\Theta(s,\sigma)$ is applied to the current quantum state $|\phi\rangle$, yielding the new state $|\phi'\rangle = \Theta(s,\sigma)|\phi\rangle$.
2. Secondly, the observable $\Delta(s,\sigma) = \mathcal{O}$ is measured on $|\phi'\rangle$. The set of possible results is $\mathcal{C} = \{c_1, \cdots, c_s\}$. According to such a quantum mechanics principle, such a measurement yields the classical outcome c_k with probability $p_k = ||P(c_k)|\phi'\rangle||^2$, and the quantum state of $\mathcal{A}$ collapses to $P(c_k)|\phi'\rangle/\sqrt{p_k}$.
3. Thirdly, the current classical state s will be changed to $\delta(s,\sigma)(c_k) = s'$.

An input word ω is assumed to be accepted (rejected) if and only if the classical state after scanning σ_{l+1} is an accepting (rejecting) state. We assume that δ is well defined so that 1QCFA $\mathcal{A}$ always accepts or rejects at the end of the computation.

Let $L \subset \Sigma^*$ and $0 \le \epsilon < 1/2$, then 1QCFA $\mathcal{A}$ recognizes L with bounded error ϵ if

1. For any $\omega \in L$, $Pr[\mathcal{A} \text{ accepts } \omega] \ge 1 - \epsilon$, and
2. For any $\omega \notin L$, $Pr[\mathcal{A} \text{ rejects } \omega] \ge 1 - \epsilon$.

3   Simulation of Other Models by 1QCFA

In this section, we prove that the following automata models can be simulated by 1QCFA: DFA, coin-tossing 1PFA, MO-1QFA, MM-1QFA and 1QFACL.

Theorem 1. *Any n states DFA $\mathcal{A} = (S, \Sigma, \delta, s_0, S_{acc})$ can be simulated by a 1QCFA $\mathcal{A}' = (Q', S', \Sigma', \Theta', \Delta', \delta', |q_0\rangle', s_0', S_{acc}', S_{rej}')$ with 1 quantum state and $n + 1$ classical states.*

Proof. Actually, if we do not use the quantum component of 1QCFA, the automaton is reduced to a DFA. Let $Q' = \{|q_0\rangle'\}$, $S' = S \cup \{s_r\}$, $\Sigma' = \Sigma$, $s_0' = s_0$, $S_{acc}' = S_{acc}$ and $S_{rej}' = \{s_r\}$. For any $s \in S$ and any $\sigma \in \Sigma$, let $\Theta(s,\sigma) = I$, $\Delta'(s,\sigma) = I$, and the classical transition function δ' is defined as follows:

$$\delta'(s,\sigma)(c) = \begin{cases} s, & \sigma = \mathcal{c}; \\ \delta(s,\sigma), & \sigma \in \Sigma, \\ s, & \sigma = \$, s \in S_{acc}'; \\ s_r, & \sigma = \$, s \notin S_{acc}'. \end{cases} \tag{18}$$

where c is the measurement result.

Theorem 2. *Any n states coin-tossing 1PFA $\mathcal{A}^1 = (S^1, \Sigma^1, \delta^1, s_1^1, S_{acc}^1)$ can be simulated by a 1QCFA $\mathcal{A}^2 = (Q^2, S^2, \Sigma^2, \Theta^2, \Delta^2, \delta^2, |q_0\rangle^2, s_0^2, S_{acc}^2, S_{rej}^2)$ with 2 quantum states and $n + 1$ classical states.*

Proof. A coin-tossing 1PFA is essentially a DFA augmented with a fair coin-flip component. In every transition, coin-tossing 1PFA can use a fair coin-flip or not freely. Using the quantum component, a 1QCFA can simulate the fair coin-flip perfectly.

Lemma 1. *A fair coin-flip can be simulate by 1QCFA $\mathcal{A}$ with two quantum states, a unitary operation and a projective measurement.*

Proof. The automaton $\mathcal{A}$ simulates a coin-flip according to the following transition functions, with $|p_0\rangle$ as the starting quantum state. We use two orthogonal basis states $|p_0\rangle$ and $|p_1\rangle$. Let a projective measurement $M = \{P_0, P_1\}$ be defined by

$$P_0 = |p_0\rangle\langle p_0|, P_1 = |p_1\rangle\langle p_1|. \tag{19}$$

The results 0 and 1 represent the results of coin-flip "head" and "tail", respectively. The corresponding unitary operation will be

$$U = \begin{pmatrix} \frac{1}{\sqrt{2}} & \frac{1}{\sqrt{2}} \\ \frac{1}{\sqrt{2}} & -\frac{1}{\sqrt{2}} \end{pmatrix}. \tag{20}$$

This operator changes the state $|p_0\rangle$ or $|p_1\rangle$ to a superposition state $|\psi\rangle$ or $|\phi\rangle$, respectively, as follows:

$$|\psi\rangle = \frac{1}{\sqrt{2}}(|p_0\rangle + |p_1\rangle), \quad |\phi\rangle = \frac{1}{\sqrt{2}}(|p_0\rangle - |p_1\rangle). \tag{21}$$

When measuring $|\psi\rangle$ or $|\phi\rangle$ with M, we will get the result 0 or 1 with probability $\frac{1}{2}$, respectively. This is similar to a coin-flip process. If the result is 0, we simulate "head" result of the coin-flip; if the result is 1, we simulate "tail" result of the coin-flip. So the Lemma is proved.

If the current state of coin-tossing 1PFA $\mathcal{A}^1$ is s and the scanning symbol is $\sigma \in \Sigma$, $\mathcal{A}^1$ makes a coin-flip. The current state of $\mathcal{A}^1$ will change to t_1 or t_2, in both cases with probability $\frac{1}{2}$. We use a 1QCFA $\mathcal{A}^2$ to simulate this step as follows:

1. Use the quantum component of 1QCFA $\mathcal{A}^2$ to simulate a fair coin-flip. We assume the outcome to be 0 or 1.
2. We define $\delta^2(s, \sigma)(0) = t_1$ and $\delta^2(s, \sigma)(1) = t_2$.

The other parts of the simulation are similar to the one described in the proof of Theorem 1.

Theorem 3. *Any n quantum states MO-1QFA $\mathcal{A}^1 = (Q^1, \Sigma^1, \Theta^1, |q_0\rangle^1, Q^1_{acc})$ can be simulated by a 1QCFA $\mathcal{A}^2 = (Q^2, S^2, \Sigma^2, \Theta^2, \Delta^2, \delta^2, |q_0\rangle^2, s_0^2, S^2_{acc}, S^2_{rej})$ with n quantum states and 3 classical states.*

Proof. We use the quantum component of 1QCFA to simulate the evolution of quantum states of MO-1QFA and use the classical states of 1QCFA to calculate the accepting probability. Let $Q^2 = Q^1$, $S^2 = \{s_0^2, s_a^2, s_r^2\}$, $\Sigma^2 = \Sigma^1$, $|q_0\rangle^2 = |q_0\rangle^1$, $S^2_{acc} = \{s_a^2\}$ and $S^2_{rej} = \{s_r^2\}$. For any current classical state s and scanning symbol σ, the quantum transition function is defined to be

$$\Theta^2(s, \sigma) = \Theta^1(\sigma). \tag{22}$$

The measurement function is defined to be

$$\Delta^2(s, \sigma) = \begin{cases} I, & \sigma \neq \$; \\ \{P_a, P_r\}, & \sigma = \$. \end{cases} \tag{23}$$

where $P_a = \sum_{q \in Q_{acc}} |q\rangle\langle q|$, $P_r = I - P_a$. If we assume the outcome to be c_a or c_r, then the classical transition function will be defined to be

$$\delta^2(s, \sigma)(c) = \begin{cases} s, & \sigma \neq \$; \\ s_a^2, & \sigma = \$, c = c_a; \\ s_r^2, & \sigma = \$, c = c_r. \end{cases} \tag{24}$$

Theorem 4. *Any n quantum states MM-1QFA $\mathcal{A}^1 = (Q^1, \Sigma^1, \Theta^1, |q_0\rangle^1, Q_{acc}^1, Q_{rej}^1)$ can be simulated by a 1QCFA $\mathcal{A}^2 = (Q^2, S^2, \Sigma^2, \Theta^2, \Delta^2, \delta^2, |q_0\rangle^2, s_0^2, S_{acc}^2, S_{rej}^2)$ with n quantum states and 3 classical states.*

Proof. We use the quantum component of 1QCFA to simulate both the evolution of quantum states of MM-1QFA and its projective measurements. We use the classical states of 1QCFA to calculate the accepting and rejecting probability. Let $Q^2 = Q^1$, $S^2 = \{s_0^2, s_a^2, s_r^2\}$, $\Sigma^2 = \Sigma^1$, $|q_0\rangle^2 = |q_0\rangle^1$, $S_{acc}^2 = \{s_a^2\}$ and $S_{rej}^2 = \{s_r^2\}$. For any current classical state s and any scanning symbol σ, the quantum transition function is defined to be

$$\Theta^2(s, \sigma) = \Theta^1(\sigma). \tag{25}$$

The measurement function is defined to be

$$\Delta^2(s, \sigma) = \{P_a, P_r, P_n\}, \tag{26}$$

where $P_a = \sum_{q \in Q_{acc}} |q\rangle\langle q|$, $P_r = \sum_{q \in Q_{rej}} |q\rangle\langle q|$ and $P_n = \sum_{q \in Q_{non}} |q\rangle\langle q|$. If we assume the classical outcomes to be c_a, c_r or c_n, then the classical transition function will be defined to be

$$\delta^2(s, \sigma)(c) = \begin{cases} s_a^2, & s = s_a^2; \\ s_r^2, & s = s_r^2; \\ s_a^2, & s = s_0^2, c = c_a; \\ s_r^2, & s = s_0^2, c = c_r; \\ s_0^2, & s = s_0^2, c = c_n, \sigma \neq \$; \\ s_r^2, & s = s_0^2, c = c_n, \sigma = \$. \end{cases} \tag{27}$$

Although 1QFACL can accept all regular languages, their behavior seems to be rather complicated. We prove that any 1QFACL can be simulated by a 1QCFA with an easy to understand behavior.

Theorem 5. *Any n quantum states 1QFACL $\mathcal{A}^1 = (Q^1, \Sigma^1, \Theta^1, |q_0\rangle^1, \mathcal{O}^1, \mathcal{L}^1)$, whose control language $\mathcal{L}^1$ can be recognized by an m states DFA $\mathcal{A} = (S, \Sigma, \delta, s_0, S_{acc})$, can be simulated by a 1QCFA $\mathcal{A}^2 = (Q^2, S^2, \Sigma^2, \Theta^2, \Delta^2, \delta^2, |q_0\rangle^2, s_0^2, S_{acc}^2, S_{rej}^2)$ with n quantum states and $m + 1$ classical states.*

Proof. We use the quantum component of 1QCFA to simulate the evolution of quantum states of 1QFACL and also its projective measurements. We use the classical states of 1QCFA to simulate DFA $\mathcal{L}^1$. Let $Q^2 = Q^1$, $S^2 = S \cup \{s_r\}$, $\Sigma^2 = \Sigma^1$, $s_0^2 = s_0$, $|q_0\rangle^2 = |q_0\rangle^1$, $S_{acc}^2 = S_{acc}$ and $S_{rej}^2 = \{s_r\}$. For any current classical state s and any scanning symbol σ, the quantum transition function will be defined to be

$$\Theta^2(s, \sigma) = \Theta^1(\sigma). \qquad (28)$$

The measurement function is defined to be

$$\Delta^2(s, \sigma) = \{P(c_i) : i = 1, \cdots, t\}, \qquad (29)$$

where $P(c_i)$ denotes the projector onto the eigenspace corresponding to c_i. We assume that the set of possible classical outcomes is $\mathcal{C} = \{c_1, \cdots, c_t\}$, where $\mathcal{C} = \Sigma$, then the classical transition function will be defined to be

$$\delta^2(s, \sigma)(c) = \begin{cases} \delta(s, c), & \sigma \neq \$; \\ \delta(s, c), & \sigma = \$, \delta(s, c) \in S_{acc}; \\ s_r, & \sigma = \$, \delta(s, c) \notin S_{acc}. \end{cases} \qquad (30)$$

4   Closure Properties of Languages Accepted by 1QCFA

For convenience, we denote by 1QCFA(ϵ) the classes of languages recognized by 1QCFA with bounded error ϵ. Moreover, let $QS(\mathcal{A})$ and $CS(\mathcal{A})$ denote the numbers of quantum states and classical states of a 1QCFA $\mathcal{A}$. We start to consider the operation of complement.

Theorem 6. *If $L \in 1QCFA(\epsilon)$, then also $L^c \in 1QCFA(\epsilon)$, where L^c is the complement of L.*

Proof. Let a 1QCFA(ϵ) $\mathcal{A} = (Q, S, \Sigma, \Theta, \Delta, \delta, |q_0\rangle, s_0, S_{acc}, S_{rej})$ accept L with a bounded error ϵ. We can construct the 1QCFA $\mathcal{A}^c$ only by exchanging the classical accepting and rejecting states in $\mathcal{A}$. That is, $\mathcal{A}^c = (Q, S, \Sigma, \Theta, \Delta, \delta, |q_0\rangle, s_0, S_{acc}^c, S_{rej}^c)$, where $S_{acc}^c = S_{rej}$, $S_{rej}^c = S_{acc}$ and the other components remain the same as those defined in $\mathcal{A}$. Afterwards we have:

1. If $\omega \in L^c$, then $\omega \notin L$. Indeed, for an input ω, $\mathcal{A}$ will enter a rejecting state with probability at least $1 - \epsilon$ at the end of the computation. With the same input ω, $\mathcal{A}^c$ will enter an accepting state with probability at least $1 - \epsilon$ at the end of the computation. Hence, $\mathcal{A}^c$ accepts ω with the probability at least $1 - \epsilon$;
2. The case $\omega \notin L^c$ is treated in a symmetric way.

Remark 1. According to the construction given above, if $QS(\mathcal{A}) = n$, $CS(\mathcal{A}) = m$, then $QS(\mathcal{A}^c) = n$, $CS(\mathcal{A}^c) = m$.

Theorem 7. *If $L_1 \in 1QCFA(\epsilon_1)$ and $L_2 \in 1QCFA(\epsilon_2)$, then $L_1 \cap L_2 \in 1QCFA(\epsilon)$, where $\epsilon = \epsilon_1 + \epsilon_2 - \epsilon_1\epsilon_2$.*

Proof. Let $\mathcal{A}^i = (Q^i, S^i, \Sigma^i, \Theta^i, \Delta^i, \delta^i, |q_0\rangle^i, s_0^i, S_{acc}^i, S_{rej}^i)$ be 1QCFA to recognize L_i with bounded error ϵ_i $(i = 1, 2)$. We construct a 1QCFA $\mathcal{A} = (Q, S, \Sigma, \Theta, \Delta, \delta, |q_0\rangle, s_0, S_{acc}, S_{rej})$ where:

1. $Q = Q^1 \otimes Q^2$,
2. $S = S^1 \times S^2$,
3. $\Sigma = \Sigma^1 \cap \Sigma^2$,
4. $s_0 = \langle s_0^1, s_0^2 \rangle$,
5. $|q_0\rangle = |q_0\rangle^1 \otimes |q_0\rangle^2$,
6. $S_{acc} = S_{acc}^1 \times S_{acc}^2$,
7. $S_{rej} = (S_{acc}^1 \times S_{rej}^2) \cup (S_{rej}^1 \times S_{acc}^2) \cup (S_{rej}^1 \times S_{rej}^2)$
8. For any classical state $s = \langle s^1, s^2 \rangle \in S$ and any $\sigma \in \Sigma$, the quantum transition function of $\mathcal{A}$ is defined to be

$$\Theta(s, \sigma) = \Theta(\langle s^1, s^2 \rangle, \sigma) = \Theta^1(s^1, \sigma) \otimes \Theta^2(s^2, \sigma). \tag{31}$$

9. For any classical state $s = \langle s^1, s^2 \rangle \in S$ and any $\sigma \in \Sigma$, the measurement function of $\mathcal{A}$ is defined to be

$$\Delta(s, \sigma) = \Delta(\langle s^1, s^2 \rangle, \sigma) = \Delta^1(s^1, \sigma) \otimes \Delta^2(s^2, \sigma). \tag{32}$$

As classical measurements outcomes are then tuples $c_{ij} = \langle c_i, c_j \rangle$.

10. For any classical state $s = \langle s^1, s^2 \rangle \in S$ and any $\sigma \in \Sigma$, the classical transition function of $\mathcal{A}$ is defined to be

$$\delta(s, \sigma)(c_{ij}) = \delta(\langle s^1, s^2 \rangle, \sigma)(\langle c_i, c_j \rangle) = \langle \delta^1(s^1, \sigma)(c_i), \delta^2(s^2, \sigma)(c_j) \rangle. \tag{33}$$

In terms of the 1QCFA $\mathcal{A}$ constructed above, for any $\omega \in \Sigma^*$, we have:

1. If $\omega \in L_1 \cap L_2$, then $\mathcal{A}$ will enter a state $\langle t_1, t_2 \rangle \in S_{acc}^1 \times S_{acc}^2$ at the end of the computation with probability at least $(1 - \epsilon_1)(1 - \epsilon_2)$. $\mathcal{A}$ accepts ω with the probability at least $(1 - \epsilon_1)(1 - \epsilon_2) = 1 - (\epsilon_1 + \epsilon_2 - \epsilon_1\epsilon_2)$.
2. If $\omega \in L_1$ but $\omega \notin L_2$, then $\mathcal{A}$ will enter a state $\langle t_1, t_2 \rangle \in S_{acc}^1 \times S_{rej}^2$ at the end of the computation with probability at least $(1 - \epsilon_1)(1 - \epsilon_2)$. $\mathcal{A}$ rejects ω with the probability at least $1 - (\epsilon_1 + \epsilon_2 - \epsilon_1\epsilon_2)$.
3. The case $\omega \notin L_1$ but $\omega \in L_2$ is symmetric to the previous one and therefore the same is the outcome.
4. If $\omega \notin L_1$ and $\omega \notin L_2$, then $\mathcal{A}$ will enter a state $\langle t_1, t_2 \rangle \in S_{rej}^1 \times S_{rej}^2$ at the end of the computation with probability at least $(1 - \epsilon_1)(1 - \epsilon_2)$. $\mathcal{A}$ rejects ω with the probability at least $1 - (\epsilon_1 + \epsilon_2 - \epsilon_1\epsilon_2)$.

So $L_1 \cap L_2 \in 1QCFA(\epsilon)$.

Remark 2. According to the construction given above, let $QS(\mathcal{A}^1) = n_1, CS(\mathcal{A}^1) = m_1, QS(\mathcal{A}^2) = n_2$ and $CS(\mathcal{A}^2) = m_2$, then $QS(\mathcal{A}) = n_1n_2, CS(\mathcal{A}) = m_1m_2$.

A similar outcome holds for the union operation.

Theorem 8. *If $L_1 \in 1QCFA(\epsilon_1)$ and $L_2 \in 1QCFA(\epsilon_2)$, then $L_1 \cup L_2 \in 1QCFA(\epsilon)$, where $\epsilon = \epsilon_1 + \epsilon_2 - \epsilon_1\epsilon_2$.*

Proof. Let $\mathcal{A}^i = (Q^i, S^i, \Sigma^i, \Theta^i, \Delta^i, \delta^i, |q_0\rangle^i, s_0^i, S_{acc}^i, S_{rej}^i)$ be 1QCFA to recognize L_i with bounded error ϵ_i $(i = 1, 2)$. The construction of the 1QCFA $\mathcal{A} = (Q, S, \Sigma, \Theta, \Delta, \delta, |q_0\rangle, s_0, S_{acc}, S_{rej})$ is the same as in the proof of Theorem 7 except for S_{acc} and S_{rej}. We define $S_{acc} = (S_{acc}^1 \times S_{rej}^2) \cup (S_{rej}^1 \times S_{acc}^2) \cup (S_{acc}^1 \times S_{acc}^2)$ and $S_{rej} = S_{rej}^1 \times S_{rej}^2$. The rest of the proof is similar to the proof in Theorem 7.

Remark 3. In the last proof the set of input symbols was defined as $\Sigma = \Sigma^1 \cap \Sigma^2$. Actually, if we take $\Sigma = \Sigma^1 \cup \Sigma^2$, the theorem still holds. In that case, we extend Σ^i to Σ by adding a rejecting classical state s_r^i to $\mathcal{A}^i$. For any classical state $s^i \in S^i$ and $\sigma^i \notin \Sigma^i$, the quantum transition function is defined to be $\Theta^i(s^i, \sigma^i) = I$, the measurement function is defined to be $\Delta^i(s^i, \sigma^i) = I$. We assume the measurement result to be c, then the classical transition function will be defined to be $\delta^i(s^i, \sigma^i)(c) = s_r^i$. For the new adding state s_r^i, we define the transition functions as follow: for any $\sigma \in \Sigma$, $\Theta^i(s_r^i, \sigma) = I$, $\Delta^i(s_r^i, \sigma) = I$, $\delta^i(s_r^i, \sigma)(c) = s_r^i$, where c is the the measurement result.

5   Succinctness Results

State complexity and succinctness results are an important research area of classical automata theory, see [31], with a variety of applications. Once quantum versions of classical automata were introduced and explored, it started to be of large interest to find out through succinctness results a relation between the power of classical and quantum automata model. This has turned out to be an area of surprising outcomes that again indicated that relations between classical and corresponding quantum automata models is intriguing. For example, it has been shown, see [2, 4, 5, 13], that for some languages 1QFA require exponentially less states that classical 1FA, but for some other languages it can be in an opposite way.

Since 1QCFA can simulate both 1FA and 1QFA, and in this way they combine the advantages of both of these models, it is of interest to explore the relation between the state complexity of languages for the case that they are accepted by 1QCFA and MM-1QFA and this we will do in this section.

The main result we obtain when considering languages $L_m = \{a^*b^* \mid |a^*b^*| = km, k = 1, 2, \cdots\}$, where m is a prime. For survey on the famous language $\{a^* \mid |a^*| = km, k = 1, 2, \cdots\}$, the reader may refer to [7].

Obviously, there exist a $2m + 2$ states DFA, depicted in Figure 1 that accepts L_m.

Lemma 2. *DFA $\mathcal{A}$ depicted in Figure 1 is minimal.*

Proof. We show that any two different state s and t are distinguishable (i.e., there exists a string z such that exactly one of the following states $\widehat{\delta}(p, z)^2$ or $\widehat{\delta}(q, z)$ is an accepting state [31]).

---

[2] For any string $x \in \Sigma^*$ and any $\sigma \in \Sigma$, $\widehat{\delta}(s, \sigma x) = \widehat{\delta}(\delta(s, \sigma), x)$; if $|x| = 0$, $\widehat{\delta}(s, x) = s$ [11].

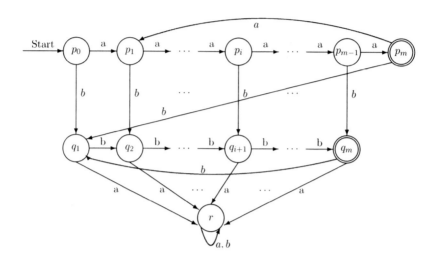

Fig. 1. DFA $\mathcal{A}$ recognizing L_m

1. For $0 \leq i \leq m$, $0 \leq j \leq m$ and $i \neq j$, we have $\widehat{\delta}(p_i, a^{m-i}) = p_m$ and $\widehat{\delta}(p_j, a^{m-i}) = p_k$, where $k \neq m$. Hence, p_i and p_j are distinguishable.
2. For $1 \leq i \leq m$, $1 \leq j \leq m$ and $i \neq j$, we have $\widehat{\delta}(q_i, b^{m-i}) = q_m$ and $\widehat{\delta}(q_j, b^{m-i}) = q_k$, where $k \neq m$. Hence, q_i and q_j are distinguishable.
3. For $0 \leq i \leq m$ and $1 \leq j \leq m$, we have $\widehat{\delta}(p_i, a^{m-i}) = p_m$ and $\widehat{\delta}(q_j, a^{m-i}) = r$. Hence, p_i and q_j are distinguishable.
4. Obviously, the state r is distinguishable from any other state s.

Therefore, the Lemma has been proved.

Lemma 3 ([2, 18]). *For any prime m, any 1PFA recognizing L_m with probability $1/2 + \epsilon$, for a fixed $\epsilon > 0$, has at least m states.*

Remark 4. The proof can be obtained by an easy modification of the proof from the paper [2] where the state complexity of the language $L_p = \{a^i \mid i$ is divisible by $p\}$ is considered.

Lemma 4 ([2]). *(Forbidden construction) Let L be a regular language, and let $\mathcal{A}$ be its minimal DFA. Assume that there is a word w such that $\mathcal{A}$ contains states s, t (a forbidden construction) satisfying:*

1. $s \neq t$,
2. $\widehat{\delta}(s, x) = t$,
3. $\widehat{\delta}(t, x) = t$ *and*
4. t *is neither "all-accepting" state, nor "all-rejecting" state (i.e., there exist strings u and v such that $\widehat{\delta}(t, u)$ is an accepting state and $\widehat{\delta}(t, v)$ is not an accepting state).*

Then L cannot be recognized by an MM-1QFA with bounded error $\frac{7}{9} + \epsilon$ for any fixed $\epsilon > 0$.

Theorem 9. *For any fixed $\epsilon > 0$, L_m cannot be recognized by an MM-1QFA with bounded error $\frac{7}{9} + \epsilon$.*

Proof. According to Lemma 4, we know that L_m cannot be accepted by any MM-1QFA with bounded error $\frac{7}{9} + \epsilon$ since its minimal DFA (see Figure 1) contains the "Forbidden construction" of Lemma 4. For example, we can take $s = p_0$, $t = p_m$, $x = a^m$, then we have $\widehat{\delta}(p_0, a^m) = p_m$, $\widehat{\delta}(p_m, a^m) = p_m$, $\widehat{\delta}(p_m, b^m) = q_m$ and $\widehat{\delta}(p_m, ba) = r$.

Let $L_1 = \{a^*b^*\}$ and $L_2 = \{w \mid w \in \{a,b\}^*, |w| = km, k = 1, 2, \cdots\}$ where m is a prime. So we have $L_m = L_1 \cap L_2$. We will show L_1 and L_2 can be recognized by 1QCFA.

Lemma 5. *The language L_1 can be recognized by a 1QCFA $\mathcal{A}^1$ with certainty with 1 quantum state and 4 classical states.*

Proof. L_1 can be accepted by a DFA $\mathcal{A}$ with 3 classical states (see Figure 2). According to Theorem 1, $\mathcal{A}$ can be simulated by a 1QCFA $\mathcal{A}^1$ with 1 quantum state and 4 classical states.

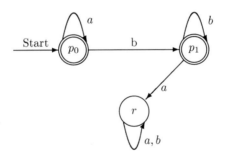

Fig. 2. A DFA recognizing the language L_1

Lemma 6 ([2]). *For any $\epsilon > 0$, there is an MM-1QFA $\mathcal{A}$ with $\mathbf{O}(\log m)$ quantum states recognizing L_2 with a bounded error ϵ.*

Lemma 7. *For any $\epsilon > 0$, there is a 1QCFA $\mathcal{A}^2$ with $\mathbf{O}(\log m)$ quantum states and 3 classical states recognizing L_2 with a bounded error ϵ.*

Proof. According to Lemma 6, there is an MM-1QFA $\mathcal{A}$ with $\mathbf{O}(\log m)$ quantum states recognizing L_2 with bounded error ϵ. According to Theorem 4, an $\mathbf{O}(\log m)$ quantum states MM-1QFA $\mathcal{A}$ can be simulated by a 1QCFA with $\mathbf{O}(\log m)$ quantum states and 3 classical states.

Theorem 10. *For any $\epsilon > 0$, L_m can be recognized by a 1QCFA with $\mathbf{O}(\log m)$ quantum states and 12 classical states with a bounded error ϵ.*

Proof. $L_m = L_1 \cap L_2$. According to Lemma 5, the language L_1 can be recognized by 1QCFA $\mathcal{A}^1$ with 1 quantum state and 4 classical states with certainty (i.e., $\epsilon_1 = 0$). According to Lemma 7, for any $\epsilon > 0$, the language L_2 can be recognized by 1QCFA $\mathcal{A}^2$ with $\mathbf{O}(\log m)$ quantum states and 3 classical states with a bounded error ϵ. According to Theorem 7, 1QCFA is closed under intersection. Hence, there is a 1QCFA $\mathcal{A}$ recognize L_m with a bounded error ϵ. Therefore $QS(\mathcal{A}^1) = 1$, $CS(\mathcal{A}^1) = 4$, $QS(\mathcal{A}^2) = \mathbf{O}(\log m)$ and $CS(\mathcal{A}^2) = 3$, so $QS(\mathcal{A}) = QS(\mathcal{A}^1) \times QS(\mathcal{A}^2) = \mathbf{O}(\log m)$, $CS(\mathcal{A}) = CS(\mathcal{A}^1) \times CS(\mathcal{A}^2) = 12$.

6   Conclusions

2QCFA were introduced by Ambainis and Watrous [3]. In this paper, we investigated the one-way version of 2QCFA, namely 1QCFA. Firstly, we gave a formal definition of 1QCFA. Secondly, we showed that DFA, coin-tossing 1PFA, MO-1QFA, MM-1QFA and 1QFACL can be simulated by 1QCFA. As we know, the behavior of 1QFACL seems to be rather complicated. However, when we used a 1QCFA to simulate a 1QFACL, the behavior of 1QCFA started to be seen as quite natural. Thirdly, we studied closure properties of languages accepted by 1QCFA, and we proved that the family of languages accepted by 1QCFA is closed under intersection, union, and complement. Fourthly, for any fixed $\epsilon_1 > 0$ and any prime m we have showed that the language $L_m = \{a^*b^* \mid |a^*b^*| = km, k = 1, 2, \cdots\}$, cannot be recognized by any MM-1QFA with bounded error $\frac{7}{9} + \epsilon_1$, and any 1PFA recognizing it has at last m states, but L_m can be recognized by a 1QCFA for any error bound $\epsilon > 0$ with $\mathbf{O}(\log m)$ quantum states and 12 classical states. Thus, 1QCFA can make use of merits of both 1FA and 1QFA.

To conclude, we would like to propose some problems for further consideration.

1. How about the state complexity of 1QCFA compared with other 1QFA for recognizing the same languages, such as one-way quantum finite automata together with classical states in [26]?
2. Are 1QCFA closed under catenation and reversal?

Acknowledgment. The authors are thankful to the anonymous referees and editor for their comments and suggestions that greatly help to improve the quality of the manuscript.

References

1. Ambainis, A., Beaudry, M., Golovkins, M., Kikusts, A., Mercer, M., Thérrien, D.: Algebraic results on quantum automata. Theory Comput. Syst. 39, 165–188 (2006)
2. Ambainis, A., Freivalds, R.: One-way quantum finite automata: strengths, weaknesses and generalizations. In: Proceedings of the 39th Annual Symposium on Foundations of Computer Science, pp. 332–341. IEEE Computer Society, Palo Alfo (1998)

3. Ambainis, A., Watrous, J.: Two-way finite automata with quantum and classical states. Theoretical Computer Science 287, 299–311 (2002)

4. Ambainis, A., Nahimovs, N.: Improved constructions of quantum automata. Theoretical Computer Science 410, 1916–1922 (2009)

5. Ambainis, A., Nayak, A., Ta-Shma, A., Vazirani, U.: Dense quantum coding and quantum automata. Journal of the ACM 49(4), 496–511 (2002)

6. Bertoni, A., Mereghetti, C., Palano, B.: Quantum Computing: 1-Way Quantum Automata. In: Ésik, Z., Fülöp, Z. (eds.) DLT 2003. LNCS, vol. 2710, pp. 1–20. Springer, Heidelberg (2003)

7. Bertoni, A., Mereghetti, C., Palano, B.: Small size quantum automata recognizing some regular languages. Theoretical Computer Science 340, 394–407 (2005)

8. Brodsky, A., Pippenger, N.: Characterizations of 1-way quantum finite automata. SIAM Journal on Computing 31, 1456–1478 (2002)

9. Gruska, J.: Quantum Computing. McGraw-Hill, London (1999)

10. Gruska, J.: Descriptional complexity issues in quantum computing. J. Automata, Languages Combin. 5, 191–218 (2000)

11. Hopcroft, J.E., Ullman, J.D.: Introduction to Automata Theory, Languages, and Computation. Addision-Wesley, New York (1979)

12. Kondacs, A., Watrous, J.: On the power of quantum finite state automata. In: Proceedings of the 38th IEEE Annual Symposium on Foundations of Computer Science, pp. 66–75 (1997)

13. Le Gall, F.: Exponential separation of quantum and classical online space complexity. In: Proceedings of SPAA 2006, pp. 67–73 (2006)

14. Li, L.Z., Qiu, D.W.: Determining the equivalence for one-way quantum finite automata. Theoretical Computer Science 403, 42–51 (2008)

15. Li, L.Z., Qiu, D.W.: A note on quantum sequential machines. Theoretical Computer Science 410, 2529–2535 (2009)

16. Li, L.Z., Qiu, D.W., Zou, X.F., Li, L.J., Wu, L.H., Mateus, P.: Characterizations of one-way general quantum finite automata. Theoretical Computer Science 419, 73–91 (2012)

17. Mereghetti, C., Palano, B.: Quantum finite automata with control language. RAIRO- Inf. Theor. Appl. 40, 315–332 (2006)

18. Mereghetti, C., Palano, B., Pighizzini, G.: Note on the Succinctness of Deterministic, Nondeterministic, Probabilistic and Quantum Finite Automata. RAIRO-Inf. Theor. Appl. 5, 477–490 (2001)

19. Monras, A., Beige, A., Wiesner, K.: Hidden Quantum Markov Models and non-adaptive read-out of many-body states. ArXiv:1002.2337 (2010)

20. Moore, C., Crutchfield, J.P.: Quantum automata and quantum grammars. Theoretical Computer Science 237, 275–306 (2000)

21. Nielsen, M.A., Chuang, I.L.: Quantum Computation and Quantum Information. Cambridge University Press, Cambridge (2000)

22. Paschen, K.: Quantum finite automata using ancilla qubits. Technical Report, University of Karlsruhe (2000)

23. Paz, A.: Introduction to Probabilistic Automata. Academic Press, New York (1971)

24. Qiu, D.W.: Some Observations on Two-Way Finite Automata with Quantum and Classical States. In: Huang, D.-S., Wunsch II, D.C., Levine, D.S., Jo, K.-H. (eds.) ICIC 2008. LNCS, vol. 5226, pp. 1–8. Springer, Heidelberg (2008)

25. Qiu, D.W., Li, L.Z., Mateus, P., Gruska, J.: Quantum Finite Automata. In: Wang, J. (ed.) Handbook of Finite State Based Models and Applications, pp. 113–144. CRC Press, Boca Raton (2012)

26. Qiu, D.W., Mateus, P., Sernadas, A.: One-way quantum finite automata together with classical states. arXiv:0909.1428
27. Qiu, D.W., Yu, S.: Hierarchy and equivalence of multi-letter quantum finite automata. Theoretical Computer Science 410, 3006–3017 (2009)
28. Yakaryilmaz, A., Cem Say, A.C.: Succinctness of two-way probabilistic and quantum finite automata. Discrete Mathematics and Theoretical Computer Science 12(4), 19–40 (2010)
29. Yakaryilmaz, A., Cem Say, A.C.: Unbounded-error quantum computation with small space bounds. Information and Computation 209, 873–892 (2011)
30. Yakaryilmaz, A., Cem Say, A.C.: Languages recognized by nondeterministic quantum finite automata. Quantum Information and Computation 10(9-10), 747–770 (2010)
31. Yu, S.: Regular Languages. In: Rozenberg, G., Salomaa, A. (eds.) Handbook of Formal Languages, pp. 41–110. Springer, Heidelberg (1998)
32. Zheng, S.G., Li, L.Z., Qiu, D.W.: Two-Tape Finite Automata with Quantum and Classical States. International Journal of Theoretical Physics 50, 1262–1281 (2011)
33. Zheng, S.G., Qiu, D.W., Li, L.Z.: Some languages recongnied by two-way finite automata with quantum and classical states. International Journal of Foundation of Computer Science. Also arXiv:1112.2844 (2011) (to appear)
34. Zheng, S.G., Qiu, D.W., Gruska, J., Li, L.Z., Mateus, P.: State succinctness of two-way finite automata with quantum and classical states. ArXiv:1202.2651 (2012)

Author Index